METHODS IN MOLECULAR BIOLOGY

T0335413

Series Editor
John M. Walker
School of Life and Medical Sciences
University of Hertfordshire
Hatfield, Hertfordshire, AL10 9AB, UK

For further volumes:
http://www.springer.com/series/7651

Cancer Drug Resistance

Overviews and Methods

Edited by

José Rueff

Centre for Toxicogenomics and Human Health, Genetics, Oncology and Human Toxicology, NOVA Medical School/Faculdade de Ciências Médicas, Universidade Nova de Lisboa, Lisbon, Portugal

António Sebastião Rodrigues

Centre for Toxicogenomics and Human Health, Genetics, Oncology and Human Toxicology, NOVA Medical School/Faculdade de Ciências Médicas, Universidade Nova de Lisboa, Lisbon, Portugal

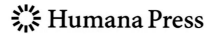

Editors
José Rueff
Centre for Toxicogenomics and Human Health,
    Genetics, Oncology and Human Toxicology
NOVA Medical School/Faculdade de
    Ciências Médicas
Universidade Nova de Lisboa
Lisbon, Portugal

António Sebastião Rodrigues
Centre for Toxicogenomics and Human Health,
    Genetics, Oncology and Human Toxicology
NOVA Medical School/Faculdade de
    Ciências Médicas
Universidade Nova de Lisboa
Lisbon, Portugal

ISSN 1064-3745   ISSN 1940-6029   (electronic)
Methods in Molecular Biology
ISBN 978-1-4939-3345-7   ISBN 978-1-4939-3347-1   (eBook)
DOI 10.1007/978-1-4939-3347-1

Library of Congress Control Number: 2016930407

Springer New York Heidelberg Dordrecht London

Cover Image: Immunolfluorescence with anti-P-glycoprotein/MDR-1 monoclonal antibody of Lymphoma MDR cell line (Chapter 6).

Printed on acid-free paper

Humana Press is a brand of Springer
Springer Science+Business Media LLC New York is part of Springer Science+Business Media (www.springer.com)

Preface

The last few decades have been witness to a far-reaching transformation in the biomedical sciences in which genetics has been one amongst the main actors. Indeed, genetics casts a new light on our understanding of genes and their action, with genomic sciences enabling the rapid acquisition of knowledge of whole genome sequences, polymorphisms, and epigenomics' mechanisms of regulation of gene expression, leading to the current era of the "omics" with all its component members: the genome, the transcriptome, the proteome, the metabolome, or the variome. Other terms could enter this lexicon reflecting the increased ability of a more accurate diagnosis and characterization of the neoplastic cell types, but not still necessarily a cure.

The present impact of the major noncommunicable diseases is startling, inasmuch as it shows a strong tendency to rise and tends to increase proportionally in low-income countries. Cancer as well as other major non-communicable diseases displays an unbridled growth both in incidence and in mortality. The global burden of cancer continues to increase largely because of the aging and the growth of the world population [1] alongside a failure of cancer therapy associated with acquired and intrinsic resistance mechanisms. Indeed, of the 7.6 million cancer deaths that occur every year worldwide, [2] many are due to cancer drug resistance.

This problem is not negligible since the number of new cases is projected to rise from the 13.3 million new cases of cancer in 2010 to 21.5 million in 2030 [3]. In the European Region alone, cancer is the most important cause of death and morbidity after cardiovascular diseases, with more than three million new cases and 1.7 million deaths each year. Overall, more than 70 % of all cancer deaths occur in low-income and middle-income regions with little or no resources for the prevention, diagnosis, and treatment of cancer. The proportion of cases diagnosed in less developed countries is projected to meagerly increase from about 56 % in 2008 to a little more than 60 % in 2030 [4]. Cancer is inextricably linked with economic wealth.

Cancer care costs are a financial burden to patients, their families, and society as a whole. The importance of the effectiveness of new drugs is also illustrated by financial figures, and the global pharmaceutical market with their approximately 30 % profit margins epitomizes the costs involved in the search for new effective drugs due to acquired and intrinsic resistance mechanisms. And those costs are growing.

Although stemming from a single account of evidence and to make a long story short, the importance of developing new cancer drugs when older ones become less effective may be well illustrated by remembering imatinib and nilotinib, both functioning as competitive inhibitors at the ATP-binding site of BCR-ABL of chronic myelogenous leukemia (CML). Although imatinib is a first-line treatment for CML and might also be of interest, for instance, for glioblastoma multiform, having generated sales of more than US$2.5 billion worldwide in 2006, its less favorable therapeutic results in CML led, in a few years, to the development of the second-generation drug nilotinib which showed a relatively more favorable safety profile and is active in imatinib-resistant CML [5]. In 2012 nilotinib generated US$998 million and a 44 % growth gaining market segment share as a potent second-generation targeted therapy for chronic myeloid leukemia (CML). Although it is a truism that cancer treatment is inextricably linked to economical factors, we live in a time to invest boldly in new ways of understanding and predicting new cancer drugs' effects and their potential to induce resistance. The new era of the "omics" is poised to face the problem, and the current book was a timely initiative of Springer, Humana Press, which intends to review and update the available knowledge and mechanisms on cancer drug resistance.

Cellular resistance to drugs can develop from a variety of mechanisms which are intended to be dealt in this book, not necessarily thoroughly, which would be an impossible task.

Resistance to a particular drug, or a class of drugs with similar mechanisms of action (multi-drug resistance [MDR]), might arise from an alteration in the drug's cellular target (e.g., a mutation in the target molecule) or by an increase in the repair of drug-induced DNA damage, or a rapid metabolic biotransformation of the drug rendering it ineffective. In the last few years, the importance of DNA repair pathways in resistance to chemotherapy has been increasingly recognized, yet translation to the clinic is residual. Since many classical cancer therapies target DNA, the influence of DNA repair systems in response to DNA damage from chemotherapy and radiotherapy is critical to cell survival. The use of inhibitors of DNA repair or DNA damage signaling pathways (NER, BER, MMR, HR, and NHEJ) provides an interesting opportunity to target the genetic differences that exist between normal and tumor tissue. On the other hand, the study of genes involved in the metabolism of drugs and xenobiotics, in particular CYPs, CYPOR, and Cytb5 which may mediate the effectiveness of drugs and also drug resistance, is central to the development of next-generation therapies.

However, the most common mechanism of resistance to cancer drugs may rely on the efflux of drugs from the cell by one or more adenosine 5′-triphosphate (ATP)–binding cassette (ABC) transporters. In healthy cells, ABC transporter proteins display a variety of roles in several organs, i.e., the liver, kidneys, gastrointestinal tract, or the nervous and reproductive systems, increasing the excretion of toxins from the body. In cancer cells, the ABC transporters work to eject chemotherapeutics from the cell to nontoxic concentrations, thus decreasing their therapeutic effects. Of the more than 48 membrane proteins that comprise the ABC transporters family, at least 15 have been associated with drug resistance. Although much progress has been made to elucidate the molecular mechanism of these resistance-conferring ABC transporters, this knowledge is not sadly at a routine stage of translational to clinical relevance. It thus seems paramount to search for an integrative view of membrane transporters as mediators of the entry, distribution, and excretion of medicines and genotoxic xenobiotics in the human organism, namely the superfamilies of membrane transporters ABC and SLC (and there are some 55 families in the human *SLC* gene superfamily) and their involvement in the membrane traffic of cancer drugs. Also, as elusive as it might still be if the Warburg effect is causal or is an effect in tumorigenesis, the fact is that cancer cells are avid for glucose and thus the two different types of membrane carrier proteins, the Na+-coupled glucose transporters (*SLC5A*/SGLT2) and the glucose transporter facilitators (*SLC2A*/GLUT1), are paramount to glucose inflow and have been shown to be upregulated in some cancers. Besides their role as main players in PET scan diagnostic procedures, they may constitute potential targets for new drugs blocking the entrance of glucose in cancer cells inasmuch as those drugs may exhibit cancer cell tropism.

Moreover, genome-wide association studies (GWAS) have been extremely successful in identifying regions of the genome that are linked to a specific trait and could also be applied in detecting the most probable marker/gene responsible for a certain resistance to a drug or patient's germ line genetic variation that may also affect drug response. Furthermore,

NGS-based approaches as applied to the exome of cancer cells may open new ways to the early identification of mutated genes whose protein products are targets to new drugs.

The study of the variome of repair genes and the levels and allocation of epigenetic regulators, in particular noncoding RNAs (e.g., microRNAs) and methylation patterns, as well as the DNA lesions and the regulation of gene expression, all have a central interest in the etiology of cancer. These aspects will contribute to the increase in the effectiveness and safety of new drugs and their therapeutic use and thus will certainly be among the major players in the future treatment of cancer.

Not more important than all the above-mentioned aspects, but still overriding, is the use of methods like proteomics which are essential in evaluating protein markers that can guide us in the search for the genomic variants or mutations responsible for cancer drug resistance.

Finally, the development of databases and in silico methodologies and their use in helping to de-emphasize individual medical hunches by supplying the criteria of evidence-based medicine will surely improve the rationale of the use of new cancer drugs and their potential resistance as well as play an interesting trade-off of individual medical ethics against the social ethics (and biopolitics) of the efficient use of scarce health resources.

Had it not been for the kind invitation of Professor John M. Walker, Professor Emeritus, School of Life and Medical Sciences, University of Hertfordshire, we would never have had the boldness of entangling ourselves in the task of organizing a book on resistance to cancer drugs. Also without the prompt and supremely competent contribution of all the prestigious authors of the various chapters who kindly accepted our invitation, the book would never be here. Our gratitude is proffered to all of them.

Lisbon, Portugal   *José Rueff*
Lisbon, Portugal   *António Sebastião Rodrigues*

References

1. Jemal A, Bray F, Center MM, Ferlay J, Ward E, Forman D. Global cancer statistics. CA Cancer J Clin. 2011;61(2):69–90.
2. WHO 2008 data; http://www.who.int/cancer/en/.
3. http://www.ispor.org/news/articles/Sept-Oct11/Global-Burden-of-Non-Communicable-Diseases.asp.
4. Jemal A., Center MM, DeSantis C, Ward EM. Global patterns of cancer incidence and mortality rates and trends. Cancer Epidemiol Biomarkers Prev. 2010;19(8):1893–1907.
5. Kantarjian H, Giles F, Wunderle L, Bhalla K, O'Brien S, Wassmann B, Tanaka C, Manley P, Rae P, Mietlowski W, Bochinski K, Hochhaus A, Griffin JD, Hoelzer D, Albitar M, Dugan M, Cortes J, Alland L, Ottmann OG. Nilotinib in imatinib-resistant CML and Philadelphia chromosome-positive ALL. N Engl J Med. 2006;354(24):2542–2551.

Contents

Preface. . . . .   *v*

Contributors. . . . .   *xi*

1   Cancer Drug Resistance: A Brief Overview from a Genetic
    Viewpoint.   1
    José Rueff and António Sebastião Rodrigues

2   Classical and Targeted Anticancer Drugs: An Appraisal
    of Mechanisms of Multidrug Resistance   19
    Bruce C. Baguley

3   In Vitro Methods for Studying the Mechanisms of Resistance
    to DNA-Damaging Therapeutic Drugs.   39
    Pasarat Khongkow, Anna K. Middleton, Jocelyn P.-M. Wong,
    Navrohit K. Kandola, Mesayamas Kongsema,
    Gabriela Nestal de Moraes, Ana R. Gomes, and Eric W.-F. Lam

4   In Vitro Approaches to Study Regulation of Hepatic
    Cytochrome P450 (CYP) 3A Expression by Paclitaxel and Rifampicin.   55
    Romi Ghose, Pankajini Mallick, Guncha Taneja,
    Chun Chu, and Bhagavatula Moorthy

5   Uptake and Permeability Studies to Delineate the Role
    of Efflux Transporters.   69
    Ramya Krishna Vadlapatla, Dhananjay Pal,
    and Ashim K. Mitra

6   Dynamics of Expression of Drug Transporters:
    Methods for Appraisal.   75
    Marta Gromicho, José Rueff, and António Sebastião Rodrigues

7   Fluorimetric Methods for Analysis of Permeability, Drug Transport
    Kinetics, and Inhibition of the ABCB1 Membrane Transporter.   87
    Ana Armada, Célia Martins, Gabriella Spengler,
    Joseph Molnar, Leonard Amaral, António Sebastião Rodrigues,
    and Miguel Viveiros

8   Resistance to Targeted Therapies in Breast Cancer.   105
    Sofia Braga

9   MicroRNAs and Cancer Drug Resistance   137
    Bruno Costa Gomes, José Rueff, and António Sebastião Rodrigues

10 The Role of MicroRNAs in Resistance to Current Pancreatic
    Cancer Treatment: Translational Studies and Basic Protocols for
    Extraction and PCR Analysis.   163
    Ingrid Garajová, Tessa Y.S. Le Large, Elisa Giovannetti,
    Geert Kazemier, Guido Biasco, and Godefridus J. Peters

11 Methods for Studying MicroRNA Expression and Their
    Targets in Formalin-Fixed, Paraffin-Embedded (FFPE)
    Breast Cancer Tissues   189
    Bruno Costa Gomes, Bruno Santos, José Rueff,
    and António Sebastião Rodrigues

12 The Regulatory Role of Long Noncoding RNAs in Cancer
    Drug Resistance   207
    Marjan E. Askarian-Amiri, Euphemia Leung, Graeme Finlay,
    and Bruce C. Baguley

13 Cancer Exosomes as Mediators of Drug Resistance   229
    Maria do Rosário André, Ana Pedro, and David Lyden

14 Isolation and Characterization of Cancer Stem Cells from Primary
    Head and Neck Squamous Cell Carcinoma Tumors.   241
    Hong S. Kim, Alexander T. Pearson, and Jacques E. Nör

15 Clinical and Molecular Methods in Drug Development:
    Neoadjuvant Systemic Therapy in Breast Cancer as a Model   251
    Sofia Braga

16 Proteomics in the Assessment of the Therapeutic Response
    of Antineoplastic Drugs: Strategies and Practical Applications   281
    Vukosava Milic Torres, Lazar Popovic, Fátima Vaz,
    and Deborah Penque

17 Managing Drug Resistance in Cancer: Role of Cancer Informatics.   299
    Ankur Gautam, Kumardeep Chaudhary, Rahul Kumar,
    Sudheer Gupta, Harinder Singh, and Gajendra P.S. Raghava

Index *313*

Contributors

LEONARD AMARAL • *Department of Medical Microbiology and Immunobiology, Faculty of Medicine, University of Szeged, Szeged, Hungary; Unidade de Medicina das Viagens, Centro de Malária e Outras Doenças Tropicais, Instituto de Higiene e Medicina Tropical, Universidade Nova de Lisboa, Lisbon, Portugal*

MARIA DO ROSÁRIO ANDRÉ • *Champalimaud Foundation, Lisbon, Portugal; Department of Genetics, Oncology and Human Toxicology, NOVA Medical School/Faculty of Medical Sciences, Nova University, Lisbon, Portugal*

ANA ARMADA • *Grupo de Micobactérias, Unidade de Ensino e Investigação de Microbiologia Médica e Centro de Malária e Outras Doenças Tropicais, Instituto de Higiene e Medicina Tropical, Nova University, Lisbon, Portugal*

MARJAN E. ASKARIAN-AMIRI • *Auckland Cancer Society Research Centre, Faculty of Medical and Health Sciences, University of Auckland, Auckland, New Zealand; Molecular Medicine and Pathology, Faculty of Medical and Health Sciences, University of Auckland, Auckland, New Zealand*

BRUCE C. BAGULEY • *Auckland Cancer Society Research Centre, Faculty of Medical and Health Sciences, University of Auckland, Auckland, New Zealand*

GUIDO BIASCO • *Department of Experimental, Diagnostic and Speciality Medicine, Sant'Orsola-Malpighi Hospital, University of Bologna, Bologna, Italy*

SOFIA BRAGA • *José de Mello Saúde, Lisbon, Portugal; Departamento de Ciências Biomédicas e Medicina, Universidade do Algarve, Algarve, Portugal*

KUMARDEEP CHAUDHARY • *Bioinformatics Centre, CSIR-Institute of Microbial Technology, Chandigarh, India*

CHUN CHU • *Department of Pediatrics, Baylor College of Medicine, Houston, TX, USA*

GABRIELA NESTAL DE MORAES • *Department of Surgery and Cancer, Imperial College London, London, UK*

GRAEME FINLAY • *Auckland Cancer Society Research Centre, Faculty of Medical and Health Sciences, University of Auckland, Auckland, New Zealand; Molecular Medicine and Pathology, Faculty of Medical and Health Sciences, University of Auckland, Auckland, New Zealand*

INGRID GARAJOVÁ • *Department of Medical Oncology, VU University Medical Center, Cancer Center Amsterdam, Amsterdam, The Netherlands; Department of Experimental, Diagnostic and Specialty Medicine, Sant'Orsola-Malpighi Hospital, University of Bologna, Bologna, Italy*

ANKUR GAUTAM • *Bioinformatics Centre, CSIR-Institute of Microbial Technology, Chandigarh, India*

ROMI GHOSE • *Department of Pharmacological and Pharmaceutical Sciences, University of Houston, Houston, TX, USA*

ELISA GIOVANNETTI • *Department of Medical Oncology, VU University Medical Center, Cancer Center Amsterdam, Amsterdam, The Netherlands; Cancer Pharmacology Lab, AIRC Start-Up Unit, University of Pisa, Pisa, Italy*

BRUNO COSTA GOMES • *Centre for Toxicogenomics and Human Health, Genetics, Oncology and Human Toxicology, NOVA Medical School/Faculdade de Ciências Médicas, Universidade Nova de Lisboa, Lisbon, Portugal*

ANA R. GOMES • *Department of Surgery and Cancer, Imperial College London, London, UK*

MARTA GROMICHO • *Centre for Toxicogenomics and Human Health, Genetics, Oncology and Human Toxicology, NOVA Medical School/Faculdade de Ciências Médicas, Universidade Nova de Lisboa, Lisbon, Portugal*

SUDHEER GUPTA • *Bioinformatics Centre, CSIR-Institute of Microbial Technology, Chandigarh, India*

NAVROHIT K. KANDOLA • *Department of Surgery and Cancer, Imperial College London, London, UK*

GEERT KAZEMIER • *Department of Surgery, VU University Medical Center, Amsterdam, The Netherlands*

PASARAT KHONGKOW • *Department of Surgery and Cancer, Imperial College London, London, UK*

HONG S. KIM • *Department of Restorative Sciences, University of Michigan School of Dentistry, Ann Arbor, MI, USA*

MESAYAMAS KONGSEMA • *Department of Surgery and Cancer, Imperial College London, London, UK*

RAHUL KUMAR • *Bioinformatics Centre, CSIR-Institute of Microbial Technology, Chandigarh, India*

ERIC W.-F. LAM • *Department of Surgery and Cancer, Imperial College London, London, UK*

TESSA Y.S. LE LARGE • *Department of Medical Oncology, VU University Medical Center, Cancer Center Amsterdam, Amsterdam, The Netherlands; Department of Surgery, VU University Medical Center, Amsterdam, The Netherlands*

EUPHEMIA LEUNG • *Auckland Cancer Society Research Centre, Faculty of Medical and Health Sciences, University of Auckland, Auckland, New Zealand; Molecular Medicine and Pathology, Faculty of Medical and Health Sciences, University of Auckland, Auckland, New Zealand*

DAVID LYDEN • *Champalimaud Foundation, Lisbon, Portugal; Department of Genetics, Oncology and Human Toxicology, Nova Medical School/Faculty of Medical Sciences, Nova University, Lisbon, Portugal*

PANKAJINI MALLICK • *Department of Pharmacological and Pharmaceutical Sciences, University of Houston, Houston, TX, USA*

CÉLIA MARTINS • *Centre for Toxicogenomics and Human Health Genetics, Oncology and Human Toxicology, NOVA Medical School/Faculdade de Ciências Médicas, Universidade Nova de Lisboa, Lisbon, Portugal*

ANNA K. MIDDLETON • *Department of Surgery and Cancer, Imperial College London, London, UK*

ASHIM K. MITRA • *Division of Pharmaceutical Sciences, School of Pharmacy, University of Missouri-Kansas City, Kansas City, MO, USA*

JOSEPH MOLNAR • *Department of Medical Microbiology and Immunobiology, Faculty of Medicine, University of Szeged, Szeged, Hungary*

BHAGAVATULA MOORTHY • *Department of Pediatrics, Baylor College of Medicine, Houston, TX, USA*

JACQUES E. NÖR • *Department of Restorative Sciences, University of Michigan School of Dentistry, Ann Arbor, MI, USA; Comprehensive Cancer Center, University of Michigan, Ann Arbor, MI, USA; Department of Biomedical Engineering, University of Michigan*

College of Engineering, Ann Arbor, MI, USA; Department of Otolaryngology, University of Michigan School of Medicine, Ann Arbor, MI, USA

DHANANJAY PAL • *Division of Pharmaceutical Sciences, School of Pharmacy, University of Missouri-Kansas City, Kansas City, MO, USA*

ALEXANDER T. PEARSON • *Division of Hematology/Oncology, Department of Internal Medicine, University of Michigan Medical Center, Ann Arbor, MI, USA; Comprehensive Cancer Center, University of Michigan, Ann Arbor, MI, USA*

ANA PEDRO • *Champalimaud Foundation, Lisbon, Portugal; Department of Genetics, Oncology and Human Toxicology, Nova Medical School/Faculty of Medical Sciences, Nova University, Lisbon, Portugal*

DEBORAH PENQUE • *Laboratório de Proteómica, Human Genetics Departament, Instituto Nacional de Saúde Dr Ricardo Jorge, Lisbon, Portugal; ToxOmics-Centre of Toxicogenomics and Human Health, Universidade Nova de Lisboa, Portugal*

GODEFRIDUS J. PETERS • *Department of Medical Oncology, VU University Medical Center, Cancer Center Amsterdam, Amsterdam, The Netherlands*

LAZAR POPOVIC • *Medical Oncology Department, Oncology Institute of Vojvodina, Sremska Kamenica, Serbia; Medical Faculty, University of Novi Sad, Novi Sad, Serbia*

GAJENDRA P.S. RAGHAVA • *Bioinformatics Centre, CSIR-Institute of Microbial Technology, Chandigarh, India*

ANTÓNIO SEBASTIÃO RODRIGUES • *Centre for Toxicogenomics and Human Health, Genetics, Oncology and Human Toxicology, NOVA Medical School/Faculdade de Ciências Médicas, Universidade Nova de Lisboa, Lisbon, Portugal*

JOSÉ RUEFF • *Centre for Toxicogenomics and Human Health, Genetics, Oncology and Human Toxicology, NOVA Medical School/Faculdade de Ciências Médicas, Universidade Nova de Lisboa, Lisbon, Portugal*

BRUNO SANTOS • *Centre for Toxicogenomics and Human Health, Genetics, Oncology and Human Toxicology, NOVA Medical School/Faculdade de Ciências Médicas, Universidade Nova de Lisboa, Lisbon, Portugal*

HARINDER SINGH • *Bioinformatics Centre, CSIR-Institute of Microbial Technology, Chandigarh, India*

GABRIELLA SPENGLER • *Department of Medical Microbiology and Immunobiology, Faculty of Medicine, University of Szeged, Szeged, Hungary*

GUNCHA TANEJA • *Department of Pharmacological and Pharmaceutical Sciences, University of Houston, Houston, TX, USA*

VUKOSAVA MILIC TORRES • *Laboratory of Proteomics, Human Genetics Department, Instituto Nacional de Saúde Dr Ricardo Jorge, Lisbon, Portugal; ToxOmics-Centre of Toxicogenomics and Human Health, Universidade Nova de Lisboa, Portugal*

RAMYA KRISHNA VADLAPATLA • *Technical Services, Mylan Pharmaceuticals Inc., Morgantown, WV, USA*

FÁTIMA VAZ • *Laboratório de Proteómica, Human Genetics Departament, Instituto Nacional de Saúde Dr Ricardo Jorge, Lisbon, Portugal; ToxOmics-Centre of Toxicogenomics and Human Health, Universidade Nova de Lisboa, Portugal*

MIGUEL VIVEIROS • *Grupo de Micobactérias, Unidade de Ensino e Investigação de Microbiologia Médica e Centro de Malária e Outras Doenças Tropicais, Instituto de Higiene e Medicina Tropical, Universidade Nova de Lisboa, Lisbon, Portugal*

JOCELYN P.-M. WONG • *Department of Surgery and Cancer, Imperial College London, London, UK*

Chapter 1

Cancer Drug Resistance: A Brief Overview from a Genetic Viewpoint

José Rueff and António Sebastião Rodrigues

Abstract

Cancer drug resistance leading to therapeutic failure in the treatment of many cancers encompasses various mechanisms and may be intrinsic relying on the patient's genetic makeup or be acquired by tumors that are initially sensitive to cancer drugs. All in all, it may be responsible for treatment failure in over 90 % of patients with metastatic cancer. Cancer drug resistance, in particular acquired resistance, may stem from the micro-clonality/micro-genetic heterogeneity of the tumors whereby, among others, the following mechanisms may entail resistance: altered expression of drug influx/efflux transporters in the tumor cells mediating lower drug uptake and/or greater efflux of the drug; altered role of DNA repair and impairment of apoptosis; role of epigenomics/epistasis by methylation, acetylation, and altered levels of microRNAs leading to alterations in upstream or downstream effectors; mutation of drug targets in targeted therapy and alterations in the cell cycle and checkpoints; and tumor microenvironment that are briefly reviewed.

Key words Intrinsic resistance and pharmacogenetics, Acquired resistance and tumor micro-heterogeneity, Acquired resistance and adaptive compensatory pathways, Uptake and efflux transporters in resistance, DNA repair and resistance, Epigenomics and resistance, Tumor microenvironment and resistance

1 Innate or Intrinsic and Acquired Resistance: Definitions and Mechanisms

Cancer drug resistance classically either stems from host factors (innate or intrinsic resistance) or is an acquired resistance of the tumor cells by means of genetic or epigenetic alterations in the cancer cells [1]. Another way of defining the types of cancer drug resistance is to consider the pharmacokinetic-based resistance and the cell-dependent resistance (for a review *see* ref. 2). For the sake of straightforwardness and from a genetic point of view, let us assume that the mechanisms involved in intrinsic resistance are by and large due to the germinal genetic makeup, and that the mechanisms responsible for acquired resistance rely on mutational or epigenetic phenomena occurring in the tumor cells leading to failure of response to therapeutics. It is well known that tumors

José Rueff and António Sebastião Rodrigues (eds.), *Cancer Drug Resistance: Overviews and Methods*, Methods in Molecular Biology, vol. 1395, DOI 10.1007/978-1-4939-3347-1_1, © Springer Science+Business Media New York 2016

exhibit micro-clonality with a high degree of genetic heterogeneity making possible the recruitment of resistant cells to continue growing in spite of the therapy.

This genetic heterogeneity may render the tumor cells particularly versatile in modifying rates of efflux of drugs, up-regulating DNA repair processes, or activating alternative survival signaling pathways [3]. That combination of genomic and epigenetic instability associated with the acquisition of a stem cell-like phenotype is probably an important part of the tumor behavior explaining drug resistance. Such a biological behavior typically characterizes the basis of acquired resistance based on genetic phenomena of tumor cells and quite independent of the constitutional germ-line genome of the patient. A caveat must, however, be alluded to by mentioning that the final therapeutic failure or success also depends on various factors like the competence and effectiveness of the immunological surveillance or the various factors involved in metastasis which are outside the scope of this brief overview, although tumor cell variants may have low immunogenicity due to genetic heterogeneity and become resistant to immune attack [4].

Whatever the mechanisms underlying cancer and its progression, drug resistance is a major problem since it is believed to cause treatment failure in over 90 % of patients with metastatic cancer [5].

2   Intrinsic Resistance

2.1   Intrinsic Resistance, Clinical Expertise, and Clinical Guidelines

In a schematic manner, intrinsic resistance might be defined as a failure of response to the initial drug (or combination of drugs), indicating that before receiving therapy the resistance mechanisms/factors were already present. Intrinsic resistance may result largely, but not only from some major factors: (1) possible pre-existence of resistant cells in the tumor that render the therapy unsuccessful causing or leading anyway to a wrong adequacy of the administered drugs(s) to that particular cancer patient; (2) a different type of unsuccessful treatment may result from the fact that the actual cancer patient has low tolerance to the drug(s) and/or their side effects are unbearable and the dose has to be lowered resulting in putative failure of treatment; or even (3) from factors involved in the ADME (absorption, distribution, metabolism, excretion) such that the drug does not attain its best pharmacokinetic profile to exert its effects on the tumor or is subject to pharmacogenetic patterns which determine different levels of availability of the active metabolite of the drug. Tamoxifen is amongst the examples of the latter.

As far as intrinsic resistance is concerned and even when the prescribed drug(s) belong to the first-line therapy for a specific tumor, there are no easy ways to predict or estimate resistance; neither an easy strategy has been found to overcome resistance, which is based on highly complex and individually variable biological mechanisms, apart from the few cases where pharmacogenomic patterns can be sought indicating germ-line patterns of low drug efficiency.

In the clinical practice, drug resistance can only be recognized during treatment. Thus, in order to try to prevail over (or tentatively overcome) intrinsic resistance one has to firstly take into consideration the far-reaching clinical competence of the medical staff on the judgement of the adequacy of the treatment and its therapeutic scheme to the particular patient under treatment, as well as the adherence to clinical guidelines, although guidelines are the expression of average patients and many patients are simply not in the average [6]. Understanding the diversity of both genetic and therapeutic factors that can determine innate patient responsiveness to anticancer drugs is thus a multiform endeavor.

2.2 Intrinsic Resistance and Pharmacogenetic Patterns

A main germ-line-determined factor of intrinsic resistance, as pointed out above, encroaching on the ADME variables, is drug metabolism and biotransformation which depends in large part on the activities of cytochromes P450 for phase I and on conjugation reactions for phase II. Cytochrome P450s (CYP) are members of a large superfamily of heme proteins, with a pivotal role in xenobiotic biotransformation, as well as in the endobiotic biosynthesis and catabolism of steroid hormones, bile acid, lipid-soluble vitamins, and fatty acids. At least 57 human microsomal CYPs have been recognized, some 15 of which are involved in drug metabolism [7] whose activities are supported by electron transfer from NADPHcytochrome P450 oxidoreductase (CYPOR) [8]. Interindividual variability in CYP-mediated xenobiotic metabolism is extensive. This type of intrinsic resistance is thus germ-line determined and has as main responsible organs primarily the liver, but also the lungs or the kidneys, among others.

As mentioned briefly above, a representative case of a pharmacogenetic CYP-dependent pattern is the pharmacokinetics of tamoxifen, a drug in clinical use for treatment and prevention of estrogen-dependent breast cancer. Tamoxifen is, however, but a pro-drug, being transformed among other metabolites to endoxifen. Endoxifen is the metabolite with the higher potent anti-estrogen effect since it has a much higher affinity for the estrogen receptor and attains higher plasma levels. However, single-nucleotide polymorphisms in the CYP2D6 gene, particularly the presence of two null alleles, predict for reduced tamoxifen metabolism and possibly poorer outcome than expected in patients with a wild-type genotype due to lower biotransformation to endoxifen. However, studies evaluating the impact of genetic polymorphisms resulting in CYP2D6 with reduced or no activity on long-term outcome of breast cancer do not still allow, by and large, a recommendation for typing of CYP2D6 polymorphisms as indicators for predictive outcome of treatment of estrogen-dependent breast cancer. It is expected that the future may bring about predictable tests to evaluate germ-line-determined pharmacogenetic phenotypes that may help in designing more effective treatments with lower relapse rates. Of course, tumor-associated factors stemming from acquired mutations and/or epimutations should also be considered on their role in the fate of the drug(s) [9].

3   Acquired Resistance

Various mechanisms may bring about acquired resistance. Some of those can be categorized according to the functions which appear modified in the tumor rendering the tumor cell more competitive for growth and metastasis and better resisting cancer drugs.

Categories of acquired drug resistance are seemingly due to secondary genetic alterations (both mutations and epimutations, the latter defined as an abnormal up-regulation of otherwise normally repressed genes, or downregulation of genes active in normal cells, or still by copy number changes), and they encompass namely (1) increased rates of drug efflux of drugs or decreased rates of drug influx into the tumor cells, mediated by transmembrane transporters of drug uptake and/or efflux (e.g., SLCs, ATP-binding cassettes (ABCs)); (2) biotransformation and drug metabolism mainly due to CYPs in the tumor; (3) altered role of DNA repair and impairment of apoptosis; (4) role of epigenomics/epistasis by methylation, acetylation, and altered levels of microRNAs leading to alterations in upstream or downstream effectors; (5) mutation of drug targets in targeted therapy and alterations in the cell cycle and checkpoints; and (6) tumor microenvironment [1].

3.1   Acquired Resistance and Tumor Micro-heterogeneity

In order to tackle the categories of acquired drug resistance, one should thus take into account the recognizable genetic heterogeneity that is present in many tumors (if not all). Indeed, cancer cells within one tumor of a patient at any given moment in time may display overwhelming heterogeneity for various traits related to tumorigenesis, such as those that may modify or modulate all the above categories of acquired resistance leading to angiogenic, invasive, and metastatic potential [10–12].

Tumors, besides turning the organism of the patient into a genetic mosaic, themselves display genetic mosaicism. Tumors are, indeed, composed of subclones, subpopulations of genetically identical cells that can be distinguished from other subclones by the mutations they harbor. Such subclones compete for biological dominance during cancer progression, and drug treatment can lead to formerly minor tumor subclones becoming dominant if they are resistant to treatment. These subclones are indeed positively selected to outgrowth and resistance to apoptosis and although representing a smaller cell population they are endowed with a rapidly growing capacity [13]. There is a crucial need to understand the mechanisms driving genomic instability so that therapeutic approaches to limit cancer diversity, adaptation, and drug resistance can be developed [14].

Besides inter-tumor heterogeneity which different patients, indeed different genotypic tumors, bear, although probably histologically classified as of the same type, intra-tumor heterogeneity is claiming nowadays our attention [15], since many if not most somatic mutations detected by exome sequencing may not be detected across every tumor region. As pointed out by Castano et al.: "The tumour 'onco-genotype', which defines the collection of disease-related mutations and that evolves over time due to inherent genomic instability, differs obviously among patients so that nearly every tumour cell population is unique, thus adding to the clinical challenges" [16]. Or, as appropriately referred to by Sharma and Settleman, "Cancer is … actually a hundred diseases masquerading as one" [17].

Intra-tumor heterogeneity may have conspicuous consequences in therapeutic failure or cancer drug resistance. The tumor "onco-genotype," which defines the collection of disease-related mutations often occurring mainly as "driver mutations" evolves over time due to inherent genomic instability, not only accumulating various "passenger mutations," but also accumulating mutations, genome rearrangements, and polisomy, involving critical genes for the tumor progression and its resistance to drugs that had previously been found to be effective in the refraining of tumor growth and metastasis. Driver and passenger mutations may change places as the tumor evolves. As such, resistance appears to select for subclones bearing mutations in the genes or pathways targeted by the drug.

But genomic instability and thus heterogeneity leading to resistance may show the way to a rising strategy to overcome this problem through the use of combinations of targeted therapies with the goal of defeating several drivers.

This may, nonetheless, involve insurmountable costs per patient. This is a central problem that should not be neglected since it raises important financial, political, and even ethical questions concerning the access and availability of those drugs that may provide, if not the cure, at least some extra time of life [18].

The very point of using targeted therapies to specific cellular oncoproteins should be traced back to the inspiring concept of "oncogene addiction" coined in 2000 by Weinstein [19–22] whereby despite the multiple genetic and epigenetic abnormalities of the cancer cells, their growth control can often be impaired by the inactivation of a single oncogene, i.e., the "Achilles heel" of the cancer cell that could reasonably be thought to be blocked/inactivated therapeutically.

This innovating cutting-edge concept was based on the assumption that a given oncogene may play a key role on the cell circuitry of signaling pathways of the cells so that they lose cell cycle control and apoptosis mechanisms leading to sustained proliferation and survival. The examples are many and some brought in the basis for the "oncogene addiction" concept, namely *MYC* (the first in supporting the concept), *RAS* genes, and the most representative activated tyrosine kinases, like the *BCR-ABL* or the ErbB receptor tyrosine kinase family [17]. Unfortunately this key concept and the profound therapeutic basis it helped to create soon uncovered new mechanism of cancer drug resistance, namely by the adaptive compensatory pathways or oncogenic bypass, as discussed below [3, 23].

Tumor micro-heterogeneity may also be linked to the epithelial–mesenchymal transition (EMT). The epithelial phenotype of cells can undergo transition to a mesenchymal phenotype, a process driven by various transcription factors that is associated with increased motility and invasive capacity as well as increased cancer drug resistance. Signaling pathways activated in EMT seem to include, in some cancers, Wnt/β catenin, Notch, PI3K/AKT, among others, leading to increased resistance to drug treatment, both chemotherapy and targeted therapy, namely resistance to EGFR inhibitors [3, 24]. Nonetheless, EMT may not occur in all tumors, like melanomas, albeit displaying phenotypes with either the expression of high MITF-M and E-cadherin with more differentiated noninvasive behavior, or expressing high N-cadherin, Slug, and Axl and with a more invasive behavior [25].

3.2 Acquired Resistance and Adaptive Compensatory Pathways

Some of the secondary genetic and epigenetic alterations occurring in tumors which may determine cellular diversity with the subsequent occurrence of subclonal heterogeneity may also coexist with adaptive nonhereditary mechanisms, in particular adaptive responses or fluctuation in protein levels downstream to the receptors to targeted therapies, leading to activation of alternative compensatory signaling pathways [3].

This type of bypass to the main pathway by which the drug is exerting its therapeutic effect is what can be called compensatory adaptation or oncogenic bypass [3, 23]. Compensation thus does not affect drug-target interaction but adapts the signaling circuitry of the tumor cell, thus escaping the growth-blocking activity of the drug.

This bypass thus lowers the dependence for tumor growth of the signal transduction pathway whose triggering receptor is being blocked by the drug through the activation of a parallel pathway which results in failure of growth control by the drug being administered. This type of transactivation by other receptor partners thus results in resistance to the target-directed first drug administered and can only be overcome by the use of combinations of targeted therapies, as mentioned above.

Most targeted chemotherapeutic drugs, indeed, block only a single cellular pathway and as consequence cancers frequently acquire resistance by up-regulating alternative compensatory pathways.

But besides multi-targeted therapies, fortunate situations exist, and probably new ones will come to be uncovered whereby some key product genes of the cell circuitry may control more than one signaling pathways. Steroid receptor coactivator-3 (SRC-3), also known as AIB1 (*a*mplified *i*n *b*reast cancer 1), is probably such an example. It is a member of the p160 steroid receptor coactivator family composed of SRC-1 (NCOA1), SRC-2 (TIF2/GRIP1/NCOA2), and SRC-3. SRC-3 coordinates multiple signaling networks, suggesting that SRC-3 inhibition offers a promising therapeutic strategy [26, 27].

4 Uptake and Efflux of Drugs Mediated by Transporters: Role in Resistance

The rates of abnormal efflux or influx of drugs to the cancer cell, as well as their abnormal biotransformation to inactive metabolites, are among the main mediator mechanisms leading to pharmacokinetics-mediated resistance, whereas proficient DNA repair, or lack of an abnormal epigenetic-controlled expression of a key gene product controlling cell cycle regulation, determines a pharmacodynamic resistance.

Multifunctional efflux transporters from the ABC gene family have been known for more than two decades to play a role in multidrug resistance (MDR) of tumor cells conferring resistance to various anticancer drugs. The human genome encodes 48 ABC transporters, organized into seven distinct subfamilies (ABCA–ABCG), and at least 15 of these members are associated with MDR [28].

ABC proteins are involved in the ATP-dependent efflux of substrates such as phospholipids, sterols, bile salts, and amphipathic drugs. While various ABC transporters have been observed to export chemotherapy drugs using in vitro experimental systems, the ones having the major involvement of drug transport seem to be ABCB1, ABCC1, and ABCG2 [1].

Tumors originating from tissues with naturally high levels of ABC transporters' expression may be intrinsically drug resistant (e.g., colon, kidney, pancreas, and liver carcinoma), whereas tumors from tissues with low expression may display an increase only after chemotherapy, acquiring resistance through up-regulation of gene expression. In both cases though, the evolving nature of the initial cancer clone will dictate whether influx/efflux membrane transporters may have a role in cancer drug resistance. Either because in the first case they may not be so much expressed in the genetically altered cancer cell, or because in the latter case cancer cells expressing high levels of resistance due to drug efflux may be selected to proliferate.

In many solid tumors over-expression of ABC transporters and drug resistance is unequivocal. Over-expression of ABCG2, in particular, is associated with resistance to a wide range of different anticancer agents including mitoxantrone, camptothecins, anthracyclines, flavopiridol, and antifolates [29], but a wider range of cancer drugs are substrates of various ABC transporters [30].

The attempts to use inhibitors of ABC transporters to circumvent ABC-mediated MDR in vivo faced, however, high toxicity observed in vivo in clinical trials, and also because clinical efficacy can only be reached with the inhibition of various transporters.

In the case of breast cancer resistance, the major efflux transporter protein is the breast cancer-resistant protein, a member of the ABCG family (BCRP/ABCG2). It is noteworthy that the c-MET downstream phosphoinositide 3-kinase (PI3K)/AKT signaling activates over-expression to BCRP/ABCG2 in a doxorubicin-resistant ovarian cancer line, thus apparently linking the cell signaling circuitry controlling the cell cycle and proliferation to the levels of expression of a drug efflux transporter showing how intertwined is the network of cancer pathways and mechanisms that more often than not render cancer drug resistance a burdensome phenomenon [28, 31].

In leukemia, it was shown that the expression and functionality of ABCB1 hampers complete remission and survival [32–34]. Acute myeloid leukemia (AML) patients who had joint expression of ABCB1 and ABCG2 had the poorest prognosis [35]. However, the role of ABCG2 as a cause of MDR in acute lymphoid leukemia (ALL) is a matter of debate [36]. In pediatric patients with ALL, ABCB1 does not seem to have a prognostic significance [37].

In chronic myeloid leukaemia (CML) patients ABCG2 gene expression levels correlated with ABCB1 and ABCC1, and interestingly there seems to exist a correlation between efflux genes and the influx gene SLC22A1 which supports the hypothesis that absolute bioavailability may also be influenced by the balance between efflux and influx transport and most of these transporters were also found over-expressed in the majority of resistant CML cell lines [38, 39].

It is worth noting that namely hematopoietic stem cells express higher levels of ABCB1 than their matured counterparts, which contributed to the concept that cancer stem cells may represent a small subset of cancer cells within a cancer that have the ability to self-renew, thus constituting a reservoir of self-sustaining cells [2], which does not set aside the concept that genetic diversification and clonal selection by the cancer drugs may simultaneously occur with a reiterative process of clonal expansion from stem-like cells in some tumors [40].

Efflux pumps of the ABC transporters' family are subject to microRNA-mediated gene regulation. As a matter of fact, it appears that ABC transporters are entrenched in a concerted microRNA-guided network of concurrently regulated proteins that mediate altered drug transport and cell survival upon defy by cancer drugs or adverse survival conditions due to exposure to environmental detrimental compounds. There is increasing evidence that microRNAs are crucially involved in coordinating and fine-tuning this complex network of proteins mediating increased drug efflux and cell survival. microRNA-93, for example, activates c-Met/PI3K/Akt pathway which in turn activates over-expression to BCRP/ABCG2, as mentioned above [41].

microRNAs play, therefore, an important epigenetic role in controlling the levels of expression of ABC transporters' genes, being thus connected with drug distribution as well as with drug resistance [42].

5   DNA Repair and Cancer Drug Resistance

Mutations in genes involved in the DNA damage response (DDR) can increase the risk of developing cancer which is plentifully illustrated by the various cancer syndromes involving mutations in genes coding for repair enzymes, from ataxia telangiectasia (ATM), or Fanconi anemia (FANC genes) to breast and ovary cancers (BRAC1/BRAC2).

It is also well established that, besides rare Mendelian gene defects of high penetrance, common variations (e.g., SNPs) in DNA repair genes of low penetrance may alter protein function and the individual's capacity to repair damaged DNA, hence increasing cancer susceptibility. However, those SNPs occurring in DNA repair genes may also possibly have an important role in cancer drug resistance [43–50].

Many cancer drugs exert their effects by and large by causing DNA damage, like epirubicin, doxorubicin, 5-fluorouracil, or cisplatin which find their place as first-line drugs for some cancers. DNA damage entails the triggering of the DDR which is a key mechanism enabling cancer cells to survive through repair of the induced DNA lesions and thereby developing resistance [51]. But lack of DDR proficiency can also definitely contribute to cancer drug resistance.

BRCA genes are involved in repairing DNA through homologous recombination following DNA strand breaks (DSB). The second hit is acquired in the tumor genome, rendering these tumors susceptible to DNA-damaging agents once they have defects in their DNA repair machinery. Moreover, some 60–80 % of breast tumors from BRCA1 mutation carriers display a triple-negative phenotype (TNBC) [52]. BRCA gene products are nonfunctional in a subset of sporadic triple-negative breast tumors (TNBCs), generally through promoter hypermethylation or other epigenetic pathways, and also by mutations occurring concurrently with tumors' progression heterogeneity. This has been termed the "BRACness" of sporadic TNBCs. BRCAness leads to a better response upon intensive exposure to alkylating agents as adjuvant chemotherapy and to hypersensitivity to DSB-inducing agents such as bifunctional alkylators and platinum salts, but not doxorubicin and docetaxel. Also the clinical responses are lower with taxane- and/or anthracycline-based neoadjuvant chemotherapy (NAC) in the case of tumors bearing BRACness [53]. Thus, up- or downregulation of DDR genes may provide tumor cells with escape mechanisms to cancer drugs and induce chemotherapy resistance.

Disabling alterations in DNA repair pathways are frequently observed in cancer. These DNA repair defects may either be mutations or epimutations and are specific to cancer cells. It is thought that these molecular defects produce a "mutator phenotype," which allows cancer cells to accumulate additional cancer-promoting mutations.

The molecular understanding of DNA repair mechanisms, namely DSB repair, has led to the development of targeted therapies to selectively trigger cancer cells that display defects in homologous recombination-mediated DNA DSB repair. These pharmacological approaches for the treatment of homologous recombination-defective tumors predominantly aim at repressing the activity of PARP1, which is crucial for base excision repair, or inhibit the nonhomologous end joining kinase DNA-PKcs (DNA-dependent protein kinase, catalytic subunit). Whereas normal cells can bypass PARP1 (poly ADP-ribose polymerase 1) inhibitor- or DNA-PKcs inhibitor-induced genotoxic lesions via homologous recombination, homologous recombination-defective cancer cells are unable to properly repair DNA DSBs, in the presence of PARP1 or DNA-PKcs inhibitors, ultimately leading to apoptotic cancer cell death [54].

The identification of genes associated with the DNA repair activity and related with individual response to chemotherapeutic agents is therefore crucial since it may allow the development of customized strategies for cancer treatment. The recent approval by the US FDA of olaparib, a (PARP) inhibitor, is a relevant move towards the class of personalized cancer drugs targeted to the blocking of DNA repair functions ultimately triggering cell death.

Indeed, the search for targeted therapies has also focused on DNA repair pathways [51], besides the ones developed for molecular players having a key role on the cell circuitry of signaling pathways of the cells so as to avoid that they lose cell cycle control and apoptosis mechanisms, as mentioned above. Efforts are now also focused in targets of DNA integrity, or, stated otherwise, not only targeted to "gatekeepers" as the genes that should be inactivated for a cell to become cancerous, but also targeted to "caretakers," the genes involved in maintaining genetic stability [55].

The biological significance of DNA repair mechanisms is highlighted by the fact that their deregulation can contribute to the initiation and progression of cancer, but on the other hand, DNA repair can confer resistance to front-line cancer treatments, might there be cancer drugs or radiotherapy which relies on the generation of DNA damage to kill cancer cells. The way cancer cells (or cancer stem cells) recognize DNA damage and undertake DNA repair is therefore a key mechanism for therapeutic resistance or recurrence [56].

6   Epigenomics and Resistance: The Role of Methylation, Acetylation, and microRNAs

Whatever the mutations involved in the initiation of a cancer, not all may end up as a clinically diagnosed cancer in all individuals. One possible important reason for this is that the outcome of a mutation can depend upon other genetic variants in the genome. This can broadly define epistatic interactions, which may increase the effects of the hypostatic gene or, conversely, alleviate its effects [57]. They can occur between different variants within the same gene or between variants in different genes. The latter might be important to consider in cancer since the wealth of "passenger mutations" in a cancer may modulate the effect of the "driver mutation," acting as putative modifier genes. For example mutations in *ERS1*, the gene coding for the estrogen receptor (ER), have been linked to treatment failure and shown to be recurrent in metastatic clinical samples playing an important role in acquired endocrine therapy resistance [58].

But epimutations may also play an important role, by up- or downregulating the expression of receptors, might them be hormonal receptors or receptors used by targeted therapies.

In estrogen-dependent breast cancer, approximately some 20 % of ER-positive tumors lose its expression during tamoxifen treatment. This loss of expression may be the result of epigenetic silencing of ER expression.

Several mechanisms have been proposed to explain the absence of ER expression. These mechanisms involve epigenetic changes such as aberrant methylation of CpG islands of the ER promoter and histone deacetylation. This fact has been used as a predictor of poor outcome and tamoxifen resistance. Other mechanisms proposed in the loss of ER expression are hypoxia, over-expression of EGFR or HER2, and MAPKs hyperactivation. Also, PI3K pathway activation confers anti-estrogen resistance. Also, altered expression of specific microRNAs has been implicated in tamoxifen resistance development predicting the outcome and therapeutic response in breast cancer [59].

Steroid receptor coactivator-3 (SRC-3) promotes numerous aspects of cancer, through its capacity as a coactivator for nuclear hormone receptors and other transcription factors, and via its ability to control multiple growth pathways simultaneously. Gene amplification and protein over-expression of Sarc3 are well established. In fact, SRC-3 is over-expressed in 60 % of breast cancers which may be implicated in tamoxifen resistance [27], namely through potentiating of E2F1 activity (a target of pRb-mediated repression). Binding of SRC to transcription factors will further recruit other chromatin modification factors, such as acetyltransferases and methyltransferases that modify the chromatin structure and alter the transcription levels of their target genes. Thus, it is conceivable that these changes may affect the expression levels of many genes. In tamoxifen-treated breast cancer patients, SRC-3 over-expression is associated with high levels of HER-2/*neu*, tamoxifen resistance, and poor disease-free survival [60].

Mechanisms involved in epigenetic-driven drug resistance encompass epigenetic changes resulting in gene transcription of drug transporters (ABCB1), pro-apoptotic genes (*DAPK, APAF-1*), DNA repair proteins (MLH1, MGMT, FANCF), and histone modifiers (KDM5A). Fortunately, treatment of drug-resistant tumor cell populations bearing epimutations with cytotoxic or targeted drugs in combination with epigenetic drugs, such as inhibitors of histone deacetylases (e.g., vorinostat, trichostatin A), DNA methyl transferases, and histone methyltransferases, may reverse a drug-resistant epigenome into a drug-sensitive epigenome, thereby rendering tumor cells sensitive to the cytotoxic or targeted drug. Indeed, the large variability in drug resistance of individual cells is to be found, maybe not primarily in cancer cells' mutations due to genetic instability of the tumor, but also and most decisively in the different transcriptional network states produced by epigenetic mechanisms in the same cancer genome [61].

Epigenetic regulation, particularly by microRNAs, besides DNA methylation or histone acetylation, plays an important role in carcinogenesis and oncotherapy. The approximately 2000 different human microRNA species identified form a intertwined network of concurrently regulated proteins that mediate cell survival upon a challenge by cancer drugs, and as already mentioned above they may control the levels of expression of ABC transporters' genes, being thus connected with drug resistance [42]. Also long noncoding RNAs (lncRNAs) are also able to regulate mRNAs' levels of expression correlated with cancer drug resistance. The fine-tuning of the ncRNA system is on the other hand also regulated by hypermethylation making the whole of the epigenetic machinery a self-regulated system whose overall implications in cancer drug resistance are yet to be fully uncovered.

7   Tumor Microenvironment and Resistance

The tumor microenvironment (TME) consists of vascular cells, fibroblasts, infiltrating immune cells, the extracellular matrix (ECM), and the signaling molecules bound to it [62]. TME has many roles in tumor progression and metastasis, including the creation of a hypoxic environment, increased angiogenesis, and invasion and changes in expression of noncoding RNAs. There is a molecular cross talk between the tumor and its microenvironment that determines tumor progression [63].

The microenvironment could be a major niche where some mechanisms of drug resistance may take place through the reduction of drug distribution throughout the tumor, therefore protecting high proportions of cells from damage induced by the drug [64]. The dissection of interactions between tumors and their microenvironment can reveal important mechanisms underlying drug resistance [64, 65].

It is noteworthy that landscaper genes seem to facilitate the growth of neoplastic lesions by creating a microenvironment that aids in unregulated cellular proliferation. Loss of components of the extracellular matrix (ECM) may lead to a microenvironment which can stimulate unregulated growth, clonal proliferation, and ultimately neoplastic lesions [66].

There are also interesting data suggesting that, at least in patients with *BRCA1/2*-related breast cancers, genomic alterations in the stroma coexist equally with alterations in the epithelium, and, thus, the genetically unstable stroma might provide for a microenvironment that functions as a landscaper that positively selects for genomic instability in the epithelium [67]. However, this might not be the case in other situations whereby the suggestion of epithelial:mesenchymal interactions remains but a possibility in the causality of malignant development [68].

8   Concluding Remarks

Substantial scientific advances over the last years have allowed us to understand the genomic landscapes and portraits of individual tumors [69, 70]. Numerous genetic alterations have been identified in individual tumors, but the number of cancer-promoting genes is considered relatively small, in the order of 100–150 [69]. Two to eight driver gene mutations can be found in tumors, but the vast majority will be passenger mutations. These driver genes can be grouped into well-known signaling pathways, the fittingly called hallmarks of cancer [71]; tumor-promoting mutations are seemingly involved in three major biological processes, cell fate, cell survival, and genome maintenance [69]. Hence, their identification has led to the concept of tailored mechanism-based targeted therapies aimed at inhibiting some of the specific oncogenic pathways mentioned above. This strategy has the advantage of only targeting tumor cells while doing little or no harm to normal tissues. The vast information garnered by the latest genome-wide sequencing studies has not yet been fully translated into the clinic, but several instances of targeted therapies have emerged from the identification of specific alterations in driver genes (i.e., those that confer a growth and survival advantage), such as protein kinases and development of small-molecule inhibitors. As discussed above, targeted therapies include epidermal growth factor receptor (EGFR) inhibitors, human epidermal growth factor receptor 2 (HER2), or breakpoint cluster region-Abl proto-oncogene 1 (BCR-ABL) inhibitors with some success [72]. Nevertheless, outcomes of these targeted therapies have revealed to be suboptimal, particularly as their usage becomes more widespread, and clinical responses are generally short-lived. Unfortunately, in most patients with solid tumors, the cancer evolves to become resistant within a few months [73]. Drug resistance to these targeted therapies arrives sooner or later. In some cases initial drug resistance can be attributed to misexpression of a number of genes, frequently occurring in refractory tumors, and responsible for cellular drug extrusion, as discussed [74]. However, in most cases drug resistance is due to multiple factors, including under- or over-expression of specific targets, mutations in target genes, and epigenetic alterations in DNA [2, 75].

To understand the reasons for the apparent inevitability of cancer drug resistance, one must focus on the knowns and the unknowns of cancer development and progression. First and foremost, the clinical detection of a tumor occurs many years, perhaps decades, after the initial oncogenic triggering event. The average time it takes for a tumor to reach detection size varies with several factors, including tissue affected, rate of tissue self-renewal, and exposure to mutagens and carcinogens. For example, it has been estimated that colorectal cancer requires about 17 years for a large benign tumor to evolve into an advanced cancer [76]. When comparing different tumors with different progression periods, the number of accumulated mutations and genomic alterations will necessarily be different. As a consequence, the response of different tumors towards chemotherapy will depend on these mutational landscapes. It follows that different tumor sensitivities will arise, and cellular adaptation to chemotherapy will necessarily be more or less effective and rapid. It also follows that resistance to chemotherapy will depend on the number of cells with sufficiently wide mutational landscapes that could allow escape from cell death. Recent studies have indicated that at the onset, tumors already possess mutated cells that could be responsible for resistance. According to Tomasetti et al. [77] more than half of somatic mutations in self-renewing tissues are already present before the onset of neoplasia, and the number of mutations correlates with the age of the patient. It is plausible that some of these mutations could drive drug resistance. Indeed, several experimental and theoretical studies have reached the conclusion that a small number of cells resistant to any targeted agent are always present in large solid tumors at the start of therapy and that these cells clonally expand once therapy is administered [78]. If this is the case, then treatment with multiple therapy could be more effective in delaying the onset of resistance. In accordance, recent mathematical modeling suggests that dual therapy results in higher long-term disease control for most patients, whereas for some patients with larger disease burdens triple therapy would be more effective [79, 80]. Nevertheless, one constraint on this approach is the expectable higher systemic toxicity with multiple drug regimens. Hence, ideally one should detect a tumor at the earliest stage possible and comparison of tumors should be performed with similar mutational landscapes, more likely correlated with age. Current efforts are beginning to address this issue, notably in the potential use of circulating tumor cells (CTCs) to detect tumors earlier [81].

Hence, it would be desirable to use a combination of genomics, proteomics, and functional assays to evaluate the mechanisms underlying drug resistance. Unfortunately, the difficulty in accessing tumors and the low amount of biological material available from high-grade tumor specimens preclude this approach. Thus, the usage of drug-resistant cell lines in vitro has been invaluable in elucidating specific resistance pathways, and shall continue to be so.

Finally, we should take into account the dynamic nature of resistance mechanisms. One of the reasons for failure to eradicate tumor cells could well lie in the successive alternation of one resistance mechanism with another, as cells proliferate in vivo and adapt to drug regimens. This would mean that current strategies to circumvent drug resistance would have to depend on continuous monitoring of patients and prescription of a cocktail of chemotherapeutic drugs, each targeting one or more of known drug resistance pathways. The feasibility of such an approach, especially in what regards toxicity and efficacy, is not predictable. Ideally, the earliest that the tumor is detected the lower the heterogeneity of tumor cells, and the more successful the therapy should be. However, in the long term, drug resistance is unfortunately probably inevitable [2].

References

1. Gottesman MM (2002) Mechanisms of cancer drug resistance. Annu Rev Med 53:615–627

2. Rodrigues AS, Dinis J, Gromicho M, Martins M, Laires A, Rueff J (2012) Genomics and cancer drug resistance. Curr Pharm Biotechnol 13:651–673

3. Holohan C, Van Schaeybroeck S, Longley DB, Johnston PG (2013) Cancer drug resistance: an evolving paradigm. Nat Rev Cancer 13:714–726

4. Kim R, Emi M, Tanabe K (2007) Cancer immunoediting from immune surveillance to immune escape. Immunology 121:1–14

5. Longley DB, Johnston PG (2005) Molecular mechanisms of drug resistance. J Pathol 205:275–292

6. Cooper RA, Straus DJ (2012) Clinical guidelines, the politics of value, and the practice of medicine: physicians at the crossroads. J Oncol Pract 8:233–235

7. Guengerich FP (2006) Cytochrome P450s and other enzymes in drug metabolism and toxicity. AAPS J 8:E101–E111

8. Palma BB, Silva ESM, Urban P, Rueff J, Kranendonk M (2013) Functional characterization of eight human CYP1A2 variants: the role of cytochrome b5. Pharmacogenet Genomics 23:41–52

9. Higgins MJ, Rae JM, Flockhart DA, Hayes DF, Stearns V (2009) Pharmacogenetics of tamoxifen: who should undergo CYP2D6 genetic testing? J Natl Compr Canc Netw 7:203–213

10. Marusyk A, Polyak K (2010) Tumor heterogeneity: causes and consequences. Biochim Biophys Acta 1805:105–117

11. Polyak K (2011) Heterogeneity in breast cancer. J Clin Invest 121:3786–3788

12. Salk JJ, Fox EJ, Loeb LA (2010) Mutational heterogeneity in human cancers: origin and consequences. Annu Rev Pathol 5:51–75

13. Baguley BC (2011) The paradox of cancer cell apoptosis. Front Biosci (Landmark Ed) 16: 1759–1767

14. Burrell RA, McGranahan N, Bartek J, Swanton C (2013) The causes and consequences of genetic heterogeneity in cancer evolution. Nature 501:338–345

15. Gerlinger M, Rowan AJ, Horswell S, Larkin J, Endesfelder D, Gronroos E, Martinez P, Matthews N, Stewart A, Tarpey P, Varela I, Phillimore B, Begum S, McDonald NQ, Butler A, Jones D, Raine K, Latimer C, Santos CR, Nohadani M, Eklund AC, Spencer-Dene B, Clark G, Pickering L, Stamp G, Gore M, Szallasi Z, Downward J, Futreal PA, Swanton C (2012) Intratumor heterogeneity and branched evolution revealed by multiregion sequencing. N Engl J Med 366:883–892

16. Castano Z, Fillmore CM, Kim CF, McAllister SS (2012) The bed and the bugs: interactions between the tumor microenvironment and cancer stem cells. Semin Cancer Biol 22:462–470

17. Sharma SV, Settleman J (2007) Oncogene addiction: setting the stage for molecularly targeted cancer therapy. Genes Dev 21:3214–3231

18. Fleck LM (2013) "Just caring": can we afford the ethical and economic costs of circumventing cancer drug resistance? J Pers Med 3:124–143

19. Weinstein IB (2000) Disorders in cell circuitry during multistage carcinogenesis: the role of homeostasis. Carcinogenesis 21:857–864

20. Weinstein IB, Joe A (2008) Oncogene addiction. Cancer Res 68:3077–3080, discussion 3080

21. Weinstein IB, Joe AK (2006) Mechanisms of disease: oncogene addiction—a rationale for molecular targeting in cancer therapy. Nat Clin Pract Oncol 3:448–457

22. Weinstein IB (2002) Cancer. Addiction to oncogenes—the Achilles heal of cancer. Science 297:63–64

23. Borst P (2012) Cancer drug pan-resistance: pumps, cancer stem cells, quiescence, epithelial to mesenchymal transition, blocked cell death pathways, persisters or what? Open Biol 2:120066

24. da Silva SD, Hier M, Mlynarek A, Kowalski LP, Alaoui-Jamali MA (2012) Recurrent oral cancer: current and emerging therapeutic approaches. Front Pharmacol 3:149

25. Kim JE, Leung E, Baguley BC, Finlay GJ (2013) Heterogeneity of expression of epithelial-mesenchymal transition markers in melanocytes and melanoma cell lines. Front Genet 4:97

26. Tien JC, Xu J (2012) Steroid receptor coactivator-3 as a potential molecular target for cancer therapy. Expert Opin Ther Targets 16:1085–1096

27. Ma G, Ren Y, Wang K, He J (2011) SRC-3 has a role in cancer other than as a nuclear receptor coactivator. Int J Biol Sci 7:664–672

28. Vadlapatla RK, Vadlapudi AD, Pal D, Mitra AK (2013) Mechanisms of drug resistance in cancer chemotherapy: coordinated role and regulation of efflux transporters and metabolizing enzymes. Curr Pharm Des 19:7126–7140

29. Robey RW, Polgar O, Deeken J, To KW, Bates SE (2007) ABCG2: determining its relevance in clinical drug resistance. Cancer Metastasis Rev 26:39–57

30. Lemos C, Jansen G, Peters GJ (2008) Drug transporters: recent advances concerning BCRP and tyrosine kinase inhibitors. Br J Cancer 98:857–862

31. Jung KA, Choi BH, Kwak MK (2015) The c-MET/PI3K signaling is associated with cancer resistance to doxorubicin and photodynamic therapy by elevating BCRP/ABCG2 expression. Mol Pharmacol 87:465–476

32. Marie JP, Zittoun R, Sikic BI (1991) Multidrug resistance (mdr1) gene expression in adult acute leukemias: correlations with treatment outcome and in vitro drug sensitivity. Blood 78:586–592

33. Legrand O, Simonin G, Perrot JY, Zittoun R, Marie JP (1998) Pgp and MRP activities using calcein-AM are prognostic factors in adult acute myeloid leukemia patients. Blood 91:4480–4488

34. Wuchter C, Leonid K, Ruppert V, Schrappe M, Buchner T, Schoch C, Haferlach T, Harbott J, Ratei R, Dorken B, Ludwig WD (2000) Clinical significance of P-glycoprotein expression and function for response to induction chemotherapy, relapse rate and overall survival in acute leukemia. Haematologica 85:711–721

35. Benderra Z, Faussat AM, Sayada L, Perrot JY, Chaoui D, Marie JP, Legrand O (2004) Breast cancer resistance protein and P-glycoprotein in 149 adult acute myeloid leukemias. Clin Cancer Res 10:7896–7902

36. Steinbach D, Legrand O (2007) ABC transporters and drug resistance in leukemia: was P-gp nothing but the first head of the Hydra? Leukemia 21:1172–1176

37. Steinbach D, Furchtbar S, Sell W, Lengemann J, Hermann J, Zintl F, Sauerbrey A (2003) Contrary to adult patients, expression of the

multidrug resistance gene (MDR1) fails to define a poor prognostic group in childhood AML. Leukemia 17:470–471

38. Gromicho M, Magalhaes M, Torres F, Dinis J, Fernandes AR, Rendeiro P, Tavares P, Laires A, Rueff J, Sebastiao RA (2013) Instability of mRNA expression signatures of drug transporters in chronic myeloid leukemia patients resistant to imatinib. Oncol Rep 29:741–750

39. Gromicho M, Dinis J, Magalhães M, Fernandes A, Tavares P, Laires A, Rueff J, Rodrigues A (2011) Development of imatinib and dasatinib resistance: dynamics of the drug transporters expression ABCB1, ABCC1, ABCG2, MVP and SLC22A1. Leuk Lymphoma 52:1980–1990

40. Greaves M, Maley CC (2012) Clonal evolution in cancer. Nature 481:306–313

41. Ohta K, Hoshino H, Wang J, Ono S, Iida Y, Hata K, Huang SK, Colquhoun S, Hoon DS (2015) MicroRNA-93 activates c-Met/PI3K/Akt pathway activity in hepatocellular carcinoma by directly inhibiting PTEN and CDKN1A. Oncotarget 6:3211–3224

42. Haenisch S, Werk AN, Cascorbi I (2014) MicroRNAs and their relevance to ABC transporters. Br J Clin Pharmacol 77:587–596

43. Silva S, Moita R, Azevedo A, Gouveia R, Manita I, Pina J, Rueff J, Gaspar J (2007) Menopausal age and XRCC1 gene polymorphisms: role in breast cancer risk. Cancer Detect Prev 31:303–309

44. Bastos HN, Antao MR, Silva SN, Azevedo AP, Manita I, Teixeira V, Pina JE, Gil OM, Ferreira TC, Limbert E, Rueff J, Gaspar JF (2009) Association of polymorphisms in genes of the homologous recombination DNA repair pathway and thyroid cancer risk. Thyroid 19:1067–1075

45. Conde J, Silva S, Azevedo A, Teixeira V, Pina J, Rueff J, Gaspar J (2009) Association of common variants in mismatch repair genes and breast cancer susceptibility: a multigene study. BMC Cancer 9:344

46. Santos LS, Gomes BC, Gouveia R, Silva SN, Azevedo AP, Camacho V, Manita I, Gil OM, Ferreira TC, Limbert E, Rueff J, Gaspar JF (2013) The role of CCNH Val270Ala (rs2230641) and other nucleotide excision repair polymorphisms in individual susceptibility to well-differentiated thyroid cancer. Oncol Rep 30:2458–2466

47. Santos LS, Branco SC, Silva SN, Azevedo AP, Gil OM, Manita I, Ferreira TC, Limbert E, Rueff J, Gaspar JF (2012) Polymorphisms in base excision repair genes and thyroid cancer risk. Oncol Rep 28:1859–1868

48. Gomes BC, Silva SN, Azevedo AP, Manita I, Gil OM, Ferreira TC, Limbert E, Rueff J, Gaspar JF (2010) The role of common variants of non-homologous end-joining repair genes XRCC4, LIG4 and Ku80 in thyroid cancer risk. Oncol Rep 24:1079–1085

49. Silva SN, Tomar M, Paulo C, Gomes BC, Azevedo AP, Teixeira V, Pina JE, Rueff J, Gaspar JF (2010) Breast cancer risk and common single nucleotide polymorphisms in homologous recombination DNA repair pathway genes XRCC2, XRCC3, NBS1 and RAD51. Cancer Epidemiol 34:85–92

50. Silva SN, Gomes BC, Rueff J, Gaspar JF (2011) DNA repair perspectives in thyroid and breast cancer: the role of DNA repair polymorphisms. In: Vengrova S (ed) DNA repair and human health. InTech. doi:10.5772/22944

51. Rodrigues AS, Gomes BC, Martins C, Gromicho M, Oliveira NG, Guerreiro PS, Rueff J (2013) DNA repair and resistance to cancer therapy. In: Chen C (ed) DNA repair and resistance to cancer therapy, new research directions in DNA repair. Intech. doi:10.5772/53952

52. Atchley DP, Albarracin CT, Lopez A, Valero V, Amos CI, Gonzalez-Angulo AM, Hortobagyi GN, Arun BK (2008) Clinical and pathologic characteristics of patients with BRCA-positive and BRCA-negative breast cancer. J Clin Oncol 26:4282–4288

53. Akashi-Tanaka S, Watanabe C, Takamaru T, Kuwayama T, Ikeda M, Ohyama H, Mori M, Yoshida R, Hashimoto R, Terumasa S, Enokido K, Hirota Y, Okuyama H, Nakamura S (2015) BRCAness predicts resistance to taxane-containing regimens in triple negative breast cancer during neoadjuvant chemotherapy. Clin Breast Cancer 15:80–85

54. Dietlein F, Reinhardt HC (2014) Molecular pathways: exploiting tumor-specific molecular defects in DNA repair pathways for precision cancer therapy. Clin Cancer Res 20:5882–5887

55. Kinzler KW, Vogelstein B (1997) Cancer-susceptibility genes. Gatekeepers and caretakers. Nature 386(761):763

56. Dexheimer TS (2013) DNA repair pathways and mechanisms. In: Mathews LA, Cabarcas SM, Hurt EM (eds) DNA repair of cancer stem cells. Springer, Dordrecht, Netherlands, pp 19–32. doi:10.1007/978-94-007-4590-2_2

57. Lehner B (2011) Molecular mechanisms of epistasis within and between genes. Trends Genet 27:323–331

58. Alluri PG, Speers C, Chinnaiyan AM (2014) Estrogen receptor mutations and their role in breast cancer progression. Breast Cancer Res 16:494

59. Garcia-Becerra R, Santos N, Diaz L, Camacho J (2012) Mechanisms of resistance to endocrine therapy in breast cancer: focus on signaling pathways, miRNAs and genetically based resistance. Int J Mol Sci 14:108–145

60. Yan J, Tsai SY, Tsai MJ (2006) SRC-3/AIB1: transcriptional coactivator in oncogenesis. Acta Pharmacol Sin 27:387–394

61. Wilting RH, Dannenberg J-H (2012) Epigenetic mechanisms in tumorigenesis, tumor cell heterogeneity and drug resistance. Drug Resist Updat 15:21–38

62. Theodora F (2014) Tumor plasticity driven by spatial micro-heterogeneity in the tumor microenvironment contributes to therapy resistance. J Carcinog Mutagen. doi:10.4172/2157-2518.S8-006

63. Catalano V, Turdo A, Di Franco S, Dieli F, Todaro M, Stassi G (2013) Tumor and its microenvironment: a synergistic interplay. Semin Cancer Biol 23:522–532

64. Brown JM, Giaccia AJ (1998) The unique physiology of solid tumors: opportunities (and problems) for cancer therapy. Cancer Res 58:1408–1416

65. Straussman R, Morikawa T, Shee K, Barzily-Rokni M, Qian ZR, Du J, Davis A, Mongare MM, Gould J, Frederick DT, Cooper ZA, Chapman PB, Solit DB, Ribas A, Lo RS, Flaherty KT, Ogino S, Wargo JA, Golub TR (2012) Tumour micro-environment elicits innate resistance to RAF inhibitors through HGF secretion. Nature 487:500–504

66. Srivastava S, Grizzle WE (2010) Biomarkers and the genetics of early neoplastic lesions. Cancer Biomark 9:41–64

67. Weber F, Shen L, Fukino K, Patocs A, Mutter GL, Caldes T, Eng C (2006) Total-genome analysis of BRCA1/2-related invasive carcinomas of the breast identifies tumor stroma as potential landscaper for neoplastic initiation. Am J Hum Genet 78:961–972

68. Playford RJ (2001) Landscaper seeks remunerative position. Gut 48:594–595

69. Vogelstein B, Papadopoulos N, Velculescu VE, Zhou S, Diaz LA Jr, Kinzler KW (2013) Cancer genome landscapes. Science 339:1546–1558

70. Cgar N (2012) Comprehensive molecular portraits of human breast tumours. Nature 490:61–70

71. Hanahan D, Weinberg RA (2011) Hallmarks of cancer: the next generation. Cell 144:646–674

72. Haber DA, Gray NS, Baselga J (2011) The evolving war on cancer. Cell 145:19–24

73. Kaiser J (2011) Combining targeted drugs to stop resistant tumors. Science 331:1542–1545

74. Gottesman MM, Fojo T, Bates SE (2002) Multidrug resistance in cancer: role of ATP-dependent transporters. Nat Rev Cancer 2:48–58

75. Ramos P, Bentires-Alj M (2015) Mechanism-based cancer therapy: resistance to therapy, therapy for resistance. Oncogene 34:3617–3626

76. Jones S, Chen WD, Parmigiani G, Diehl F, Beerenwinkel N, Antal T, Traulsen A, Nowak MA, Siegel C, Velculescu VE, Kinzler KW, Vogelstein B, Willis J, Markowitz SD (2008) Comparative lesion sequencing provides insights into tumor evolution. Proc Natl Acad Sci U S A 105:4283–4288

77. Tomasetti C, Vogelstein B, Parmigiani G (2013) Half or more of the somatic mutations in cancers of self-renewing tissues originate prior to tumor initiation. Proc Natl Acad Sci U S A 110:1999–2004

78. Diaz LA Jr, Williams RT, Wu J, Kinde I, Hecht JR, Berlin J, Allen B, Bozic I, Reiter JG, Nowak MA, Kinzler KW, Oliner KS, Vogelstein B (2012) The molecular evolution of acquired resistance to targeted EGFR blockade in colorectal cancers. Nature 486:537–540

79. Bozic I, Reiter JG, Allen B, Antal T, Chatterjee K, Shah P, Moon YS, Yaqubie A, Kelly N, Le DT, Lipson EJ, Chapman PB, Diaz LA Jr, Vogelstein B, Nowak MA (2013) Evolutionary dynamics of cancer in response to targeted combination therapy. eLife 2:e00747

80. Komarova NL, Katouli AA, Wodarz D (2009) Combination of two but not three current targeted drugs can improve therapy of chronic myeloid leukemia. PLoS One 4:e4423

81. Polzer B, Medoro G, Pasch S, Fontana F, Zorzino L, Pestka A, Andergassen U, Meier-Stiegen F, Czyz ZT, Alberter B, Treitschke S, Schamberger T, Sergio M, Bregola G, Doffini A, Gianni S, Calanca A, Signorini G, Bolognesi C, Hartmann A, Fasching PA, Sandri MT, Rack B, Fehm T, Giorgini G, Manaresi N, Klein CA (2014) Molecular profiling of single circulating tumor cells with diagnostic intention. EMBO Mol Med 6(11):1371–1386

Chapter 2

Classical and Targeted Anticancer Drugs: An Appraisal of Mechanisms of Multidrug Resistance

Bruce C. Baguley

Abstract

The mechanisms by which tumor cells resist the action of multiple anticancer drugs, often with widely different chemical structures, have been pursued for more than 30 years. The identification of P-glycoprotein (P-gp), a drug efflux transporter protein with affinity for multiple therapeutic drugs, provided an important potential mechanism and further work, which identified other members of ATP-binding cassette (ABC) family that act as drug transporters. Several observations, including results of clinical trials with pharmacological inhibitors of P-gp, have suggested that mechanisms other than efflux transporters should be considered as contributors to resistance, and in this review mechanisms of anticancer drug resistance are considered more broadly. Cells in human tumors exist is a state of continuous turnover, allowing ongoing selection and "survival of the fittest." Tumor cells die not only as a consequence of drug therapy but also by apoptosis induced by their microenvironment. Cell death can be mediated by host immune mechanisms and by nonimmune cells acting on so-called death receptors. The tumor cell proliferation rate is also important because it controls tumor regeneration. Resistance to therapy might therefore be considered to arise from a reduction of several distinct cell death mechanisms, as well as from an increased ability to regenerate. This review provides a perspective on these mechanisms, together with brief descriptions of some of the methods that can be used to investigate them in a clinical situation.

Key words P-glycoprotein, Cytokinetics, Apoptosis, Repopulation, Immune checkpoints

1 Introduction

The appearance of resistance to cancer therapy is hugely distressing for cancer patients; it may occur at the outset of drug treatment, as is frequently the case with tumors such as glioblastoma and pancreatic cancer, or may develop following initial response to successful first-line therapy. A common clinical experience is that the chance of response to a further drug or drug combination decreases with each relapse. In some cases, the mechanisms of resistance can be identified in molecular terms; for example, resistance to the cytotoxic drug temozolomide may be a consequence of expression of the DNA repair enzyme MGMT [1], and resistance to a targeted drug acting on a mutant BRAF protein may be a consequence of

José Rueff and António Sebastião Rodrigues (eds.), *Cancer Drug Resistance: Overviews and Methods*, Methods in Molecular Biology, vol. 1395, DOI 10.1007/978-1-4939-3347-1_2, © Springer Science+Business Media New York 2016

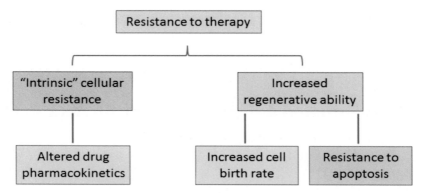

Fig. 1 Examples of resistance to therapy. (*Left-hand side*) Tumor cells may be non-responsive to either individual drugs or to groups of drugs because of a lack of expression of appropriate drug targets, or because of expression of cellular transport mechanisms that restrict access of the drug to the target. Tumor cells can also reside in "pharmacological sanctuaries" where diffusion limits access of the drug to the tumor. (*Right-hand side*) Tumor cells may be damaged by therapy but have a reduced rate of cell death. Alternatively they may be killed but surviving cells regenerate more effectively

expression of alternative signaling proteins in the RAF pathway [2]. In most clinical cases, the basis of resistance is not clearly defined and tumor progression often appears to be accompanied by resistance to all available drugs. The term "multidrug resistance" (MDR), which is also applied to multidrug-resistant microbial infections [3], has often been used to describe this situation. An enormous amount of work on cancer resistance is currently being undertaken, with over 1000 new publications each month, and this review can provide only a perspective on the field.

There are many possible reasons for resistance to cancer treatment and, as summarized in Fig. 1, they can be divided into two broad (and partially overlapping) categories. The first, which we have called "intrinsic resistance," involves a decreased ability of a therapeutic agent to induce cellular damage that is potentially cytostatic or cytotoxic to cancer cells. The second category reflects a dynamic response, i.e., the life and death responses of cancer cells that govern the repopulation of the tumor following therapy. Two hypothetical examples of resistance are illustrated diagrammatically in Fig. 2; here hypothetical tumor populations have potential population-doubling times of either 14 days (Fig. 2a, b) or 7 days (Fig. 2c). For the sensitive tumor population (Fig. 2a), each cycle of treatment (administered weekly in this example) reduces the viable population by 90 %, and the surviving population cannot completely regenerate in the interval between successive therapies. Thus, after five cycles, the surviving population is reduced by 99.9 %. For an intrinsically resistant tumor population (Fig. 2b), each cycle of treatment reduces the population by 30 % rather than 90 %; now, the surviving population can regenerate during the interval between successive cycles of treatment and no lasting

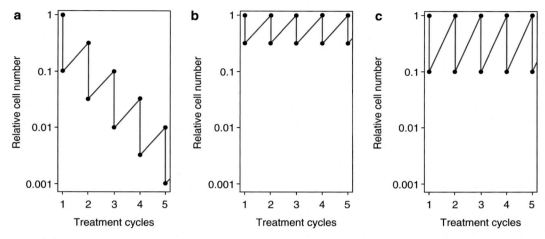

Fig. 2 Models of multiple drug resistance. (**a**) Tumor is sensitive to therapy and multiple applications result in the progressive reduction of the tumor population. (**b**) Tumor is partially resistant to therapy but the tumor population can regenerate in the time between successive therapeutic doses. (**c**) Tumor is sensitive to therapy but has an increased ability to regenerate between successive therapeutic doses, negating the therapeutic effect. See text for details

therapeutic effect is observed. For a dynamically resistant population (Fig. 2c), the cancer cell population remains intrinsically sensitive to therapy, as in Fig. 2a, but the tumor population can repopulate more effectively because the population-doubling time is 7 days rather than 14 days. Thus, the tumor cell population can again regenerate during the interval between successive treatment cycles and no lasting therapeutic effect is observed (Fig. 2c). These concepts have been discussed previously [4, 5].

Most previous reviews on multidrug resistance have focused on resistance of tumor cells to cytotoxic drugs. With the development of targeted cancer therapies, it is important to examine a broader range of resistance mechanisms and to determine which are most relevant to human cancer. This review commences with a description and discussion of the resistance mechanisms and continues by examining some of the experimental protocols that can be used to study these mechanisms.

2 Resistance Involving Altered Drug Pharmacokinetics

Resistance can also occur because of increased drug clearance, or by decreased diffusion, in both cases leading to a reduced amount of drug entering the cell, although resistance may apply only to one or a small number of anticancer drugs. An early finding was that patients with acute myeloblastic leukemia who failed to respond to the drug cytosine arabinoside had a shorter plasma half-life, apparently because of increased expression of the drug-metabolizing enzyme cytidine deaminase [6]. This led to the consideration of pharmacological factors in the optimization of treatment with this drug [7]. The same principle has been applied to patients treated with the drug cisplatin; individual variations in the pharmacokinetics of this drug have been allowed for in treatment protocols by basing dose on the area under the plasma concentration-time curve (AUC) [8]. Tumor pharmacokinetics of anticancer drugs in solid tumors are highly dependent on both drug diffusion rates and drug diffusion distances; these can be modeled experimentally using a three-dimensional matrix based on solid tumor imaging [9]. Diffusion barriers within tumor tissue, caused for instance by a low vascular density, can give rise to a pharmacokinetic "sanctuary" and to pockets of resistant cancer cells.

3   Intrinsic Multidrug Resistance Mechanisms

3.1   Resistance Mediated by P-gp

An important early step towards our understanding of drug resistance was made with the identification of a single protein, over-expression of which was accompanied by increased resistance to a variety of structurally unrelated anticancer drugs [10–12]. This protein was given the term P-glycoprotein (P-gp; MDR1; gene *ABCB1*) because chemical analysis showed it to contain multiple oligosaccharides attached to the protein. The development of an antibody to P-gp allowed the distribution of protein, both within single cells and in different organs of the body, to be studied. P-gp was initially found to be associated with the plasma membrane of resistant cultured cancer cells and further structural and biochemical studies led to the formulation of a molecular model where the protein actively transported a variety of drugs and other molecules, typically those containing hydrophobic and basically charged features, out of the cell. The P-gp transporter was embedded within the plasma membrane with the polysaccharide chains on the external cell surface and ATP-binding protein domains on the cytoplasmic side. P-gp had a tandemly duplicated structure with each half containing six potential lipophilic transmembrane domains and one nucleotide-binding site [13]. Molecular structural studies indicated that the protein could adopt two main conformations, a looser "open" form and a tightly twisted "closed" form, with the transition to the tightly twisted form driven by ATP hydrolysis. P-gp, like the protein albumin, has the ability to bind to a variety of small-molecular-weight molecules, which are generally categorized by the presence of lipophilic and/or basically charged motifs. A simple model for the action of P-gp in the membrane suggests that it functions a little like a floor mop; the open form, with its multiple transmembrane regions, can bind to a range of molecules, but a twist in conformation leads to a closed form that lacks these binding sites, allowing the protein to "squeeze out" attached molecules and

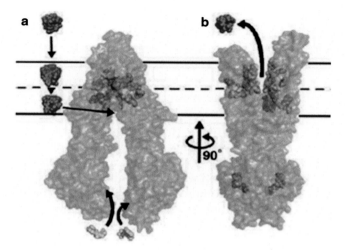

Fig. 3 Model for the action of P-gp, reproduced from the original article (Aller et al. [14]), with the permission of the publisher. P-gp protein is embedded in a membrane (either the plasma membrane or an organelle membrane) and has two conformations. In the "open" form it can interact and trap a variety of substrate molecules that enter from membrane sites. In the "closed" form it excludes these substrate molecules and the transition from the open to the closed form requires energy, which is derived by ATP hydrolysis. See text for further details

discharge them from the membrane surface. A molecular model of P-gp, as previously proposed [14], is shown in Fig. 3.

Subsequent research has indicated that P-gp is not always located on the plasma membrane; in some cell lines it is found in intracellular organelles including lysosomes, the Golgi apparatus, and the nuclear envelope [15, 16]. Here, P-gp acts on substrates to transport them from the cytoplasm to the lumen of an organelle, meaning that resistance is mediated by sequestration of drug into vesicles rather than direct outward transport. Sequestered drug can in turn be released from the cell by exocytosis.

A number of "classical" anticancer drugs, including anthracyclines such as doxorubicin, epirubicin, and daunorubicin; epipodophyllotoxins such as etoposide and teniposide; and taxanes such as paclitaxel and docetaxel, are substrates for P-gp-mediated transport [13]. The development of targeted anticancer therapies such as inhibitors of the epidermal growth factor receptor tyrosine kinase [17] raises the question of whether the efficacy of these drugs is also affected by expression of P-gp. Here the situation is complex because while such drugs can be substrates for P-gp [18] they may also antagonize P-gp function [19]. In one study, brain tissue AUC values for the drug erlotinib were found to increase by 3.8-fold in mice lacking expression of P-gp [20], consistent with a role of P-gp in intrinsic sensitivity.

3.2 Resistance Mediated by Other Transporters

The discovery of P-gp was followed by the identification of a second transporter, designated MRP1 (ABCC1), and its relative MRP2 (ABCC2), which were also associated with the MDR phenotype [21, 22]. As with P-gp, the subcellular distribution of these transporters has been found to extend to cellular organelles as well as the plasma membrane. MRP1 and MRP2 can be distinguished from P-gp in being able to couple drug transport to glutathione transport [22]. The action of these transporters is also coordinated with another type of resistance mechanism whereby a variety of cytotoxic agents, usually lipophilic, are metabolized by conjugation with hydrophilic molecules such as glucuronic acid [23]. Such conjugates not only have reduced activity as a cytotoxic species, but also have increased affinity for the transporter.

Subsequent studies on ABC transporters have identified the breast cancer resistance protein BCRP (ABCG2) as a further transporter associated with drug resistance. BCRP differs from P-gp, MRP1, and MRP2 in having two subunits rather than the tandemly repeated form of the other transporters. A superfamily of proteins, designated as ATP-binding cassette ("ABC") transporters, has now been identified; it comprises the products of 48 genes and encompasses a broad variety of molecular structures [24]. Subfamily A includes P-gp, subfamily B includes MRP1 and MRP2, and subfamily G includes BCRP. Members of the ABC transporter family carry out a wide variety of functions, many essential, in normal tissue [22, 24] and only a few family members are well characterized in terms of mediating anticancer drug resistance. Molecules binding to MRP1 and MRP2 include a variety of natural products, such as vinblastine, vincristine, paclitaxel, docetaxel, doxorubicin, daunorubicin, epirubicin, and etoposide as well as the synthetic cytotoxic anticancer drugs mitoxantrone and methotrexate [13]. Targeted anticancer drugs can also be substrates for MRP1 and MRP2 [19] and cells transfected with the gene for MRP2 were found to exhibit 6.4-fold resistance to sorafenib but showed no change in susceptibility to the structurally related drug sunitinib [25]. In one study, brain tissue accumulation of erlotinib was found to be reduced in mice over-expressing BCRP [26].

3.3 Pharmacokinetic Consequences of Expression of ABC Transporters

While many studies on ABC transporters have been carried out using cultured cells, it is important to consider transporter action in the context of tumor tissue. Tumors generally have a multicellular organization in which the majority of cells are not adjacent to the vascular endothelium. Drugs must diffuse from the bloodstream to tumor cells either through the extracellular compartment of tumor tissue or by uptake/efflux by cells comprising the tumor tissue. Drugs can cross the plasma membranes of normal and tumor cells within the tumor tissue by either transporter-mediated or passive diffusion. ABC transporters can act on intracellular drug molecules either by exporting them out of the cell again or by sequestering them into vesicles, the contents of which are subsequently released by exocytosis into the extracellular compartment. If exocytosis occurs in a polarized fashion in a direction that is distal to the vascular supply, it will effectively promote distribution of drugs to other cells within the tumor tissue. Thus, depending on their subcellular distribution, ABC transporters can either increase or decrease the distribution of a drug in tumor tissue.

A pharmacological study of SN 28049, a new DNA-binding topoisomerase II poison [27], illustrates how drug transporters might influence tissue pharmacokinetics. Tumor tissue AUC values were evaluated in two murine tumors and three human melanoma xenografts and found to vary by over two orders of magnitude, with the murine colon 38 (MCA38) tumor showing the highest value [28]. Cultured colon 38 tumor cells showed strongly positive staining for MRP1 expression in cytoplasmic bodies [29] and although other explanations are possible, MRP1-mediated sequestration of SN 28049 in cytoplasmic vesicles may contribute to the high AUC value and long tumor tissue half-life in colon 38 tumor tissue. The vesicles could constitute a depot form, slow release which enhances the overall activity of SN 28049 against the colon 38 tumor [29]. Among the murine tumors and human melanoma xenografts tested, antitumor activity was related to the observed tumor tissue pharmacokinetics, consistent with this hypothesis [28].

3.4 Use of Inhibitors of ABC Transporters in Combination Chemotherapy

Since the cellular action of ABC drug transporters involves the ATP-dependent transport of these drugs out of the cytoplasm either to the cell exterior or into subcellular vesicles, the concept of inhibiting the action of P-gp in order to increase the cytoplasmic (and nuclear) concentration of a substrate anticancer drug presented a promising therapeutic strategy. The concept was first suggested more than 30 years ago [30] and initial studies were carried out in rodents using drugs such as verapamil (a Ca^{2+} channel blocker), cyclosporine (an immunosuppressive agent), tamoxifen (a steroid receptor antagonist), and calmodulin antagonists in conjunction with cytotoxic agents [31]. Most preclinical studies were carried out using cultured cells but some in vivo studies were reported [32]. Clinical trials identified a number of problems including alteration of the pharmacokinetics of administered cytotoxic drugs and consequent increases in toxicity. Consequent changes in dose made the efficacy of the co-administered transporter inhibitor difficult to assess. Subsequent studies aimed at increasing affinity of the inhibitor for the transporter and increasing its dose potency led to so-called second-generation inhibitors such as dexverapamil (an analogue of verapamil), valspodar (an analogue of cyclosporine), and biricodar (a pipecolinate derivative). Further development sought to minimize interaction with cytochrome P450 and to optimize individual transporters, and led to "third-generation" inhibitors such as elacridar, tariquidar, zosuquidar, and laniquidar [5, 33]. Many clinical trials have been carried out with ABC transporter inhibitors but as yet no definitively increased therapeutic benefit has been demonstrated. A small Phase II clinical study in breast cancer patients showed that co-administration of tariquidar showed limited ability to increase response to doxorubicin, paclitaxel, or docetaxel [34]. A larger Phase II clinical study in breast cancer patients treated with docetaxel with or without zosuquidar concluded that there were no differences in progression-free survival, overall survival, or response rate between the two groups of patients [35]. Ongoing difficulties include selection of appropriate tumors and measurement of the effects of the ABC transporter inhibitor on the pharmacokinetics of the cytotoxic drug.

3.5 Multidrug Resistance Mechanisms not Mediated by ABC Transporters

Several additional classes of resistance to multiple anticancer drugs of differing structures have been defined. One involves the modification of the enzyme DNA topoisomerase II, a target protein for cytotoxic action. Cells were identified that lacked ABC transporter expression and yet were resistant to the drugs doxorubicin, etoposide, and amsacrine, which have widely differing structures. The phenomenon was termed "atypical" multidrug resistance and the cause was traced to reduced activity of DNA topoisomerase II and consequent induction of DNA damage, which was essential for the cytotoxic activity of these drugs [36].

A second class of resistance involves the increased expression of a DNA repair enzyme which attenuates the cytotoxic activity of multiple agents. The discovery of the antitumor activity of nitrogen mustard (mechlorethamine) in 1945 led to testing of a large number of clinical anticancer agents whose activity depends mainly on the O^6-alkylation of the DNA constituent guanine [37]. These include melphalan, cyclophosphamide, ifosfamide, dacarbazine (DTIC), temozolomide, carmustine (BCNU), lomustine (CCNU), 1-(2-chloroethyl)-3-(4-methylcyclohexyl)-1-nitrosourea (MeCCNU), 1-(4-amino-2-methyl-pyrimidinyl)methyl-3(2-chloroethyl)-3-nitrosourea (ACNU); N-methyl-N-nitrosourea (MNU), N-ethyl-N-nitrosourea (ENU), procarbazine, and streptozotocin. A single enzyme, O^6-alkylguanine-DNA alkyltransferase (AGT), also known as methylguanine transferase (MGMT), acts to repair some of the DNA lesions induced by these drugs. An approach to overcoming this resistance is to co-administer an inhibitor of DNA O^6-alkylation; trials of one such inhibitor, O^6-benzylguanine, are currently under way [37, 38].

A third class of resistance involves activation of an alternate signaling pathway for cell proliferation and survival. Most examples are found in the use of targeted therapies; one example is provided in the MAP kinase pathway. Some tumors express a mutant form of the BRAF enzyme, one of the components of the MAP kinase pathway, and since cells become dependent on signaling by this overactive enzyme, their proliferation and survival are compromised by inhibitors of the mutant enzyme [39]. Resistance to multiple agents, including vemurafenib and dabrafenib, can then be mediated by up-regulation of CRAF, which provides an alternative pathway for proliferation and survival [40].

4 Resistance Involving Altered Tumor Cytokinetics

Tumor cells in a solid cancer generally grow in a latticelike "cage" of blood vessels, which contains not only tumor cells but also host cells, typically fibroblasts, tumor-associated macrophages, other cells, and stromal/capsular components [41]. The volume-doubling times of the vascular cages reflect those of the tumor itself, which from imaging studies in human cancers cover a broad range with a median of about 4 months [42]. The vascular cage expands by a number of mechanisms including the generation of new vascular endothelial cells (angiogenesis), the co-option of existing blood vessels of normal tissue by tumor cells, the recruitment of circulating endothelial precursor cells into the vascular, and the phenotypic conversion (vasculogenic mimicry) of tumor cells to a vascular phenotype [43]. Because the potential doubling times of human tumor cells comprising the tumor, typically about 6 days [44], are much shorter than that of the tumor itself, tumor cells within the vascular cage exist in a constant stage of turnover; on average, for every 100 tumor cells dividing, approximately 90 cells are lost [45]. Resistance to multiple agents, or perhaps to all agents, can occur by modulation of the dynamic balance between tumor cell birth and death.

In vivo potential doubling times of human tumor cells vary over quite a wide range [44], raising the question of why the most rapidly growing cells are not selected for during tumor evolution. However, in a solid tumor microenvironment with extensive cell turnover, cell death mechanisms dominate the selection of tumor cells that are least susceptible to cell death mechanisms and that will therefore have a survival advantage. Cell loss from the tumors, even in the absence of therapy, involves a diverse variety of mechanisms as shown in Fig. 4, and there are a corresponding number of control mechanisms. In contrast, increases in the tumor cell population occur almost exclusively by cell division, apart from a small number of tumor cells migrating from other sites (Fig. 4).

Apoptosis is likely to be the dominant mechanism for tumor cell loss and tumors can be characterized by their "apoptotic index" [46]. Tumor cells express so-called death receptors, such as Fas, DR4, and DR5 and interaction with the corresponding ligands (TRAIL/Apo2L and FasL), which are also expressed in both tumor cells, leads to apoptosis [47]; cell death may thus due to cell-cell proximity and thus occur as a consequence of crowding. Host immune mechanisms make an important potential

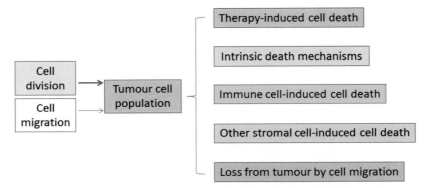

Fig. 4 Human tumors exhibit a high rate of cell turnover, which drives tumor evolution and also the development of resistance. One main mechanism (cell division) drives cell population increases but several mechanisms drive population decreases. Decreases in any of these mechanisms can therefore contribute to resistance. See text for further details

contribution to tumor cell loss. Cells may be killed by host T-lymphocytes in a complex mechanism that combines the release of FasL to activate death receptors and the release of cytotoxic granules that are taken up by the target cells [48]. Tumor-associated macrophages and dendritic cells also play important roles in immune cell-mediated tumor cell loss [49]. Tumor cells may die of other programmed cell death mechanisms [50] and may also be lost by migration out of the tumor by coupling with macrophages and export along collagen fibers into the bloodstream [51].

4.1 Resistance Arising from an Increased Rate of Tumor Cell Proliferation

A high tumor cell proliferation rate is important for resistance because it allows more efficient regeneration in the intervals between successive cycles of therapy (Fig. 2). This argument applies to radiotherapy and even surgery in addition to cytotoxic or targeted therapy. Clinically, more rapid proliferation is indicated by a shorter Tpot value, as determined in vivo [44, 52], and there is clinical evidence that higher proliferation rates are associated with shorter survival [53], particularly in the radiotherapy of head and neck tumors [54]. Cytokinetic data can also be obtained using an in vitro approach where surgical cancer samples are cultured for a short time (1 week) and the estimated cell proliferation rates are compared to clinical outcome. In two studies, one of ovarian cancer and one of glioma, decreased patient survival was significantly related to shorter culture cell cycle time [55].

4.2 Resistance to Apoptosis

Resistance to apoptosis has been described as one of the hallmarks of cancer [56] and increased resistance to the induction of apoptosis is an obvious mechanism that could apply to both cytotoxic and targeted anticancer therapeutic agents. Early experiments with the Lewis lung transplantable murine carcinoma (3LL) sought to clarify the role of apoptosis in resistance. Two variants of this tumor are one which had been maintained in vivo and recently grown in culture, and one that had been adapted to culture conditions over a considerable period. When grown to high cell density in culture, the first variant maintained a high proportion of S-phase cells and showed a high rate of cell loss, while the second variant entered a state of reduced proliferation with a low proportion of S-phase cells. These properties were echoed by those shown in vivo when the lines were grown as subcutaneous tumors; the tumors containing the more slowly growing cells were resistant to the cytotoxic drugs tested [57, 58].

There are several clinical examples where a low proliferation rate is associated with resistance to cancer chemotherapy [59–61]. This seems to conflict with reports that a higher proliferation rate is associated with reduced survival [53, 54]. However, it is quite possible that both occur because they operate on different time scales; a lower proliferation rate can be associated with resistance to apoptosis while a higher proliferation rate is associated with increased tumor regeneration. These considerations may help to explain why there is no consistent reported relationship between tumor cell proliferation and sensitivity to therapy.

4.3 Resistance Because of Loss of Host Immunity Mechanisms

As shown in Fig. 4, host immunity contributes to mechanisms of cell loss within tumors. The extent of this contribution in individual tumors is still not clear but if it is a major contribution, then its loss will have major significance for the outcome of cancer therapy. Put another way, loss of tumor immunity can lead to tumor progression. The potential importance of antitumor tumor immunity in a murine system was illustrated by a study of tumor responses to the cytotoxic drug gemcitabine [62]. Here, the response of a series of murine tumors to this drug was found to be not related to the intrinsic sensitivity of tumor-derived cultured cell line to gemcitabine, but rather to biomarkers for the immunogenicity of the tumor. This study suggested not only that host immunity played a major role in tumor regression, but also that gemcitabine itself might trigger host immune responses. Subsequent work has indicated that a number of anticancer agents including gemcitabine, doxorubicin, cisplatin, and cyclophosphamide stimulate host immunity [63, 64], further supporting the hypothesis that immune mechanisms need to be considered in the context of chemotherapy and resistance [65].

More recently, clinical studies have highlighted the role of immune checkpoints in cancer immunology. T-lymphocytes have clearly delineated mechanisms by which they can kill tumor cells, but their potential to kill normal cells in autoimmune reactions must also be carefully regulated. Some of the main mechanisms of regulation involve the so-called immune checkpoints, where cell-surface proteins such as CTLA-4 or PD-1 interact with CD80/86 or PD-L1/PD-L2, respectively, to suppress T-cell responses. The significance of these processes has been highlighted by recent studies with immune checkpoint inhibitors such as ipilimumab and nivolumab, which are engineered antibodies against CTLA-4 and PD-1, respectively, as therapeutic agents aimed at combatting resistance caused by reduced T-cell responses [66].

4.4 Tumor Tissue Heterogeneity and Resistance

Heterogeneity is a hallmark of human tumors [67] and is well illustrated by the analysis of renal cell carcinoma, where tumor subpopulations with distinct gene expression profiles, and consequently different predictions of clinical outcome, are obtained from different biopsies of the same tumor [68]. Tumor heterogeneity can also be discerned in established tumor cell lines, as shown for MCF-7, a typical human breast cancer line. Growth of this line in the absence of estrogen signaling causes an immediate cessation of culture growth, followed several months later by the outgrowth of hormone-resistant cell lines. Surprisingly, the G_1-phase DNA content, median cell volume, and proliferation rates of the emerging variant lines were not the same as those of the parental cell line, strongly suggesting that they arose from expansion of pre-existing minor populations, rather than by metabolic adaptation of the parental population [69]. Changes in chromosome numbers, as well as chromosome translocations, fusions, and alterations caused by recombination events, are likely to lead to the continuous generation of genetically distinct variants during culture. These changes have been described specifically for the MCF-7 line [70].

As well as undergoing genetic variation, tumor cells can undergo reversible phenotypic switches that presumably arise from changes in the regulation of gene transcription. Two main categories of phenotypic switch have been described. The first reflects a change in the expression of stem cell characteristics and can be measured, for instance, by an increased ability of the cell population to proliferate indefinitely. The second switch, called the epithelial to mesenchymal transition (EMT), reflects a change towards more mesenchymal behavior and includes increased migratory and invasive potential. Within each category there may be differences in intrinsic cellular drug resistance [71, 72]. An important feature of these phenotypic switches is that they may be associated with multiple changes in resistance properties [71, 73].

5   General Protocols for Studying Resistance to Multiple Anticancer Drugs

The field of drug resistance is very broad and it would be impossible in a limited space to recommend protocols for aspects of resistance. The approach taken here is to review the general approaches that can be used to study resistance. Since the starting point should always be the cancer patient, discussion is directed towards protocols that might be applicable to clinical studies.

5.1 Assessment of Transport-Mediated Multidrug Resistance

The expression of ABC transporter proteins in biopsies of human tumor material can be investigated using standard histological techniques, but it should be kept in mind that staining intensity does not accurately reflect the activity of these proteins. Moreover, the subcellular location of these ABC transporters, as well as regulation by other signaling pathways, will have an effect on activity. The major challenge for the future is to develop robust in vivo methods to assess transport-related multiple drug resistance. In one approach [34], a technetium-labeled P-gp substrate, ^{99m}Tc-sestamibi (25–30 mCi per patient), was injected intravenously. Planar scintigraphic images of known tumor sites were taken after 10 min and 2 h to determine the rate of clearance. The same procedure was repeated after administration of the tariquidar, an inhibitor of ABC transporters, in order to determine its effect on ^{99m}Tc-sestamibi uptake. The tumor-to-background ratios were calculated for all tumor sites by measurement of sestamibi uptake within the visualized portion of the tumor, and comparing it with that of adjacent tissues that were without tumor involvement [34].

5.2 Assessment of Drug Resistance of Cultured Tumor Cells

Human tumor cell lines have provided the basis for a very large number of published studies on resistance mechanisms. It is important to realize that from the time they are isolated from surgical samples subjected to culture, tumor cells are subjected to severe selective pressures from their new environment, and can change their characteristics accordingly. One of the largest selective pressures is for rapid proliferation rate; the initial doubling times of surgical samples cultured from solid tumors have been measured over the first week of culture [55, 74] and cover a broad range (3 days to more than 2 months), which is similar to that of measurements of potential doubling times in vivo [44]. Development of cell lines from these surgical samples has been reported to be accompanied by a two- to threefold decrease in doubling times [45] consistent with selection of more rapidly growing variants. Another selective pressure is the presence in cultures of atmospheric oxygen concentrations; this leads to increased concentrations of reactive oxygen and consequent toxicity, but can be counteracted by the use of low-oxygen incubators in the derivation and maintenance of cell lines [75].

Not all surgical samples grow well in culture. Early studies showed that partially disaggregated samples of human metastatic melanoma grew with a high success rate in 96-well culture plates that had been coated with a thin layer of agarose to prevent proliferation of fibroblasts; the culture medium was supplemented with fetal bovine serum, insulin, transferrin, and selenite and cells were grown under 5 % oxygen [75]. Subsequent work showed that glioma cells and a range of carcinoma cells could be grown with moderate-to-good success rates using the culture medium, sometimes supplemented with growth factors. Tumor cell proliferation was assessed by uptake of ^{3}H-labeled thymidine into DNA of proliferating cells. This technique has the advantage that these cultures have a variable number of host cells, which can distort the evaluation of drug effects on the total cell population [74].

A further approach, which has also used 96-well culture plate technology and has been reported to have higher success rates with carcinoma samples [76], is to utilize cultured fibroblasts as feeder cells [77]. In this case proliferation was assessed by counting cell density. Like the technique described in the previous paragraph, this method can be used to screen for activity of both conventional cytotoxic agents and targeted therapeutics.

5.3 Transplanted Tumors in Animals

As with culture systems, the majority of reported studies have growth-established cell lines, sometimes drug-resistant cell lines as xenografts in immunodeficient mice in order to gain an understanding of in vivo resistance. However, some early studies have utilized samples of surgically removed tumor material to establish xenografts [78] and more recent work has extended this to a number of genetically characterized tumor types. Samples representing 18 distinct cancer pathologies were implanted within 24 h of surgical resection and implanted into immune-compromised nude mice with an overall take rate of 27 %. Tumors were found to retain their differentiation patterns and supporting stromal elements were preserved. Genes downregulated specifically in the tumor xenografts were enriched for pathways involved in host immune response, consistent with the immune deficiency status of the host [79].

One of the problems of this approach, as it is with cell lines, is that there is competition for survival among the tumor cells and that the most rapidly growing cells are likely to dominate. Furthermore, since first-generation xenografts will generally have to be transplanted into further mice to provide a sufficient number of tumors for measurement of resistance to multiple drugs, further selection for a proliferation rate will be made. Because these experiments are carried out in immunosuppressed mice, possible contributions of immune cell-mediated killing cannot be assessed.

5.4 Contribution of Host Immune Mechanisms in Individual Human Tumors

There is great current interest in the clinical evaluation of immune checkpoint inhibitors and most current studies are using survival or other clinical parameters as the main index for patient comparison [80]. However, there is a need for robust assays of the contribution of T-lymphocytes or of other immune mechanisms to clinical outcome. Clinical studies are still at an early stage, but the formulation of suitable assays could lead to their use to assess immune cell activity in tumor tissue before and after therapy, providing an approach to estimate the contribution of immune effects to response and thus to resistance.

6   Conclusions

The last 30 years has seen progressive change in our appreciation of the diversity of mechanisms of drug resistance, particularly of multiple drug resistance, in human cancer. However we still do not know, for any individual patient, whether a lack of observed response to therapy is due to drug-specific resistance mechanisms, to selection of tumor cells that are resistant to the induction of apoptosis by both conventional and targeted therapies, or to a generalized breakdown of tumor immunity; perhaps all three mechanisms contribute. What we do know is that tumor heterogeneity and growth kinetics are of great importance in the transitions towards resistance. Heterogeneity will be generated in tumor populations by chromosomal instability, errors in chromosome partitioning, and other factors, and individual cells are likely to vary in cell division rate, degree of resistance to apoptosis, and susceptibility to immune responses. Since potential tumor population doubling times can be as short as 3.2 days [44], a minor population (5 %) can under conditions appropriate for selective survival become a major population (80 %) in as little as 13 days. It is easy to underestimate the potential of tumor cells that are resistant to chemotherapy or to specific immune responses to be selected on this basis. There are many reports showing that the presence of an oncogenic mutations leads to resistance to apoptosis, for instance that for c-kit mutation in leukemia [81]. Because of tumor cell turnover, such resistant cells are eventually likely to dominate because of natural selection. The way in which tumor heterogeneity can so rapidly lead to resistance perhaps paints a rather bleak picture, but it should be remembered that the emergence of a particular phenotype by such selection can also lead to an opportunity for selective chemotherapy. A major challenge for the future is to develop methods to identify such phenotypes in the course of clinical treatment, so that individualized treatment can be based on appropriate biomarkers.

References

1. Wick W, Weller M, van den Bent M, Sanson M, Weiler M, von Deimling A, Plass C, Hegi M, Platten M, Reifenberger G (2014) MGMT testing--the challenges for biomarker-based glioma treatment. Nat Rev Neurol 10: 372–385

2. Spagnolo F, Ghiorzo P, Queirolo P (2014) Overcoming resistance to BRAF inhibition in BRAF-mutated metastatic melanoma. Oncotarget 5:10206–10221

3. Hughes D (2014) Selection and evolution of resistance to antimicrobial drugs. IUBMB Life 66:521–529

4. Durand RE (1993) Cell kinetics and repopulation during multifraction irradiation of spheroids: implications for clinical radiotherapy. Semin Radiat Oncol 3:105–114

5. Baguley BC (2010) Multidrug resistance in cancer. Methods Mol Biol 596:1–14

6. Baguley BC, Falkenhaug EM (1971) Plasma half-life of cytosine arabinoside (NSC-63878) in patients treated for acute myeloblastic leukemia. Cancer Chemother Rep 1(55):291–298

7. Momparler RL (2013) Optimization of cytarabine (ARA-C) therapy for acute myeloid leukemia. Exp Hematol Oncol 2:20

8. Newell DR, Eeles RA, Gumbrell LA, Boxall FE, Horwich A, Calvert AH (1989) Carboplatin and etoposide pharmacokinetics in patients with testicular teratoma. Cancer Chemother Pharmacol 23:367–372

9. Hicks KO, Pruijn FB, Secomb TW, Hay MP, Hsu R, Brown JM, Denny WA, Dewhirst MW, Wilson WR (2006) Use of three-dimensional tissue cultures to model extravascular transport and predict in vivo activity of hypoxia-targeted anticancer drugs. J Natl Cancer Inst 98:1118–1128

10. Juliano RL, Ling V (1976) A surface glycoprotein modulating drug permeability in Chinese hamster ovary cell mutants. Biochim Biophys Acta 455:152–162

11. Ling V (1997) Multidrug resistance: molecular mechanisms and clinical relevance. Cancer Chemother Pharmacol 40(Suppl):S3–S8

12. Gottesman MM, Ling V (2006) The molecular basis of multidrug resistance in cancer: the early years of P-glycoprotein research. FEBS Lett 580:998–1009

13. Endicott JA, Ling V (1989) The biochemistry of P-glycoprotein-mediated multidrug resistance. Annu Rev Biochem 58:137–171

14. Aller SG, Yu J, Ward A, Weng Y, Chittaboina S, Zhuo R, Harrell PM, Trinh YT, Zhang Q, Urbatsch IL, Chang G (2009) Structure of P-glycoprotein reveals a molecular basis for poly-specific drug binding. Science 323:1718–1722

15. Yamagishi T, Sahni S, Sharp DM, Arvind A, Jansson PJ, Richardson DR (2013) P-glycoprotein mediates drug resistance via a novel mechanism involving lysosomal sequestration. J Biol Chem 288:31761–31771

16. Molinari A, Calcabrini A, Meschini S, Stringaro A, Crateri P, Toccacieli L, Marra M, Colone M, Cianfriglia M, Arancia G (2002) Subcellular detection and localization of the drug transporter P-glycoprotein in cultured tumor cells. Curr Protein Pept Sci 3:653–670

17. Roskoski R Jr (2014) The ErbB/HER family of protein-tyrosine kinases and cancer. Pharmacol Res 79:34–74

18. Agarwal S, Sane R, Gallardo JL, Ohlfest JR, Elmquist WF (2010) Distribution of gefitinib to the brain is limited by P-glycoprotein (ABCB1) and breast cancer resistance protein (ABCG2)-mediated active efflux. J Pharmacol Exp Ther 334:147–155

19. Lainey E, Sebert M, Thepot S, Scoazec M, Bouteloup C, Leroy C, De Botton S, Galluzzi L, Fenaux P, Kroemer G (2012) Erlotinib antagonizes ABC transporters in acute myeloid leukemia. Cell Cycle 11:4079–4092

20. de Vries NA, Buckle T, Zhao J, Beijnen JH, Schellens JH, van Tellingen O (2012) Restricted brain penetration of the tyrosine kinase inhibitor erlotinib due to the drug transporters P-gp and BCRP. Invest New Drugs 30:443–449

21. Cole SP, Chanda ER, Dicke FP, Gerlach JH, Mirski SE (1991) Non-P-glycoprotein-mediated multidrug resistance in a small cell lung cancer cell line: evidence for decreased susceptibility to drug-induced DNA damage and reduced levels of topoisomerase II. Cancer Res 51:3345–3352

22. Cole SP (2014) Targeting multidrug resistance protein 1 (MRP1, ABCC1): past, present, and future. Annu Rev Pharmacol Toxicol 54:95–117

23. Tukey RH, Strassburg CP (2000) Human UDP-glucuronosyltransferases: metabolism, expression, and disease. Annu Rev Pharmacol Toxicol 40:581–616

24. ter Beek J, Guskov A, Slotboom DJ (2014) Structural diversity of ABC transporters. J Gen Physiol 143:419–435

25. Shibayama Y, Nakano K, Maeda H, Taguchi M, Ikeda R, Sugawara M, Iseki K, Takeda Y, Yamada K (2011) Multidrug resistance protein 2 implicates anticancer drug-resistance to sorafenib. Biol Pharm Bull 34:433–435

26. Elmeliegy MA, Carcaboso AM, Tagen M, Bai F, Stewart CF (2011) Role of ATP-binding cassette and solute carrier transporters in erlotinib CNS penetration and intracellular accumulation. Clin Cancer Res 17:89–99

27. Deady LW, Rodemann T, Zhuang L, Baguley BC, Denny WA (2003) Synthesis and cytotoxic activity of carboxamide derivatives of benzo[b][1,6]naphthyridines. J Med Chem 46:1049–1054

28. Lukka PB, Chen YY, Finlay GJ, Joseph WR, Richardson E, Paxton JW, Baguley BC (2013) Tumour tissue selectivity in the uptake and retention of SN 28049, a new topoisomerase II-directed anticancer agent. Cancer Chemother Pharmacol 72:1013–1022

29. Chen YY, Lukka PB, Joseph WR, Finlay GJ, Paxton JW, McKeage MJ, Baguley BC (2014) Selective cellular uptake and retention of SN 28049, a new DNA-binding topoisomerase II-directed antitumor agent. Cancer Chemother Pharmacol 74:25–35

30. Tsuruo T, Iida H, Tsukagoshi S, Sakurai Y (1981) Overcoming of vincristine resistance in P388 leukemia in vivo and in vitro through enhanced cytotoxicity of vincristine and vinblastine by verapamil. Cancer Res 41:1967–1972

31. Beck WT (1991) Modulators of P-glycoprotein-associated multidrug resistance. Cancer Treat Res 57:151–170

32. Boesch D, Gaveriaux C, Jachez B, Pourtier-Manzanedo A, Bollinger P, Loor F (1991) In

vivo circumvention of P-glycoprotein-mediated multidrug resistance of tumor cells with SDZ PSC 833. Cancer Res 51:4226–4233

33. Avendano C, Menendez JC (2002) Inhibitors of multidrug resistance to antitumor agents (MDR). Curr Med Chem 9:159–193

34. Pusztai L, Wagner P, Ibrahim N, Rivera E, Theriault R, Booser D, Symmans FW, Wong F, Blumenschein G, Fleming DR, Rouzier R, Boniface G, Hortobagyi GN (2005) Phase II study of tariquidar, a selective P-glycoprotein inhibitor, in patients with chemotherapy-resistant, advanced breast carcinoma. Cancer 104:682–691

35. Ruff P, Vorobiof DA, Jordaan JP, Demetriou GS, Moodley SD, Nosworthy AL, Werner ID, Raats J, Burgess LJ (2009) A randomized, placebo-controlled, double-blind phase 2 study of docetaxel compared to docetaxel plus zosuquidar (LY335979) in women with metastatic or locally recurrent breast cancer who have received one prior chemotherapy regimen. Cancer Chemother Pharmacol 64:763–768

36. Danks MK, Yalowich JC, Beck WT (1987) Atypical multiple drug resistance in a human leukemic cell line selected for resistance to teniposide (VM-26). Cancer Res 47:1297–1301

37. Pegg AE (2000) Repair of O(6)-alkylguanine by alkyltransferases. Mutat Res 462:83–100

38. Preuss I, Thust R, Kaina B (1996) Protective effect of O6-methylguanine-DNA methyltransferase (MGMT) on the cytotoxic and recombinogenic activity of different antineoplastic drugs. Int J Cancer 65:506–512

39. Wan PT, Garnett MJ, Roe SM, Lee S, Niculescu-Duvaz D, Good VM, Jones CM, Marshall CJ, Springer CJ, Barford D, Marais R (2004) Mechanism of activation of the RAF-ERK signaling pathway by oncogenic mutations of B-RAF. Cell 116:855–867

40. Montagut C, Sharma SV, Shioda T, McDermott U, Ulman M, Ulkus LE, Dias-Santagata D, Stubbs H, Lee DY, Singh A, Drew L, Haber DA, Settleman J (2008) Elevated CRAF as a potential mechanism of acquired resistance to BRAF inhibition in melanoma. Cancer Res 68:4853–4861

41. Zhang J, Liu J (2013) Tumor stroma as targets for cancer therapy. Pharmacol Ther 137:200–215

42. Friberg S, Mattson S (1997) On the growth rates of human malignant tumors: implications for medical decision making. J Surg Oncol 65:284–297

43. Gasparini G, Longo R, Toi M, Ferrara N (2005) Angiogenic inhibitors: a new therapeutic strategy in oncology. Nat Clin Pract Oncol 2:562–577

44. Wilson GD, McNally NJ, Dische S, Saunders MI, Des Rochers C, Lewis AA, Bennett MH (1988) Measurement of cell kinetics in human tumours in vivo using bromodeoxyuridine incorporation and flow cytometry. Br J Cancer 58:423–431

45. Baguley BC (2011) The paradox of cancer cell apoptosis. Front Biosci (Landmark Ed) 16:1759–1767

46. Diaz D, Prieto A, Reyes E, Barcenilla H, Monserrat J, Alvarez-Mon M (2008) Flow cytometry enumeration of apoptotic cancer cells by apoptotic rate. Methods Mol Biol 414:23–33

47. Micheau O, Shirley S, Dufour F (2013) Death receptors as targets in cancer. Br J Pharmacol 169:1723–1744

48. Fan Z, Zhang Q (2005) Molecular mechanisms of lymphocyte-mediated cytotoxicity. Cell Mol Immunol 2:259–264

49. Franklin RA, Liao W, Sarkar A, Kim MV, Bivona MR, Liu K, Pamer EG, Li MO (2014) The cellular and molecular origin of tumor-associated macrophages. Science 344:921–925

50. Ouyang L, Shi Z, Zhao S, Wang FT, Zhou TT, Liu B, Bao JK (2012) Programmed cell death pathways in cancer: a review of apoptosis, autophagy and programmed necrosis. Cell Prolif 45:487–498

51. Condeelis J, Pollard JW (2006) Macrophages: obligate partners for tumor cell migration, invasion, and metastasis. Cell 124:263–266

52. Meyer JS, He W (1993) Cell proliferation measurements by bromodeoxyuridine or thymidine incorporation: clinical correlates. Semin Radiat Oncol 3:126–134

53. Laing JH, Wilson GD, Martindale CA (2003) Proliferation rates in human malignant melanoma: relationship to clinicopathological features and outcome. Melanoma Res 13:271–277

54. Begg AC, Haustermans K, Hart AA, Dische S, Saunders M, Zackrisson B, Gustaffson H, Coucke P, Paschoud N, Hoyer M, Overgaard J, Antognoni P, Richetti A, Bourhis J, Bartelink H, Horiot JC, Corvo R, Giaretti W, Awwad H, Shouman T, Jouffroy T, Maciorowski Z, Dobrowsky W, Struikmans H, Wilson GD et al (1999) The value of pretreatment cell kinetic parameters as predictors for radiotherapy outcome in head and neck cancer: a multicenter analysis. Radiother oncol 50:13–23

55. Furneaux CE, Marshall ES, Yeoh K, Monteith SJ, Mews PJ, Sansur CA, Oskouian RJ, Sharples KJ, Baguley BC (2008) Cell cycle times of

short-term cultures of brain cancers as predictors of survival. Br J Cancer 99:1678–1683

56. Hanahan D, Weinberg RA (2011) Hallmarks of cancer: the next generation. Cell 144: 646–674

57. Baguley BC, Finlay GJ, Wilson WR (1986) Cytokinetic resistance of Lewis lung carcinoma to cyclophosphamide and the amsacrine derivative CI-921. Prog Clin Biol Res 223:47–61

58. Finlay GJ, Wilson WR, Baguley BC (1987) Cytokinetic factors in drug resistance of Lewis lung carcinoma: comparison of cells freshly isolated from tumours with cells from exponential and plateau-phase cultures. Br J Cancer 56:755–762

59. Itamochi H, Kigawa J, Sugiyama T, Kikuchi Y, Suzuki M, Terakawa N (2002) Low proliferation activity may be associated with chemoresistance in clear cell carcinoma of the ovary. Obstet Gynecol 100:281–287

60. Bonetti A, Zaninelli M, Rodella S, Molino A, Sperotto L, Piubello Q, Bonetti F, Nortilli R, Turazza M, Cetto GL (1996) Tumor proliferative activity and response to first-line chemotherapy in advanced breast carcinoma. Breast Cancer Res Treat 38:289–297

61. Anjomshoaa A, Lin YH, Black MA, McCall JL, Humar B, Song S, Fukuzawa R, Yoon HS, Holzmann B, Friederichs J, van Rij A, Thompson-Fawcett M, Reeve AE (2008) Reduced expression of a gene proliferation signature is associated with enhanced malignancy in colon cancer. Br J Cancer 99:966–973

62. Suzuki E, Sun J, Kapoor V, Jassar AS, Albelda SM (2007) Gemcitabine has significant immunomodulatory activity in murine tumor models independent of its cytotoxic effects. Cancer Biol Ther 6:880–885

63. de Biasi AR, Villena-Vargas J, Adusumilli PS (2014) Cisplatin-induced antitumor immunomodulation: a review of preclinical and clinical evidence. Clin Cancer Res 20:5384–5391

64. Bracci L, Schiavoni G, Sistigu A, Belardelli F (2014) Immune-based mechanisms of cytotoxic chemotherapy: implications for the design of novel and rationale-based combined treatments against cancer. Cell Death Differ 21:15–25

65. Zitvogel L, Apetoh L, Ghiringhelli F, Andre F, Tesniere A, Kroemer G (2008) The anticancer immune response: indispensable for therapeutic success? J Clin Invest 118:1991–2001

66. Mellman I, Coukos G, Dranoff G (2011) Cancer immunotherapy comes of age. Nature 480:480–489

67. Gerdes MJ, Sood A, Sevinsky C, Pris AD, Zavodszky MI, Ginty F (2014) Emerging understanding of multiscale tumor heterogeneity. Front Oncol 4:366

68. Gerlinger M, Rowan AJ, Horswell S, Larkin J, Endesfelder D, Gronroos E, Martinez P, Matthews N, Stewart A, Tarpey P, Varela I, Phillimore B, Begum S, McDonald NQ, Butler A, Jones D, Raine K, Latimer C, Santos CR, Nohadani M, Eklund AC, Spencer-Dene B, Clark G, Pickering L, Stamp G, Gore M, Szallasi Z, Downward J, Futreal PA, Swanton C (2012) Intratumor heterogeneity and branched evolution revealed by multiregion sequencing. N Engl J Med 366:883–892

69. Leung E, Kannan N, Krissansen GW, Findlay MP, Baguley BC (2010) MCF-7 breast cancer cells selected for tamoxifen resistance acquire new phenotypes differing in DNA content, phospho-HER2 and PAX2 expression, and rapamycin sensitivity. Cancer Biol Ther 9:717–724

70. Hampton OA, Den Hollander P, Miller CA, Delgado DA, Li J, Coarfa C, Harris RA, Richards S, Scherer SE, Muzny DM, Gibbs RA, Lee AV, Milosavljevic A (2009) A sequence-level map of chromosomal breakpoints in the MCF-7 breast cancer cell line yields insights into the evolution of a cancer genome. Genome Res 19:167–177

71. Kemper K, de Goeje PL, Peeper DS, van Amerongen R (2014) Phenotype switching: tumor cell plasticity as a resistance mechanism and target for therapy. Cancer Res 74:5937–5941

72. Vidal SJ, Rodriguez-Bravo V, Galsky M, Cordon-Cardo C, Domingo-Domenech J (2014) Targeting cancer stem cells to suppress acquired chemotherapy resistance. Oncogene 33:4451–4463

73. Hong IS, Lee HY, Nam JS (2015) Cancer stem cells: the 'Achilles heel' of chemo-resistant tumors. Recent Pat Anticancer Drug Discov 10:2–22

74. Marshall ES, Baguley BC, Matthews JH, Jose CC, Furneaux CE, Shaw JH, Kirker JA, Morton RP, White JB, Rice ML, Isaacs RJ, Coutts R, Whittaker JR (2004) Estimation of radiation-induced interphase cell death in cultures of human tumor material and in cell lines. Oncol Res 14:297–304

75. Marshall ES, Finlay GJ, Matthews JH, Shaw JH, Nixon J, Baguley BC (1992) Microculture-based chemosensitivity testing: a feasibility study comparing freshly explanted human melanoma cells with human melanoma cell lines. J Natl Cancer Inst 84:340–345

76. Crystal AS, Shaw AT, Sequist LV, Friboulet L, Niederst MJ, Lockerman EL, Frias RL, Gainor JF, Amzallag A, Greninger P, Lee D, Kalsy A, Gomez-Caraballo M, Elamine L, Howe E, Hur W, Lifshits E, Robinson HE, Katayama R, Faber AC, Awad MM, Ramaswamy S,

Mino-Kenudson M, Iafrate AJ, Benes CH, Engelman JA (2014) Patient-derived models of acquired resistance can identify effective drug combinations for cancer. Science 346: 1480–1486

77. Liu X, Ory V, Chapman S, Yuan H, Albanese C, Kallakury B, Timofeeva OA, Nealon C, Dakic A, Simic V, Haddad BR, Rhim JS, Dritschilo A, Riegel A, McBride A, Schlegel R (2012) ROCK inhibitor and feeder cells induce the conditional reprogramming of epithelial cells. Am J Pathol 180:599–607

78. Houghton JA, Taylor DM (1978) Maintenance of biological and biochemical characteristics of human colorectal tumours during serial passage in immune-deprived mice. Br J Cancer 37:199–212

79. Monsma DJ, Monks NR, Cherba DM, Dylewski D, Eugster E, Jahn H, Srikanth S, Scott SB, Richardson PJ, Everts RE, Ishkin A, Nikolsky Y, Resau JH, Sigler R, Nickoloff BJ, Webb CP (2012) Genomic characterization of explant tumorgraft models derived from fresh patient tumor tissue. J Transl Med 10:125

80. Wargo JA, Cooper ZA, Flaherty KT (2014) Universes collide: combining immunotherapy with targeted therapy for cancer. Cancer discov 4:1377–1386

81. Selimoglu-Buet D, Gallais I, Denis N, Guillouf C, Moreau-Gachelin F (2012) Oncogenic kit triggers Shp2/Erk1/2 pathway to down-regulate the pro-apoptotic protein Bim and to promote apoptosis resistance in leukemic cells. PLoS One 7, e49052

Chapter 3

In Vitro Methods for Studying the Mechanisms of Resistance to DNA-Damaging Therapeutic Drugs

Pasarat Khongkow, Anna K. Middleton, Jocelyn P.-M. Wong, Navrohit K. Kandola, Mesayamas Kongsema, Gabriela Nestal de Moraes, Ana R. Gomes, and Eric W.-F. Lam

Abstract

Most commonly used anticancer drugs exert their effects mainly by causing DNA damage. The enhancement in DNA damage response (DDR) is considered a key mechanism that enables cancer cells to survive through eliminating the damaged DNA lesions and thereby developing resistance to DNA-damaging agents. This chapter describes the four experimental approaches for studying DDR and genotoxic drug resistance, including the use of γ-H2AX and comet assays to monitor DNA damage and repair capacity as well as the use of clonogenic and β-galactosidase staining assays to assess long-term cell fate after DNA-damaging treatment. Finally, we also present examples of these methods currently used in our laboratory for studying the role of FOXM1 in DNA damage-induced senescence and epirubicin resistance.

Key words γ-H2AX, Comet assay, Clonogenic assay, β-Galactosidase staining, DNA damage, Resistance

1 Introduction

Genotoxic chemotherapy is one of the principal modes of cancer treatment. Most commonly used anticancer drugs, such as epirubicin, doxorubicin, 5-fluorouracil, and cisplatin, target genomic DNA. They primarily function as DNA intercalators, blocking DNA synthesis and inducing DNA double-strand breaks (DSBs) leading to cancer cell death [1]. Clinically, these DNA-damaging agents effectively inhibit the growth and spread of cancer cells. However, the majority of these treatments will eventually fail due to the development of drug resistance. Elucidation of the molecular mechanisms underlying chemoresistance is therefore needed to aid disease management and improve patient survival. As many chemotherapeutic agents mainly exert their effects by causing DNA damage, the enhancement in DNA damage response (DDR) signaling is considered as a key mechanism that can enable cancer

José Rueff and António Sebastião Rodrigues (eds.), *Cancer Drug Resistance: Overviews and Methods*, Methods in Molecular Biology, vol. 1395, DOI 10.1007/978-1-4939-3347-1_3, © Springer Science+Business Media New York 2016

cells to survive through eliminating the induced DNA lesions and thereby developing resistance to these genotoxic agents.

At the molecular level, cells respond to DNA damage by activating the so-called DDR, a complex molecular mechanism developed to repair DNA damage and maintain genome integrity. Initially, the formation of DSBs triggers activation of ATM and phosphorylation of serine 139 on histone variant H2AX at the site of the DNA break to form "foci." This process plays a key role in DDR and is required for the recruitment of DNA repair proteins, such as 53BP1, NBS1, and MDC1, to the sites of damage as well as for the activation of checkpoint proteins which arrest the cell cycle progression [2]. Once DNA lesions are completely repaired, DDR foci are disassembled and cells quickly resume normal proliferation. In contrast, severe or irreparable DNA damage induces more protracted DDR signaling and increases gamma(γ)-H2AX foci spreading and eventually cells may undergo apoptosis or cellular senescence to prevent transmission of the lesions to the daughter cells upon cell division [3]. The factors responsible for this differential outcome are still unclear, but the dose and duration of exposure to DNA-damaging agents as well as the cell types are likely to be crucial determinants [3, 4].

In a recent study in our lab, we have found that breast cancer cell lines as well as immortalized MEFs can be induced to undergo senescence by DNA-damaging agents including epirubicin and γ-irradiation. Interestingly, the dose of DNA-damaging agents required for triggering senescent phenotypes is a much lower dose than that required for induction of apoptosis [5, 6]. Therefore, the emerging knowledge about the pathways and the patterns of gene expression related to DNA damage-induced senescence in tumor cells may lead to more effective treatments for cancer patients with fewer side effects.

In this chapter, we describe the four standard approaches for studying DDR and their related drug resistance mechanisms, including the use of γ-H2AX and comet assays to monitor DNA damage and repair capacity as well as the use of clonogenic and senescence-associated β-gal staining assays to assess long-term cell survival following DNA-damaging treatment. Finally, we also present examples of applying these methods to study the role of FOXM1 in DNA damage-induced senescence and epirubicin response.

2 Materials

2.1 Chemicals, Buffers, and Equipment to Be Used in Comet Assay

1. Single-frosted glass microscope slides 70×26 mm and cover slips 24×40 mm.
2. LMA: Agarose, Low Melting Point, Analytical Grade (Promega).
3. NMA: Agarose, LE, Analytical Grade (Promega).

4. Lysis buffer: 100 mM EDTA-Na$_2$, 2.5 M NaCl, 10 mM Tris–HCl, 250 mM NaOH, pH 10. Store at room temperature (RT). Before use, add immediately 1 % v/v Triton X-100 and 10 % v/v DMSO and store at 4 °C.

5. Electrophoresis buffer: 250 mM NaOH, 1.5 mM EDTA-Na$_2$, pH 13. Store at 4 °C.

6. Neutralization buffer: 400 mM Tris–HCl, pH 7.5. Sterilize by autoclaving and store at RT.

7. Phosphate-buffered saline (PBS), pH 7.4.

8. Vectashield mounting solution with DAPI (Vector Laboratories, Cat.#H-1200).

9. Large-bed gel electrophoresis boxes and power pack.

10. Fluorescence microscope.

11. Comet image analysis software such as CometScore software (Tritek Corp., USA) or Kinetic Imaging Komet assay software (Kinetic imaging, Liverpool, UK).

2.2 Chemicals, Buffers, and Equipment to Be Used in γ-H2AX Immunofluorescent Staining

1. BD Falcon culture slides are used with four chambers each.

2. Glass cover slips.

3. PBS, pH 7.4.

4. 4 % Paraformaldehyde (PFA) in PBS.

5. 0.2 % Triton in PBS.

6. Blocking solution: 5 % goat serum. This should be kept on ice once made.

7. Primary antibodies: Rabbit monoclonal anti-γ-H2AX Ser139 diluted in 0.2 % goat serum in PBS.

8. Secondary antibodies: Goat anti-rabbit Alexa Flour 488 (Molecular Probes, Cat.# A11034) diluted in PBS.

9. Vectashield mounting solution with DAPI (Vector Laboratories, Cat.#H-1200).

10. Clear nail polish.

11. Leica TCS SP5 confocal microscope, equipped with a 63× oil immersion objective.

12. Image analysis software such as ImageJ [7], FociCounter [8], or CellProfiler [9].

2.3 Chemicals, Buffers, and Equipment to Be Used for Clonogenic Assay

1. PBS.

2. 4 % (v/v) paraformaldehyde (PFA) in PBS.

3. 0.5 % (w/v) crystal violet in dH$_2$O. Store in the dark at RT.

4. 33 % Acetic acid.

5. 6-well plates.

6. Pipettes.

7. Tubes for dilution.

8. Hemocytometer.

9. Digital camera.

10. Tecan Sunrise 96-well Microplate Reader.

2.4 Chemicals, Buffers, and Equipment to Be Used in β-Galactosidase Staining

1. PBS.

2. Fixative solution: 2 % (v/v) formaldehyde and 0.2 % (v/v) glutaraldehyde in PBS.

3. 0.1 M Citric acid ($C_6H_8O_7 \cdot H_2O$) in dH_2O.

4. 0.2 M Sodium phosphate (Na_2HPO_4) in dH_2O.

5. 0.2 M Citric acid/sodium phosphate solution (100 ml): Mix 36.85 ml of 0.1 M citric acid solution with 63.15 ml of 0.2 M sodium phosphate solution. The pH of citric acid/sodium phosphate solution must be adjusted to 6.0 by adding either citric acid or sodium phosphate solution.

6. 20 mg/ml X-gal solution in *N*-*N*-dimethylformamide (DMF): This solution should be prepared freshly or can be stored at −20 °C for short term and should be protected from light. Always prepare and store in polypropylene or glass tubes.

7. 50 mM Potassium ferrocyanide and 50 mM potassium ferricyanide, these solutions should be stored at 4 °C in the dark (cover the solution tubes with aluminum foil to protect from light).

8. 5 M NaCl.

9. 1 M $MgCl_2$.

10. SA-β-gal staining solution: 40 mM Citric acid/sodium phosphate solution, 150 mM NaCl, 2 mM $MgCl_2$, 5 mM potassium ferrocyanide, 5 mM potassium ferricyanide, and 1 mg/ml X-gal in dH_2O. This solution must be prepared freshly just before staining.

11. 6-Well plates.

2.5 Chemicals, Buffers to Be Used in Cell Cultures

1. Cell lines: Human breast carcinoma MCF-7 cells and epirubicin-resistant MCF-7 breast carcinoma (MCF-7-Epi^R) cells.

2. Complete medium (Dulbecco's modified Eagle's medium (DMEM), 10 % fetal calf serum (FCS), 2 mM l-glutamine-PenStrep). Store at 4 °C and warm to 37 °C prior to use.

3. Trypsin-EDTA.

4. PBS, pH 7.4. Store at RT after autoclaving.

5. DNA-damaging agent: Epirubicin (100 μM stock solution in the complete culture medium, store in the dark at 4 °C).

3   Methods

3.1   Measurement of Drug-Induced DNA Breaks Using Comet Assay

The comet assay, also known as the single-cell gel electrophoresis assay, is a sensitive and simple method for detecting DNA strand breaks in individual eukaryotic cells. It is widely used in a broad variety of applications including bio-monitoring, genotoxicity assessment, and as a tool to investigate DNA damage and repair capacity in a wide range of tumor cells in response to DNA-damaging agents. In this assay, cells embedded in low-melting-point agarose on a microscope slide are lysed with a lysis buffer that removes cell membranes, cytoplasm, and most of the cellular proteins. When the leftover nucleoids are treated with high-alkaline solution, the DNA supercoils start to unwind. During electrophoresis, the relaxed DNA breaks migrate towards the anode, thereby forming a "comet"-like tail. The relative intensity of the tail reflects the frequency of damaged DNA breaks [10–12].

3.1.1   Preparation of Cell Cultures and DNA-Damaging Treatment

Seed 2×10^6 of MCF-7 cells in 10 cm plates. 24 h after, the cells are treated with a DNA-damaging agent by replacing with the medium containing the desired concentration of the agent. Leave the plate in an incubator for an appropriate exposure time (*see* **Note 1**).

3.1.2   Slide Preparation

1. Prepare 1 % NMA agarose in dH$_2$O and 0.7 % LMP agarose in PBS. Microwave until the agarose is fully melted. Place the 1 % NMA agarose in a 55 °C water bath and the LMP agarose in 37 °C water bath.

2. Merge the microscope slides in a vertical staining jar containing melted 1 % NMA agarose in water and wipe one side clean. Then, allow the slides to dry overnight at room temperature. The slides must be dry before use.

3. Prepare the cell suspension for the second layer. After suitable drug exposure time, pellet cells by centrifuging at $500 \times g$ for 5 min at 4 °C (*see* **Note 2**). Resuspend cells to final concentration of ~2×10^6 cells/ml in complete medium and maintain at 4 °C. Pipet 10 µl of resuspended cells (containing ~2×10^4 cells) onto the center of the slide. Then add 85 µl of 0.7 % LMA on top of the cells and place a cover slip on top to spread the cells (avoid bubbles). The cells need to be well spread in order to measure the length of tail. Allow second LMA layer to set by placing the slides in a tray on ice for 10 min.

4. Prepare the third LMA layer. Carefully remove the cover slip, add another 85 µl of 0.7 % LMA directly onto the cells, and place the new cover slip on top. Again place the slide in a tray on ice for 10 min (*see* **Note 3**).

3.1.3 Lysis

Carefully remove the cover slip and merge the control and treated slides in a vertical staining jar containing complete lysis mixture at 4 °C. Incubate for 1 h in the dark by covering with foil ensuring that all slides are sufficiently covered.

3.1.4 Alkaline Treatment

Prior to electrophoresis, take the slides from the lysis buffer and remove all excess of buffer from slides by carefully blotting on tissue. Then, gently place slides to the gel shelf in the electrophoresis tank containing the prepared cold alkaline electrophoresis buffer (pH 13) to unwind the DNA supercoils. Ensure that all slides are sufficiently covered. Slides should be placed close together and the remaining free space should be filled with empty slides to prevent any movement. Incubate for 40 min in the dark (*see* **Note 4**).

3.1.5 Electrophoresis

Start applying the electric current at 25 V, ~300 mA, for 20 min. This must be carried out on ice.

3.1.6 Neutralization

To stop electrophoresis, carefully remove slides from the electrophoresis tank and place on a horizontal slide rack. Gently rinse with neutralization buffer for 5 min, drain, and repeat twice.

3.1.7 DAPI Staining

Add 30 μl of mounting media with DAPI (Vector Laboratories, Cat.#H-1200) onto the surface of agarose of each slide and cover with a cover slip, avoiding bubbles. Slides should be analyzed as soon as possible. For storage, keep slides in a humid dark chamber at 4 °C until they are visualized (*see* **Note 5**).

3.1.8 Analysis

Identify random fields containing comets under the microscope and score using computer image analysis. Programs are designed to differentiate comet head from tail and to measure a variety of parameters including tail length, % DNA in tail, and tail moment. These are calculated in different ways but essentially represent the product of tail length and relative tail intensity (Fig. 1).

3.2 Measuring DNA Damage Using γ-H2AX Immunofluorescent Staining

Phosphorylation of histone 2A (γ-H2AX) is one of the early events following DNA double-strand breaks (DSB). ATR, ATM, and DNA-PK are related DNA-activated kinase responsible for this phosphorylation [13]. γ-H2AX immunofluorescent staining is a method for quantification of γ-H2AX foci using a γ-H2AX-specific antibody [14]. This method can be used for in vivo studies monitoring patient response to irradiation, DNA damage-including chemotherapeutics, and for in vitro experiments with mammalian cells to study DNA damage and repair [5, 6].

3.2.1 Preparation of Cell Cultures and DNA-Damaging Treatment

Seed cells in 4-well chamber slides at 20,000–40,000 cells/ well. After 24 h, the cells are treated with a DNA-damaging agent by replacing with complete medium containing the desired concentration of the agent. Leave the chamber slide in an incubator at 37 °C for an appropriate exposure time (*see* **Note 1**).

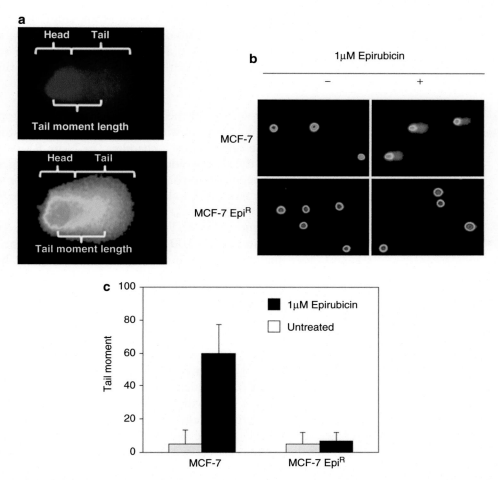

Fig. 1 Representations of comet images from MCF-7 and MCF-7 Epi[R] cells. (**a**) An example of MCF-7 cells after treatment with epirubicin, stained with DAPI, and visualized by fluorescence microscopy before (*top panel*) and after (*lower panel*) the application of comet assay software. (**b**) MCF-7 and MCF-7 Epi[R] cells were treated with or without 1 μM of epirubicin for 6 h and assayed for comet assay. The DNA damage was quantified using the tail moment (*right panel*). (**c**) Represented data are average of two independent experiments (100 comets were measured in each experiment)

3.2.2 Fixation, Permeabilization, and Blocking

At the end of the incubation time, aspire the media and wash cells once with PBS. Fix cells by adding 250 μl of 4 % PFA in each well and leave for 15 min at RT (*see* **Notes 6** and **7**). Remove fixative solution and wash three times with PBS. Permeabilize with 250 μl of 0.2 % Triton X-100 in PBS for 10 min at RT. Wash samples three times with PBS. Block the slides with 250 μl of 5 % goat serum in PBS for 60 min at RT.

3.2.3 Immunostaining

Add 250 μl of primary antibody anti-γ-H2AX Ser139 (rabbit, monoclonal, 1:250 dilution) in 0.2 % goat serum in PBS and incubate in a humidified chamber overnight at 4 °C (or 60 min at RT). After incubation, wash samples three times with PBS.

Add secondary antibody anti-rabbit Alexa Flour 488 (Molecular Probes, Cat.# A11034) at 1:500 dilution in PBS and incubate in the dark for 60 min at RT. Avoid exposing the secondary antibody to the light. After incubation, wash sample three times with PBS.

3.2.4 Counterstaining and Mounting

Detach the chamber wells from the glass slide and add one drop of Vectashield mounting medium with DAPI (Vector Laboratories Cat.#H-1200) to the center of each chamber (*see* **Note 8**). Gently place a cover slip on the slide avoiding air bubbles. Seal the edges of each cover slip with regular transparent nail polish and allow for it to dry for 5 min. Cover slides with foil and store at 4 °C (*see* **Note 9**).

3.2.5 Imaging and Image Analysis

Capture image using a Leica TCS SP5 confocal microscope. Capture images from at least 100 cells and count the number of γ-H2AX foci manually by using Image J counting analysis. Alternatively, specific image analysis software can be used such as FociCounter [8], or CellProfiler [9] cell image analysis software (*see* **Notes 10** and **11**) (Fig. 2).

3.3 Clonogenic Cell Survival Assay

Clonogenic assay is a useful long-term survival assay in cancer research laboratories. It is used to determine the ability of a single cell to proliferate and form a colony. The colony is defined as a group of at least 50 cells, which can be counted under a microscope. This method is now widely used to examine the effectiveness of anticancer cytotoxic agents on colony-forming ability in several cancer cell lines. Before and after exposing to the treatment, a small number of cells are seeded onto plates containing different concentrations of the drug and are allowed to form colonies for 14–21 days. A final cell survival rate is referred to a relationship between the dose of the agent used to produce an insult and the fraction of cells retaining their ability to proliferate [15, 16].

3.3.1 Plating

1. Trypsinize and collect cells in medium containing 10 % fetal calf serum. Centrifuge the cell suspension at 1200 rpm for 4 min to pellet cells and resuspend in fresh medium.

2. Count the cells by using a hemocytometer (*see* **Note 12**).

3. Dilute the cell suspension into the appropriate seeding concentration (low densities of cells) and seed equally into 6-well plate in a total volume of 2 ml medium, at least in duplicate (*see* **Note 12**). Number of cells seeded per well depends on cell type and its growth rate as well as the toxicity of the treatment. For example: MCF-7 cells are seeded at a density of 1000–2000 cells/well in a 6-well plate. Cells are then incubated at 37 °C and 5 % CO_2 for 24 h to allow the cells to attach to the plate prior to drug treatment.

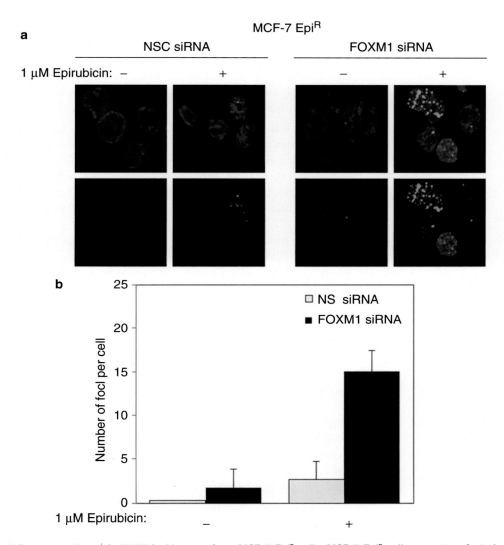

Fig. 2 Representations of γH2AX foci images from MCF-7 Epi^R cells. MCF-7 Epi^R cells were transfected with NS siRNA (non-targeting control) or FOXM1 siRNA. Twenty-four hours after transfection, cells cultured on chamber slides were either untreated or treated with 1 μM epirubicin for 24 h. Cells were then fixed and immunostained for γH2AX foci (*green*). Nuclei were counterstained with 4′-6-diamidino-2-phenylindole (DAPI; *blue*). Images were acquired with Leica TCS SP5 (×63 magnification). For each time point, images of at least 100 cells were captured and used for quantification of γH2AX foci number

3.3.2 Treatment with DNA- Damaging Agents

1. After attachment of the cells to the plate surface, they are treated with different concentrations of DNA-damaging agent. The drug concentration range should be varied and include higher concentrations that kill most of the cells as well as the lowest concentration that kills none of the cells. Leave the plate in an incubator at 5 % CO_2 for the appropriate exposure time,

which depends on the cell type and the toxicity of the treatment. For example, MCF-7 cells are treated with epirubicin (concentration range from 0 to 10 nM) for 48 h (*see* **Note 1**).

2. After the appropriate exposure time, DNA-damaging agent is removed by rinsing the cells with PBS and replacing with 2 ml fresh medium.

3. Plates are left to incubate at 37 °C in 5 % CO_2 for 7–20 days or until cells in control (untreated) plates have formed sufficiently large clones consisting of 50 or more cells. Monitoring is essential to prevent the fusion of colonies (*see* **Note 13**). In the example, the control wells for MCF-7 cells requires 9 days to form sufficiently large clones consisting of 50 or more cells.

3.3.3 Fixation and Staining

1. Following incubation, remove the medium from the cells and wash with PBS.

2. Fix the cells in each well with 1 ml of 4 % PFA for 15 min at RT.

3. Wash three times with PBS.

4. Add 1 ml of 0.5 % (w/v) crystal violet (diluted in dH_2O) for staining and incubate for 1 h at RT.

5. Remove crystal violet gently and immerse the plates in tap water to remove excess crystal violet.

6. Leave the plates with colonies to dry overnight at RT.

3.3.4 Colony Counting and Assay Analysis

1. Digital images of the colonies (at least five random fields) are obtained using a camera. Adjust the microscope and lighting to avoid shadowing on the edges of the images. Do not change the focus and light settings because the image should have the same relative background throughout (Fig. 3a).

2. Colonies are counted using imaging analysis software packages such as ImageJ. Average the colony counts (five random fields) for each condition and divide the mean by the number of cell seeded. This will give the plating efficiency (PE) [15–17]:

$$PE = \frac{Number\ of\ colonies\ counted}{Number\ of\ cells\ seeded} \times 100\%$$

Following determination of PE, the surviving fraction (SF) [16] can be determined by

$$SF = \frac{PE\ of\ treated\ sample}{PE\ of\ control} \times 100\%$$

3. Alternatively, colonies can be quantified by adding 1 ml of 33 % (w/v) acetic acid to solubilize the bound crystal violet and the

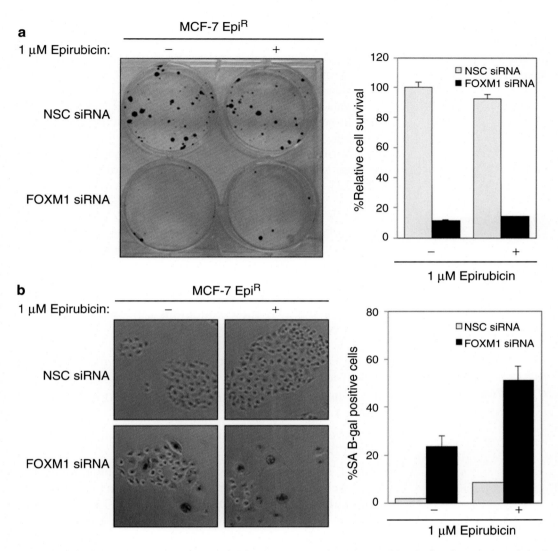

Fig. 3 Clonogenic assay and β-gal staining performed in 6-well plates (**a**) MCF-7 Epi^R cells were transfected with NSC (non-targeting control) siRNA or FOXM1 siRNA. Twenty-four hours after transfection, 2000 cells were seeded in 6-well plates, treated with 1 μM epirubicin, grown for 15 days, and then stained with crystal violet (*left panel*). In parallel, (**b**) the transfected cells were seeded in 6-well plates, treated with epirubicin. Five days after treatment, cells were stained for SAβ-gal activity. The graph shows the percentage of SAβ-gal-positive cells as measured from five different fields from two independent experiments

amount of the dye can be determined colorimetrically at a wavelength of 592 nm in a microplate reader (Tecan, Medford, MA, USA):

$$\text{Relative survival rate} = \frac{\text{Intensity of treated sample}}{\text{Intensity of control}} \times 100\%$$

3.4 Senescence-Associated β-Galactosidase Staining

Cellular senescence is a permanent proliferative arrest, which is induced by activation of the cellular DNA damage response [18]. Senescence-associate β-galactosidase (SA-β-gal) activity is the most widely employed biomarker to identify senescent cells. Eukaryotic β-galactosidase, a hydrolase normally confined to the lysosome, cleaves β-galactose residues in β-d-galactosides [19]. This enzymatic activity can be detected in cells under normal conditions at an optimal pH 3–5, while in senescent cells this activity is detectable at a higher pH of 6.0 [20]. In cells undergoing senescence the lysosomal mass is raised, generating a higher level of β-galactosidase [19, 21, 22].

3.4.1 Cell Culture and DNA-Damaging Treatment

Seed cells at a density of 20,000–40,000 cells/well on a 6-well plate in medium containing 10 % fetal calf serum. After 24 h, the cells are treated with a DNA-damaging agent by replacing with the medium containing the desired concentration of the agent. Leave the plate in an incubator for an appropriate exposure time. Note: Avoid over-confluency of the cells, or cells that have undergone too many passages, as these conditions can cause false-positive results (*see* **Note 1**).

3.4.2 Fixation and SA-β-Galactosidase Staining of Cultured Cells

1. At the end of the incubation time, aspire the media and wash cells twice with PBS.

2. Fix cells with 1 ml of fixative solution for 10 min at RT (*see* **Note 14**).

3. Remove the fixative solution and wash the fixed cells twice with PBS.

4. Add 1 ml of freshly prepared SA-β-gal staining solution to each well. Incubate cells with staining solution in a dry incubator (without supplying CO_2) at 37 °C for 12–16 h. To prevent solvent evaporation during the incubation period, wrap the plate with parafilm (*see* **Note 15**). Avoid incubating samples in a CO_2 incubator as the presence of CO_2 will lower the pH of the buffer.

5. While maintaining the SA-β-gal staining solution on the plate, visualize the cells under a microscope at 200× magnification for the development of blue coloration (positive cells) (Fig. 3b). Calculate the percentage of the SA-β-gal-positive cells by counting the cells in five random fields. The staining is easier to recognize under bright field than with phase contrast.

6. For long-term storage of the stained plates, overlay the cells with 70 % glycerol after removing the β-galactosidase staining solution and store at 4 °C.

4 Notes

1. The drug concentration and exposure time are dependent on the type of DNA-damaging agents and cell type.

2. For the cell preparation, trypsinize cells as quickly as possible to avoid any additional DNA damage.

3. Work quickly as the agarose sets quickly at RT.

4. The volume and temperature of electrophoresis buffer should be consistent in each experiment. Also, the alkali treatment time (unwinding time) should be the same.

5. Alternatively, other reagents which can be used instead of DAPI are propidium iodide (2.5 µg/ml) and Hoechst 33258 (0.5 µg/ml). Caution: Propidium iodide is a mutagenic agent, so wear a protective glove.

6. PBS and PFA need to be warmed to 37 °C prior to use.

7. Before permeabilization, samples can be left overnight in the fridge with PBS.

8. Excess PBS should be removed before adding mounting media as dilution may reduce the antifade effect in mounting media.

9. If samples need to be stored for longer than 2 weeks, they should be kept at –20 °C.

10. The dose of DNA-damaging agent necessary for γ-H2AX induction analysis depends on the endogenous levels of foci.

11. Generation of ssDNA in the S phase of cell cycle can also potentially lead to non-DSB γ-H2AX formation.

12. The accurate number of cells seeded should always be accurate in order to obtain the correct data analysis. All wells must contain the same number of cells at the start of experiment. Remember to resuspend the cells well before diluting and plating. Also, ensure an even distribution in the plate by moving plates backward and forward, and then right and left; repeat same motions three times. Leave plates in the tissue culture hood for 5 min before placing into incubator. Avoid swirling the plate to mix the cells because this causes the cells accumulating in the middle of each well.

13. The incubation time for colony formation varies according to the doubling time of the cell lines used but usually ranges from 1 to 3 weeks and cells should never be too confluent during the incubation period. It is crucial to check cells every day after treatment to ensure that colonies are not overlapping.

14. Longer fixation time may affect the SA-β-gal activity. Adjust the concentration of fixative solution depending on the cell type.

15. Independent of senescence density, induced SA-β-gal staining may occur in some cells, such as cultured fibroblasts. In this case, a blue color develops within 2 h, but with less color intensity than that observed in senescent cells and this will disappear within 2 days after the confluent cells are re-plated [20]. In general, 12–16 h is the optimal staining period for staining of senescence cells.

16. All reagents used in SA-β-gal staining should be kept under dark conditions. Avoid prolonged exposure to light during the assay [23].

17. Formaldehyde and glutaraldehyde are toxic. Wear protective groves and always handle them under a fume hood [24].

References

1. Cheung-Ong K, Giaever G, Nislow C (2013) DNA-damaging agents in cancer chemotherapy: serendipity and chemical biology. Chem Biol 20:648–659

2. Podhorecka M, Skladanowski A, Bozko P (2010) H2AX phosphorylation: its role in DNA damage response and cancer therapy. J Nucleic Acids 2010

3. d'Adda di Fagagna F (2008) Living on a break: cellular senescence as a DNA-damage response. Nat Rev Cancer 8:512–522

4. Kuilman T, Michaloglou C, Mooi WJ, Peeper DS (2010) The essence of senescence. Genes Dev 24:2463–2479

5. Khongkow P, Karunarathna U, Khongkow M, Gong C, Gomes AR, Yague E, Monteiro LJ, Kongsema M, Zona S, Man EP, Tsang JW, Coombes RC, Wu KJ, Khoo US, Medema RH, Freire R, Lam EW (2014) FOXM1 targets NBS1 to regulate DNA damage-induced senescence and epirubicin resistance. Oncogene 33:4144–4155

6. Monteiro LJ, Khongkow P, Kongsema M, Morris JR, Man C, Weekes D, Koo CY, Gomes AR, Pinto PH, Varghese V, Kenny LM, Charles Coombes R, Freire R, Medema RH, Lam EWF (2013) The Forkhead Box M1 protein regulates BRIP1 expression and DNA damage repair in epirubicin treatment. Oncogene 32:4634–4645

7. Schneider CA, Rasband WS, Eliceiri KW (2012) NIH image to ImageJ: 25 years of image analysis. Nat Methods 9:671–675

8. Jucha A, Wegierek-Ciuk A, Koza Z, Lisowska H, Wojcik A, Wojewodzka M, Lankoff A (2010) FociCounter: a freely available PC programme for quantitative and qualitative analysis of gamma-H2AX foci. Mutat Res 696:16–20

9. Carpenter A, Jones T, Lamprecht M, Clarke C, Kang I, Friman O, Guertin D, Chang J, Lindquist R, Moffat J, Golland P, Sabatini D (2006) Cell Profiler: image analysis software for identifying and quantifying cell phenotypes. Genome Biol 7:R100

10. Liao W, McNutt MA, Zhu W-G (2009) The comet assay: a sensitive method for detecting DNA damage in individual cells. Methods 48:46–53

11. Spanswick V, Hartley J, Hartley J (2010) Measurement of DNA interstrand crosslinking in individual cells using the single cell gel electrophoresis (Comet) assay. In: Fox KR (ed) Drug-DNA interaction protocols, vol 613, Methods in molecular biology. Humana Press, New York, NY, pp 267–282. doi:10.1007/978-1-60327-418-0_17

12. Clingen P, Lowe J, Green ML (2000) Measurement of DNA damage and repair capacity as a function of age using the comet assay. In: Barnett Y, Barnett C (eds) Aging methods and protocols, vol 38, Methods in molecular medicine. Humana Press, New York, NY, pp 143–157. doi:10.1385/1-59259-070-5:143

13. Paull TT, Rogakou EP, Yamazaki V, Kirchgessner CU, Gellert M, Bonner WM (2014) A critical role for histone H2AX in recruitment of repair factors to nuclear foci after DNA damage. Curr Biol 10:886–895

14. Sharma A, Singh K, Almasan A (2012) Histone H2AX phosphorylation: a marker for DNA damage. In: Bjergbæk L (ed) DNA repair protocols, vol 920, Methods in molecular biology. Humana Press, New York, NY, pp 613–626. doi:10.1007/978-1-61779-998-3_40

15. Franken NAP, Rodermond HM, Stap J, Haveman J, van Bree C (2006) Clonogenic assay of cells in vitro. Nat Protoc 1:2315–2319

16. Munshi A, Hobbs M, Meyn R (2005) Clonogenic cell survival assay. In: Blumenthal R (ed) Chemosensitivity, vol 110, Methods in molecular medicine™. Humana Press, New York, NY, pp 21–28. doi:10.1385/ 1-59259-869-2:021

17. Plumb J (2004) Cell sensitivity assays: clonogenic assay. In: Langdon S (ed) Cancer cell culture, vol 88, Methods in molecular medicine™. Humana Press, New York, NY, pp 159–164. doi:10.1385/1-59259-406-9:159

18. Campisi J (2001) Cellular senescence as a tumor-suppressor mechanism. Trends Cell Biol 11:S27–S31

19. Kurz DJ, Decary S, Hong Y, Erusalimsky JD (2000) Senescence-associated (beta)-galactosidase reflects an increase in lysosomal mass during replicative ageing of human endothelial cells. J Cell Sci 113(Pt 20):3613–3622

20. Dimri GP, Lee X, Basile G, Acosta M, Scott G, Roskelley C, Medrano EE, Linskens M, Rubelj I, Pereira-Smith O et al (1995) A biomarker that identifies senescent human cells in culture and in aging skin in vivo. Proc Natl Acad Sci U S A 92:9363–9367

21. Debacq-Chainiaux F, Erusalimsky JD, Campisi J, Toussaint O (2009) Protocols to detect senescence-associated beta-galactosidase (SA-[beta] gal) activity, a biomarker of senescent cells in culture and in vivo. Nat Protoc 4:1798–1806

22. Itahana K, Itahana Y, Dimri GP (2013) Colorimetric detection of senescence-associated beta galactosidase. Methods Mol Biol 965:143–15623

23. Burn SF (2012) Detection of β-galactosidase activity: X-gal staining. Methods Mol Biol 886: 241–250

24. Sun HW, Feigal RJ, Messer HH (1990) Cytotoxicity of glutaraldehyde and formaldehyde in relation to time of exposure and concentration. Pediatr Dent 12:303–307

Chapter 4

In Vitro Approaches to Study Regulation of Hepatic Cytochrome P450 (CYP) 3A Expression by Paclitaxel and Rifampicin

Romi Ghose, Pankajini Mallick, Guncha Taneja, Chun Chu, and Bhagavatula Moorthy

Abstract

Cancer is the second leading cause of mortality worldwide; however the response rate to chemotherapy treatment remains slow, mainly due to narrow therapeutic index and multidrug resistance. Paclitaxel (taxol) has a superior outcome in terms of response rates and progression-free survival. However, numerous cancer patients are resistant to this drug. In this investigation, we tested the hypothesis that induction of cytochrome P450 (*Cyp*)*3a11* gene by paclitaxel is downregulated by the inflammatory mediator, lipopolysaccharide (LPS), and that the pro-inflammatory cytokine, tumor necrosis factor (TNF)-α, attenuates human *CYP3A4* gene induction by rifampicin. Primary mouse hepatocytes were pretreated with LPS (1 μg/ml) for 10 min, followed by paclitaxel (20 μM) or vehicle for 24 h. RNA was extracted from the cells by trizol method followed by cDNA synthesis and analysis by real-time PCR. Paclitaxel significantly induced gene expression of *Cyp3a11* (~30-fold) and this induction was attenuated in LPS-treated samples. Induction and subsequent downregulation of CYP3A enzyme can impact paclitaxel treatment in cancer patients where inflammatory mediators are activated. It has been shown that the nuclear receptor, pregnane X receptor (PXR), plays a role in the induction of CYP enzymes. In order to understand the mechanisms of regulation of human *CYP3A4* gene, we co-transfected HepG2 cells (human liver cell line) with CYP3A4-luciferase construct and a PXR expression plasmid. The cells were then treated with the pro-inflammatory cytokine, TNFα, followed by the prototype CYP3A inducer rifampicin. It is well established that rifampicin activates PXR, leading to *CYP3A4* induction. We found that induction of CYP3A4-luciferase activity by rifampicin was significantly attenuated by TNFα. In conclusion, we describe herein several in vitro approaches entailing primary and cultured hepatocytes, real-time PCR, and transcriptional activation (transfection) assays to investigate the molecular regulation of CYP3A, which plays a pivotal role in the metabolism of numerous chemotherapeutic drugs. Genetic or drug-induced variation in CYP3A and/or PXR expression could contribute to drug resistance to chemotherapeutic agents in cancer patients.

Key words CYP3A, Chemotherapy, Inflammation, Hepatocytes, HepG2

José Rueff and António Sebastião Rodrigues (eds.), *Cancer Drug Resistance: Overviews and Methods*, Methods in Molecular Biology, vol. 1395, DOI 10.1007/978-1-4939-3347-1_4, © Springer Science+Business Media New York 2016

1   Introduction

Chemoresistance is a major obstacle to the successful treatment of cancer by chemotherapeutic agents. Paclitaxel (Taxol) is an effective anticancer agent that stabilizes microtubules and prevents them from depolymerizing. It is used to treat various cancers including ovarian, breast, lung, and head and neck cancer [1–4]. Despite good response rate in treating a number of solid malignancies, several cancer types are resistant to paclitaxel, thereby limiting its therapeutic applications. Acquired resistance to paclitaxel causes relapse and reduced survival rate, leaving few treatment options. However, the mechanism of paclitaxel chemoresistance is not well established. One potential mechanism involves activation of paclitaxel by cytochrome P450 (CYP) enzymes, particularly CYP2C8 and CYP3A4 [5, 6].

Human CYP3A4 is one of the most important drug-metabolizing enzymes (DME) that metabolizes >50 % of known drugs [7]. CYP3A enzymes are involved in the metabolism of several anticancer drugs (e.g., vinca alkaloids, docetaxel, irinotecan, teniposide, etoposide, and ifosfamide) [8–13]. Since variation in CYP3A activity can impact disposition of chemotherapeutic drugs, this enzyme plays an important role as a determinant of cancer therapy. For instance, an in vitro study showed that metabolism of vinblastine by CYP3A4 resulted in less toxic but inactive compounds, causing resistance to vinca alkaloids [10]. Furthermore, there are several reports showing the impact of CYP3A variation on cancer therapy. (1) Patients with *CYP3A4*1B* allele have high activity that results in increased docetaxel clearance; (2) enhanced CYP3A4 expression in breast tumor tissue has shown to predict therapeutic response to docetaxel [14]; (3) CYP3A4 polymorphism results in reduced metabolic activation of cyclophosphamide and shorter overall survival [15]; (4) paclitaxel induces its own metabolism through CYP3A4 induction, in vitro [16–18]. These observations suggest that patients with enhanced CYP3A4 activity due to polymorphism or due to co-administered drugs would lead to reduced concentration of paclitaxel within the cell. This can cause chemoresistance, thereby leading to poor therapeutic outcome.

CYP3A4 gene expression is regulated by basal transcription factors and nuclear receptors (NRs), including the xenobiotic NR, pregnane X receptor (PXR) [19–21]. PXR regulates *CYP3A4* expression by forming heterodimer with retinoid X receptor (RXR) and binding to distinct motifs within the promoter area of CYP3A4 [22]. PXR activation is one of the major mechanisms behind drug–drug interactions due to induction of CYP3A4 by anticancer drugs. Due to its ligand promiscuity, PXR can be activated by many anticancer drugs, including cyclophosphamide, tamoxifen, paclitaxel, vincristine, and vinblastine [23, 24]. A recent study demonstrated that the reduced chemosensitivity of colorectal cancer cells to irinotecan was reversed by the PXR antagonist sulforaphane, while the activation of PXR decreased the effectiveness of this drug [25].

As mentioned above paclitaxel is known to induce *CYP3A* gene expression [26, 27]. However, in cancer patients, this induction likely occurs in an environment where inflammatory mediators are activated. Our previous studies have shown that treatment of mice with the bacterial endotoxin induces inflammation, leading to down-regulation of *Cyp3a11* (the murine ortholog of CYP3A4) gene expression [28, 29]. Thus, it is likely that the microenvironment of inflammation in cancer patients causes attenuation of CYP3A induction by paclitaxel. This can increase overall paclitaxel concentration in cancer patients and can improve chemoresistance, but can have serious implications in terms of increased toxicity. We determined the role of lipopolysaccharide (LPS) in paclitaxel-mediated *Cyp3a11* induction in primary mouse hepatocytes. In order to further elucidate the mechanism of induction and downregulation of *CYP3A* gene expression, we examined the role of the pro-inflammatory cytokine, tumor necrosis factor (TNF)α, on human *CYP3A4* induction by the classical PXR activator, rifampicin.

Cultured primary hepatocytes are considered the "gold standard" model for induction studies. This in vitro system contains full machinery for enzyme regulation required for CYP induction studies, which includes hepatic enzymes (CYPs), hepatobiliary uptake/efflux transporter network, and cofactors (NADPH, UDPGA, GSH, PAPS, etc.). In our primary mouse hepatocyte study, we found that paclitaxel significantly induced expression of *Cyp3a11* (~30-fold), whereas the induction by paclitaxel was ~15-fold in the presence of LPS (Fig. 1). Thus, LPS markedly attenuated *Cyp3a11* induction by paclitaxel.

Similarly, HepG2 cells are widely used for transfection studies and our results show that PXR-mediated induction of CYP3A4-luciferase activity was significantly attenuated in the presence of the pro-inflammatory cytokine, TNFα. When transfected HepG2 cells were treated with 10 μM rifampicin, CYP3A4 luciferase activity was induced up to four folds and this induction was 50 % blocked upon co-treatment with TNFα (Fig. 2). Thus, our in vitro approaches utilizing primary hepatocytes and HepG2 cells demonstrate that induction of *CYP3A* gene expression by paclitaxel or rifampicin is significantly attenuated by inflammatory mediators. These findings indicate novel mechanisms of activation of CYP3A4 expression which can have implications in chemotherapy treatment.

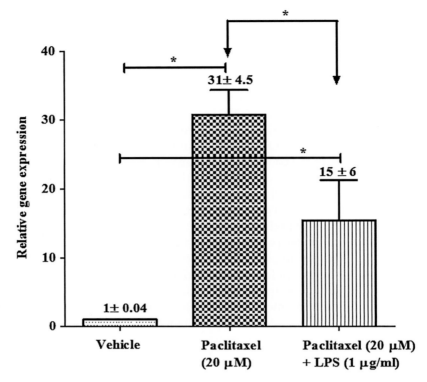

Fig. 1 Induction of *Cyp3a11* gene expression by paclitaxel is attenuated during inflammation in vitro. Primary mouse hepatocytes from C57BL/6 mice were treated with 20 μM paclitaxel in the presence and absence of 1 μg/ml LPS or 0.1 % DMSO (vehicle) for 24 h prior to RNA isolation. Relative fold changes in gene expression for *Cyp3a11* are indicated, normalized to vehicle levels. Data are shown as mean ± S.D. * indicates statistical significance at $p < 0.05$ when compared to the vehicle group

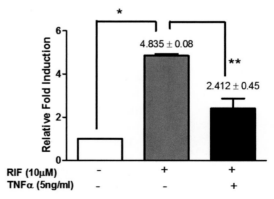

Fig. 2 CYP3A4 induction by rifampicin is attenuated by TNFα. HepG2 cells are cultured and transfected with *CYP3A4* luciferase reporter plasmid and PXR plasmid. After 24 h, they were pretreated with 10 μM rifampicin (RIF) or 0.1 % DMSO followed by 5 ng/ml TNFα or saline for 24 h prior to luciferase assay. Relative fold induction of CYP3A4 is indicated, normalized to DMSO vehicle level (*left most bar*). The data represents means ± SD of duplicate transfections for each treatment

2   Materials

2.1   Primary Mouse Hepatocyte Isolation and Treatment

1. Adult (~6–8 weeks), C57BL/6 male mice (Jackson Laboratory, BarHarbor, Maine).

2. Perfusion buffer 1: Phosphate buffer saline (PBS) without Ca/Mg: 500 ml (VWR International), EGTA: 95 mg (Sigma-Aldrich), HEPES: 0.01 M. pH adjusted to 7.4 and stored at 4 °C.

3. Perfusion buffer 2: PBS with Ca/Mg: 62.5 ml (VWR), Collagenase Type IV: 20 mg (Sigma-Aldrich).

4. Washing/plating medium: Williams medium E: 500 ml (Invitrogen), penicillin/streptomycin solution: 10,000 U/ml (Invitrogen), gentamicin-glutamine solution: 200 mM L (S/C)-glutamine and 5 mg gentamicin per ml (Sigma-Aldrich), insulin-transferrin-sodium selenite (ITS): 5 mg/ml (Sigma-Aldrich), glucagon: 4 ng/ml (Sigma-Aldrich), fetal bovine serum (FBS): 50 ml (Invitrogen). Filter sterilize and store at 4 °C.

5. Treatment medium: Williams medium E: 500 ml (Invitrogen), penicillin/streptomycin solution: 10,000 U/ml (Invitrogen), gentamicin-glutamine solution: 200 mM l-glutamine and 5 mg gentamicin per ml (Sigma-Aldrich). Filter sterilize and store at 4 °C.

6. 70 μm Cell strainer (Fisher).

7. Thread (Henry Schein).

8. Percoll solution (Sigma-Aldrich).

9. Ethanol 70 % (v/v).

10. Isoflurane.

11. Trypan blue (Sigma-Aldrich).

12. 6-Well plates BD Primaria culture dish (BD PharMingen).

13. Water bath.

14. Peristaltic pump (VWR International Mini-pump).

15. Catheter (VWR International).

16. Hemocytometer.

17. Paclitaxel (Sigma-Aldrich).

2.2   Reverse Transcription-Polymerase Chain Reaction Analysis

2.2.1   RNA Isolation

1. Trizol reagent (Sigma-Aldrich).

2. Chloroform.

3. Isopropyl alcohol.

4. UV-vis spectrophotometer (Beckman Coulter, DU800).

2.2.2   cDNA Synthesis

1. High Capacity Reverse Transcription Kit (Applied Biosystems).

2. RNase Inhibitor (Applied Biosystems).

2.2.3 Real-Time PCR	1. Probes and primers (Sigma-Aldrich).
	2. Roche PCR Master Mix (Roche Diagnostics).
	3. 96-Well PCR plate (Axygen Scientific).

2.3 HepG2 Cell Transfection and Treatment

1. HepG2 cells (ATCC).
2. SuperFect Transfection Reagent: 3 mg/ml (QIAGEN).
3. pCYP3A4-pGL3B plasmid was a generous gift from Tirona et al. [30], hPXR-pSG5 plasmid was a generous gift from Kliewer et al. [21], pRL-TK (Promega).
4. Growth medium—Dulbecco's modified Eagle's medium, high glucose: 500 ml (Life Technologies), penicillin/streptomycin solution: 10,000 U/ml (Invitrogen), fetal bovine serum (FBS): 50 ml (Invitrogen).
5. Serum-free medium—Dulbecco's modified Eagle's medium, high glucose: 500 ml (Life Technologies).
6. 96-Well, black, tissue culture-treated plates (BD Falcon).
7. Rifampicin (Sigma Aldrich).
8. Recombinant human TNFα (R&D Systems)

2.4 Luciferase Assay

1. 1× PBS without Ca/Mg: 500 ml (VWR).
2. Dual Luciferase Reporter Assay System (Promega).
3. VWR Microplate Shaker.
4. SpectraMax Microplate Reader/Luminometer.

3 Methods

3.1 Isolation of Primary Mouse Hepatocytes

1. Two-step collagenase perfusion technique (*see* **Note 1**) is followed.
2. Anesthetize mice with isoflurane. Stretch the limbs and pin them using syringe needles. Spray 70 % alcohol to wet the abdomen.
3. Make an incision through the skin at the lower portion of abdomen.
4. Once the incision is complete, push away organs with cotton tips, and expose inferior vena cava (IVC) and portal vein (PV).
5. Using curved hemostat, gently push under the IVC and tie a loose knot before inserting catheter.
6. At an angle (just inferior to the junction with renal veins), gently place the catheter in the IVC (put it in the middle area, away from the muscles) (*see* **Note 2**).
7. Carefully take out the inner needle, and make sure that blood comes out through white part.

8. Secure the knot around the catheter. Attach the tubing of peristaltic pump to the catheter, quickly cut the portal vein, and turn on the pump to pass perfusion buffer-1 for 5–6 min at 4–5 ml/min (*see* **Note 3**).

9. After that, switch to perfusion buffer-2 and perfuse for 3–4 min (~25–30 ml) until the liver begins dissociating under Glisson's capsule.

10. When liver is adequately digested it should leave a small indentation when slightly POKED WITH CURVED FORCEPS. Turn off pump, and take catheter out.

11. Remove liver (whole liver), place in 50 ml conical containing 20 ml of cold plating media, and shake vigorously enough to break up the liver as too much/harsh shaking will lower the cell viability.

12. Pour the suspension through the 70 μm cell strainer; collect in a fresh 50 ml conical tube. Follow each transfer with a small amount of media from an additional conical containing 20 ml plating media to carry through any cells left behind.

13. Place the crude 40 ml cell suspension in a centrifuge set to 4 °C and spin at 50 g for 2 min.

14. Aspirate off the supernatant. Bring pellet to 27 ml on tube with plating media, then add 9 ml of 100 % Percoll, invert to mix, spin at 100 g for 12 min (brake off: acceleration 9; deceleration 3), and aspirate off the supernatant (*see* **Note 3**).

15. Wash the preparation twice as follows: Aspirate off the supernatant, resuspend the pellet in 30 ml of cold plating media, and spin at 50 g for 2 min.

16. Aspirate suspension and resuspend pellet in 10 ml plating media.

17. Measure viability using trypan exclusion method and a hemocytometer (*see* **Note 4**).

3.2 Hepatocyte Culture and Treatment

1. Seed each well with a cell density of 500,000 cells per 2 ml. Place the plate in an incubator with 5 % CO_2 at 37 °C for 4 h.

2. After 4 h, the cells should be attached and form a monolayer (*see* **Note 5**).

3. Replace the media with fresh 2 ml plating media. Cells are maintained for 48 h with daily change of plating medium.

4. On the day of treatment, cells are incubated with treatment media 2 h prior to pretreatment with 1 μg/ml LPS for 10 min, followed by 20 μM of paclitaxel prepared in 100 % dimethyl sulfoxide (DMSO) for 24 h (*see* **Note 6**).

5. At the end of 24 h, pipette out the treatment media, wash the cells twice with 1 ml of 1× PBS, and finally add 0.25 ml of cold TRIzol reagent to each well. Store at –80 °C until RNA preparation for real-time PCR analysis.

3.3 Assessment of Induction of Cyp3a11 by Paclitaxel by Real-Time PCR Analysis

3.3.1 RNA Isolation

1. Total RNA is isolated from mouse liver using TRIzol reagent according to the manufacturer's protocol.

2. Scrape cells with cell scraper, pipette down 10–15 times to break the cell wall, transfer to microcentrifuge tubes, and incubate for 5 min at room temperature.

3. Add 0.2 ml of cold chloroform.

4. Cap tubes securely, shake vigorously by hand for 15 s, and incubate at room temperature/15 min.

5. Centrifuge the samples at 13,400g/15 min/4 °C.

6. After centrifugation, the mixture separates into a lower red, phenol-chloroform phase, an interphase, and a colorless upper aqueous phase.

7. Transfer aqueous phase into a clean tube. Add 0.5 ml of isopropyl alcohol, mix by vigorous shaking, and incubate at room temperature/10 min.

8. Centrifuge at 13,400g/10 min/4 °C.

9. After centrifugation, the RNA precipitates into whitish gel-like pellet at the bottom of the microcentrifuge tube.

10. Remove the supernatant. Add 1 ml of 75 % ethanol to the pellet, mix the sample by vortexing gently, and centrifuge at 5000g/5 min/4 °C (*see* **Note 7**).

11. Air-dry the RNA pellet in the hood for 1–2 min and dissolve in 30–35 μl of RNase-free water (*see* **Note 8**).

12. Measure the total RNA concentration using a UV-vis spectrophotometer at 260 and 280 nm wavelengths (*see* **Note 9**).

3.3.2 cDNA Synthesis

1. cDNA is synthesized using the High Capacity Reverse Transcription Kit.

2. Allow kit components to thaw on ice.

3. Keep work bench RNase free.

4. Prepare the 2× reverse transcription master mix (per 20 μl reaction) using the kit components. Referring to Table 1, calculate the volume of components needed to prepare the required number of reactions (*see* **Note 10**).

5. Prepare 6 μg RNA in a volume of 20 μl by adding RNase free in 0.2 ml thin-coated PCR tubes on ice. To this tube now add 20 μl of 2× reverse transcription master mix. Total volume is 40 μl.

6. Pulse, mix gently, pulse, and load on thermal cycler with the following conditions (Table 2):

7. After PCR, store the samples at −80 °C (*see* **Note 11**).

3.3.3 RTPCR

1. RT-PCR is performed using an ABI PRISM 7300 Sequence System instrument and software (Applied Biosystems, Foster City, CA).

Table 1
Components used in preparing the reverse transcription master mix

Component	Volume/reaction (μl) 1×
10× Reverse transcription buffer	4
25× dNTP mix (100 mM)	1.6
10× Random primers	4
Multiscribe reverse transcriptase (50 U/μl)	2
RNase inhibitor, 20 U/μl	2
RNase-free water	6.4
Total per reaction	20

Table 2
Thermal cycler conditions

	Step 1	Step 2	Step 3	Step 4
Temperature (°C)	25	37	85	4
Time	10 min	120 min	5 s	∞

2. PCR reaction mixture of 25 μl contains 15 μl of PCR mix reagent and 10 μl of 100 ng of cDNA.

3. The contents of the 15 μl of PCR mix reagent are prepared by referring to the table below. Calculate the volume of components needed to prepare the required number of reactions.

Reagent	Vol/1 rxn (μl)
Roche PCR master mix (2×)	11.25
Forward primer (100 μM)	0.075
Reverse primer (100 μM)	0.075
TaqMan probe (100 μM)	0.05
RNase-free water	3.55
Total volume	15 μl

4. Set up the experiment with the following PCR program on ABI Prism 7300 (Table 3).
   - Stage-1: 50 °C for 2 min.
   - Stage-2: 95 °C for 10 min.
   - Stage-3: denaturation—95 °C for 15 s and annealing—60 °C for 1 min, 50 cycles.

5. Cyclophilin is used as housekeeping gene for normalization.

Table 3
Primers and conditions used for RT-PCR

Gene name	Primer sequence
Cyclophilin	Forward primer: GGCCGATGACGAGCCC Reverse primer: TGTCTTTGGAACTTTGTCTGCA Probe sequence: 6TGGGCCGCGTCTCCTTCGA0
Cyp3a11	Forward primer: GGATGAGATCGATGAGG CTCTG Reverse primer: CAGGTATTCCATCTCCATCACAGT Probe sequence: CCAACAAGGCACCTCCCACGTATGA

3.4 Transient Transfection of HepG2 Cells

The following protocol is for transient transfection of HepG2 cells in a 96-well plate. The quantities mentioned below represent the amount needed for one well.

1. The day before transfection, seed 1×10^4 cells/well in 100 µl growth medium. Incubate cells at 37 °C and 5 % CO_2 (*see* **Notes 12** and **13**).

2. After 24 h, i.e., the day of transfection, dilute 2.5 µl of Superfect transfection reagent in 30 µl serum-free medium. Add all plasmid DNAs (pCYP3A4-pGL3B, hPXR-pSG5, and pRL-TK) such that total DNA per well is 0.5 µg. Mix by vortexing for 10 s (*see* **Notes 14** and **15**).

3. Incubate the cocktail mix for 10 min at room temperature.

4. While waiting, gently aspirate the growth medium from 96-well plate.

5. After 10 min, add 150 µl growth medium to the cocktail mix. Mix by inverting the tube three times and immediately transfer to 96-well plate.

6. Incubate cells with transfection complexes for 5 h (*see* **Note 16**).

7. Remove medium containing the remaining complexes from the cells by gentle aspiration.

8. Wash cells once with 100 µl PBS.

9. Add 100 µl fresh cell growth medium/well.

10. Incubate for 24 h after transfection to obtain maximal levels of gene expression (*see* **Note 17**).

3.5 HepG2 Culture and Treatment

1. Pretreat cells at a final concentration of 10 µM rifampicin (use DMSO as solvent) for 30 min or 100 % DMSO as control (*see* **Note 18**).

2. Treat cells at a final concentration of 5 ng/ml TNFα for 24 h or saline as control.

3.6 Dual-Luciferase Reporter Assay Protocol

1. Gently remove growth medium from cultured cells.

2. Add 100 μl PBS to wash the cells once. Briefly swirl the plate to remove detached cells and residual growth medium. Completely remove PBS from the wells.

3. Add 20 μl 1× passive lysis buffer.

4. Place the plate on a shaker/rocker with gentle shaking for 15 min at room temperature (*see* **Note 19**).

5. While the plate is on the shaker, set up the luminometer to record the luminescence reading.

6. Dispense 100 μl Luciferase Assay Reagent II (LAR II) directly to wells using multichannel pipette. Gently tap the plate to mix the reagent and immediately measure firefly luciferase activity (*see* **Note 20**).

7. Add 100 μl 1× Stop and Glo Reagent directly to wells using multichannel pipette. Again, tap and immediately measure Renilla Luciferase activity (*see* **Note 21**).

4 Notes

1. Mouse should not die before the infusion starts as it will lead to blotchy liver due to impaired perfusion and result in poor yield and viability.

2. Puncture of inferior vena cava (IVC) with catheter should be done carefully and in first attempt. Improper puncture will cause blood loss, poor liver perfusion, and digestion, making second attempt impossible.

3. Collagenase performance is variable between lots. Therefore, first test hepatocyte preparation quality and quantity with several lots of collagenase, and then buy larger bottles of good lot numbers. Low viability problem can be solved by incorporating following changes: decrease Percol percentage; reduce collagenase amount; reduce flow rate of perfusion buffers and digestion time; and double the initial centrifugation speed and time. The hepatocyte quality depends on the species and strain, so optimization has to be done accordingly.

4. Measurement of viability of isolated hepatocytes using trypan exclusion method: (a) Count on hemocytometer (10 μl cells + 80 μl PBS + 10 μl trypan blue); (b) count the four quadrants: $(x/4) \times 10^4 \times 10$ cells/ml (usually about 2×10^6 cells/ml). Avoid using hepatocyte preparation with <90 % viability as such preparation will not form confluency and result in dedifferen-

tiation of hepatocytes. Typical yield is between 1.5 and 2 million cells per ml.

5. The isolated hepatocytes should be plated as soon as possible to avoid hypoxia.

6. Concentrations of organic solvents must be kept low (<1 %) to avoid any effects of the organic solvents on the hepatocyte attachment.

7. Take out all the ethyl alcohol by pipetting; otherwise it will reduce the RNA solubility in water and hence result in low yield and stability.

8. Do not let the RNA pellet dry for long time as it will reduce the solubility and stability of RNA. Incubate for 10 min at 55 °C if there is difficulty in dissolving the RNA pellet in RNase-free water. The amount of water used to resuspend will depend on the size of pellet.

9. A ratio of absorbance at 260/280 is used to assess the purity and quantitatively determine concentration of RNA. Generally a ratio of >1.8 is generally accepted as pure RNA. Low ratio indicates impure RNA due to protein contamination.

10. Include additional reactions in the calculations to provide excess volume for the loss that occurs during reagent transfers. Air bubbles may interfere with the assay; therefore eliminate by brief (10 s) centrifugation.

11. Store at −80 °C to prevent degradation for long-term storage of RNA.

12. The optimal confluency at the time of transfection complex addition is 40–80 % (0.5–2.0×10^4 cells/well), although it is preferable to seed less cells (~50 % confluency). High cell density leads to insufficient uptake of transfection complexes, thereby showing decreased expression of the gene of interest. Moreover, overgrown cultures are often more resistant to complete lysis by passive lysis buffer.

13. Ensure that cells with low passage number (<50 splitting cycles) are used. As passage number increases, transfection efficiency decreases.

14. It is important to use serum-free media as FBS and antibiotics present during this step interfere with complex formation and decrease transfection efficiency.

15. Trans effects: When cells are co-transfected with firefly and Renilla vectors, they can potentially affect each other's reporter gene expression especially if one or both contain very strong promoter elements. Therefore, it is necessary to first optimize both the amount of experimental plasmid (luciferase plasmid) and the co-reporter plasmid (Renilla plasmid). Ratio of 10:1 or 5:1 for Firefly:Renilla vector is usually feasible.

16. Incubation with Superfect-DNA complexes for 3–5 h yields optimal results. Although if excessive cell death is observed, decrease the exposure time.

17. Time to obtain maximum gene expression is again dependent on the cell type and gene of interest. It is preferable to incubate the cells for a minimum of 16 h.

18. The percentage of DMSO should not exceed 0.5 % of total cell culture volume/well. High concentration of organic solvent is extremely toxic to HepG2 cells.

19. Ensure that cell monolayer is evenly covered with 1× passive lysis buffer for complete lysis.

20. Never thaw LAR II in water bath as it is heat labile. Thaw at room temperature and mix before use.

21. Firefly luciferase and Renilla activity readings should be at least ten times the baseline reading.

Acknowledgements

This work was supported by NIH grants R21-DA035751 to RG and R01-ES-009132, R01-HL-112516, and R01-ES-019689 to BM.

References

1. McGuire WP, Hoskins WJ, Brady MF, Kucera PR, Partridge EE, Look KY, Clarke-Pearson DL, Davidson M (1996) Cyclophosphamide and cisplatin compared with paclitaxel and cisplatin in patients with stage III and stage IV ovarian cancer. N Engl J Med 334:1–6

2. Brockstein B, Haraf DJ, Stenson K, Fasanmade A, Stupp R, Glisson B, Lippman SM, Ratain MJ, Sulzen L, Klepsch A, Weichselbaum RR, Vokes EE (1998) Phase I study of concomitant chemoradiotherapy with paclitaxel, fluorouracil, and hydroxyurea with granulocyte colony-stimulating factor support for patients with poor-prognosis cancer of the head and neck. J Clin Oncol 16:735–744

3. McGuire WP, Blessing JA, Moore D, Lentz SS, Photopulos G (1996) Paclitaxel has moderate activity in squamous cervix cancer. A Gynecologic Oncology Group study. J Clin Oncol 14:792–795

4. Johnson DH, Paul DM, Hande KR, Shyr Y, Blanke C, Murphy B, Lewis M, De Vore RF III (1996) Paclitaxel plus carboplatin in advanced non-small-cell lung cancer: a phase II trial. J Clin Oncol 14:2054–2060

5. Bolis G, Scarfone G, Polverino G, Raspagliesi F, Tateo S, Richiardi G, Melpignano M, Franchi M, Mangili G, Presti M, Villa A, Conta E, Guarnerio P, Cipriani S, Parazzini F (2004) Paclitaxel 175 or 225 mg per meters squared with carboplatin in advanced ovarian cancer: a randomized trial. J Clin Oncol 22:686–690

6. Monsarrat B, Chatelut E, Royer I, Alvinerie P, Dubois J, Dezeuse A, Roche H, Cros S, Wright M, Canal P (1998) Modification of paclitaxel metabolism in a cancer patient by induction of cytochrome P450 3A4. Drug Metab Dispos 26:229–233

7. Li AP, Kaminski DL, Rasmussen A (1995) Substrates of human hepatic cytochrome P450 3A4. Toxicology 104:1–8

8. Walker D, Flinois JP, Monkman SC, Beloc C, Boddy AV, Cholerton S, Daly AK, Lind MJ, Pearson AD, Beaune PH et al (1994) Identification of the major human hepatic cytochrome P450 involved in activation and N-dechloroethylation of ifosfamide. Biochem Pharmacol 47:1157–1163

9. Marre F, Sanderink GJ, de Sousa G, Gaillard C, Martinet M, Rahmani R (1996) Hepatic biotransformation of docetaxel (Taxotere) in vitro: involvement of the CYP3A subfamily in humans. Cancer Res 56:1296–1302

10. Yao D, Ding S, Burchell B, Wolf CR, Friedberg T (2000) Detoxication of vinca

alkaloids by human P450 CYP3A4-mediated metabolism: implications for the development of drug resistance. J Pharmacol Exp Ther 294:387–395

11. Mathijssen RH, van Alphen RJ, Verweij J, Loos WJ, Nooter K, Stoter G, Sparreboom A (2001) Clinical pharmacokinetics and metabolism of irinotecan (CPT-11). Clin Cancer Res 7:2182–2194

12. Relling MV, Nemec J, Schuetz EG, Schuetz JD, Gonzalez FJ, Korzekwa KR (1994) O-demethylation of epipodophyllotoxins is catalyzed by human cytochrome P450 3A4. Mol Pharmacol 45:352–358

13. Rodriguez-Antona C, Ingelman-Sundberg M (2006) Cytochrome P450 pharmacogenetics and cancer. Oncogene 25:1679–1691

14. Miyoshi Y, Taguchi T, Kim SJ, Tamaki Y, Noguchi S (2005) Prediction of response to docetaxel by immunohistochemical analysis of CYP3A4 expression in human breast cancers. Breast Cancer 12:11–15

15. Petros WP, Hopkins PJ, Spruill S, Broadwater G, Vredenburgh JJ, Colvin OM, Peters WP, Jones RB, Hall J, Marks JR (2005) Associations between drug metabolism genotype, chemotherapy pharmacokinetics, and overall survival in patients with breast cancer. J Clin Oncol 23:6117–6125

16. Kostrubsky VE, Lewis LD, Strom SC, Wood SG, Schuetz EG, Schuetz JD, Sinclair PR, Wrighton SA, Sinclair JF (1998) Induction of cytochrome P4503A by taxol in primary cultures of human hepatocytes. Arch Biochem Biophys 355:131–136

17. Bahadur N, Leathart JB, Mutch E, Steimel-Crespi D, Dunn SA, Gilissen R, Houdt JV, Hendrickx J, Mannens G, Bohets H, Williams FM, Armstrong M, Crespi CL, Daly AK (2002) CYP2C8 polymorphisms in Caucasians and their relationship with paclitaxel 6alpha-hydroxylase activity in human liver microsomes. Biochem Pharmacol 64:1579–1589

18. Taniguchi R, Kumai T, Matsumoto N, Watanabe M, Kamio K, Suzuki S, Kobayashi S (2005) Utilization of human liver microsomes to explain individual differences in paclitaxel metabolism by CYP2C8 and CYP3A4. J Pharmacol Sci 97:83–90

19. Harmsen S, Meijerman I, Beijnen JH, Schellens JH (2007) The role of nuclear receptors in pharmacokinetic drug-drug interactions in oncology. Cancer Treat Rev 33:369–380

20. Bertilsson G, Heidrich J, Svensson K, Asman M, Jendeberg L, Sydow-Backman M, Ohlsson R, Postlind H, Blomquist P, Berkenstam A (1998) Identification of a human nuclear receptor defines a new signaling pathway for CYP3A induction. Proc Natl Acad Sci U S A 95:12208–12213

21. Lehmann JM, McKee DD, Watson MA, Willson TM, Moore JT, Kliewer SA (1998) The human orphan nuclear receptor PXR is activated by compounds that regulate CYP3A4 gene expression and cause drug interactions. J Clin Invest 102:1016–1023

22. Goodwin B, Hodgson E, Liddle C (1999) The orphan human pregnane X receptor mediates the transcriptional activation of CYP3A4 by rifampicin through a distal enhancer module. Mol Pharmacol 56:1329–1339

23. Poso A, Honkakoski P (2006) Ligand recognition by drug-activated nuclear receptors PXR and CAR: structural, site-directed mutagenesis and molecular modeling studies. Mini Rev Med Chem 6:937–947

24. Synold TW, Dussault I, Forman BM (2001) The orphan nuclear receptor SXR coordinately regulates drug metabolism and efflux. Nat Med 7:584–590

25. Raynal C, Pascussi JM, Leguelinel G, Breuker C, Kantar J, Lallement B, Poujol S, Bonnans C, Joubert D, Hollande F, Lumbroso S, Brouillet JP, Evrard A (2010) Pregnane X Receptor (PXR) expression in colorectal cancer cells restricts irinotecan chemosensitivity through enhanced SN-38 glucuronidation. Mol Cancer 9:46

26. Nallani SC, Goodwin B, Maglich JM, Buckley DJ, Buckley AR, Desai PB (2003) Induction of cytochrome P450 3A by paclitaxel in mice: pivotal role of the nuclear xenobiotic receptor, pregnane X receptor. Drug Metab Dispos 31:681–684

27. Ding X, Staudinger JL (2005) Induction of drug metabolism by forskolin: the role of the pregnane X receptor and the protein kinase a signal transduction pathway. J Pharmacol Exp Ther 312:849–856

28. Ghose R, White D, Guo T, Vallejo J, Karpen SJ (2008) Regulation of hepatic drug-metabolizing enzyme genes by Toll-like receptor 4 signaling is independent of Toll-interleukin 1 receptor domain-containing adaptor protein. Drug Metab Dispos 36:95–101

29. Ghose R, Guo T, Vallejo JG, Gandhi A (2011) Differential role of Toll-interleukin 1 receptor domain-containing adaptor protein in Toll-like receptor 2-mediated regulation of gene expression of hepatic cytokines and drug-metabolizing enzymes. Drug Metab Dispos 39:874–881

30. Tirona RG, Lee W, Leake BF, Lan LB, Cline CB, Lamba V, Parviz F, Duncan SA, Inoue Y, Gonzalez FJ, Schuetz EG, Kim RB (2003) The orphan nuclear receptor HNF4alpha determines PXR- and CAR-mediated xenobiotic induction of CYP3A4. Nat Med 9:220–224

Chapter 5

Uptake and Permeability Studies to Delineate the Role of Efflux Transporters

Ramya Krishna Vadlapatla, Dhananjay Pal, and Ashim K. Mitra

Abstract

Chemotherapy is one of the major therapeutic interventions in oncology. Despite numerous advances and intensive research, a large number of patients acquire multidrug resistance (MDR) and no longer respond to chemotherapy. Efflux transporters play a predominant role in mediating MDR. Cellular accumulation (uptake) and permeability studies serve as invaluable methods to detect drug efflux/transport mechanism. These methods are generally performed on transfected cells (e.g., MDCKII-MDR1, MDCKII-MRP2, and MDCKII-BCRP) or cells expressing high amount of intrinsic efflux transporters (Caco-2) utilizing specific inhibitors as positive controls. This chapter presents a method of performing uptake and permeability studies, including the preparation of various buffers required for the study.

Key words Efflux transporters, Uptake, Bidirectional permeability, Transwells, DPBS buffer, Stop and lysis solutions, TEER values

1 Introduction

Cancer remains a major health problem in the USA accounting for nearly one of every four deaths. In 2014, a total of 1,665,540 new cancer cases and 585,720 cancer deaths are projected to occur [1]. Chemotherapy is one of the major therapeutic interventions in oncology. Unlike localized treatments such as radiation and surgery, chemotherapy is considered a systemic treatment. Depending on the type/stage of cancer progression, chemotherapy can be administered alone or in combination with other therapy. Despite numerous advances and intensive research, a large number of patients acquire multidrug resistance (MDR) and no longer respond to chemotherapy. MDR is a clinical outcome where resistance develops to functionally and structurally unrelated drugs. This resistance, either inherent or acquired, represents a major obstacle to successful chemotherapy. Several mechanisms of MDR have been suggested, which can be categorized into

José Rueff and António Sebastião Rodrigues (eds.), *Cancer Drug Resistance: Overviews and Methods*, Methods in Molecular Biology, vol. 1395, DOI 10.1007/978-1-4939-3347-1_5, © Springer Science+Business Media New York 2016

pharmacokinetic (leading to reduced intratumor drug levels) and pharmacodynamic (changes in microenvironment or biology of the cancer cells) pathways [2]. These mechanisms include (1) decreased influx, (2) increased efflux, (3) activation of detoxifying systems, (4) drug sequestration, (5) alteration of specific drug targets, (6) activation of DNA repair, (7) modulation of cellular death pathways, and (8) impact of microenvironment. The first four mechanisms account for pharmacokinetic resistance while the last four mechanisms lead to pharmacodynamic resistance. These mechanisms may function independently or synergistically, leading to treatment failure and rise in cancer deaths [3]. Although the pathways can act independently, they are more often intertwined. Of the various mechanisms involved, up-regulation of efflux transporters and drug-metabolizing enzymes (DMEs) constitute a major resistance phenotype.

Efflux transporters and Phase I DMEs orchestrate a defensive system against various chemotherapeutics. ATP-binding cassette (ABC) efflux transporters are ubiquitous and identified in both prokaryotes and eukaryotes. ABC efflux transporters include a large family of transmembrane proteins involved in translocation of diverse compounds including amino acids, bile salts, drug molecules, metal ions, metabolites, nucleotides, peptides, sterols, and sugars. Synergistic action leading to MDR could possibly arise due to two different factors—overlapping substrate specificity and coordinated regulation of their expression. Anticancer drugs such as docetaxel, doxorubicin, etoposide, imatinib, paclitaxel, teniposide, vinblastine, and vincristine are substrates for both efflux transporters and metabolism enzymes. This close overlap of substrates and tissue distribution may lead to development of MDR [4, 5]. Also, the inducible expression of efflux transporters and DMEs is governed by molecular mechanisms involving nuclear receptors (NRs). Pregnane X receptor (PXR) and constitutive androstane/activated receptor (CAR) are the master orphan NRs involved in xenobiotic metabolism and elimination [6, 7]. Certain chemotherapeutics can activate NRs, leading to increased expression of both efflux transporters and metabolizing enzymes [8, 9]. This untoward activation can lead to reduced oral absorption in the intestine, increased excretion from the liver, and diminished uptake of anticancer agents by the tumor cells. These factors substantially increase the severity of drug resistance.

Several studies including cellular accumulation, drug permeability, gene/protein expression of efflux transporters, and NRs may be considered for analysis of drug resistance mediated by efflux transporters. In this chapter we mainly focus on presenting the methods of delineating uptake and permeability of drug molecules.

2   Materials

Prepare all solutions using ultrapure water (prepared by purifying distilled deionized water to attain a sensitivity of 18 MΩ cm at 25 °C) and analytical/biological grade reagents. Add each ingredient to the stirring water in the beaker and ensure that the solution is thoroughly mixed. Prepare and store all the buffers at 4 °C (unless indicated otherwise). Meticulously follow all waste disposal regulations (as specified by your institute) when disposing waste materials.

1. Dulbecco's phosphate-buffered saline (DPBS) buffer: 130 mM NaCl, 0.03 mM KCl, 7.5 mM Na_2HPO_4, 1.5 mM KH_2PO_4, 1 mM $CaCl_2$, 0.5 mM $MgSO_4$, 20 mM HEPES, and 5 mM glucose. Adjust the solution pH to 7.4 (physiological) using 1 N NaOH or 1 N HCl. pH can be altered based on the study design.

2. Ice-cold stop solution: 200 mM KCl and 2 mM HEPES.

3. Lysis solution: 0.1 % v/v Triton X-100 and 0.3 N NaOH. Once prepared, store the stop buffer solution at room temperature.

3   Methods

3.1   Cellular Drug Accumulation Study

1. Perform cellular drug accumulation studies on confluent and healthy cells cultured in 12-well polystyrene plates (3.8 cm² growth area per well).

2. Prior to experimentation, warm DPBS and drug solutions at 37 °C.

3. Aspirate the culture medium and add 1 mL of DPBS. Incubate the cells at 37 °C for 10 min. Remove the washed buffer and add 1 mL of fresh DPBS again. Perform this washing step for a total of three times.

4. Remove the DPBS completely.

5. Initiate the uptake study by adding 1 mL of radioactive* drug solution into each well and incubate the cells at 37 °C for 30 min (time of incubation can be adjusted based on specific study). Concentration of radioactive drugs may be adjusted to 0.5–2.0 μCi/mL based on specific activity. Nonradioactive drugs appropriately solubilized in DPBS can also be utilized for cellular uptake studies. In that case, the sensitivity of analytical methods is crucial in quantifying the drug amount.

   *Only authorized workers should use radioactive compounds. Authorization can be obtained from institutional radiation safety office.

6. After the incubation period, remove the drug solutions and add 1 mL of ice-cold stop solution. After 3 min, remove the stop solution and add fresh stop solution again. Perform this washing step for a total of three times, 3 min each.

7. Remove the stop solution completely and add 1 mL of lysis solution. Keep the cells aside at room temperature overnight.

8. Next day, perform cell lysis by mixing vigorously using a disposable plastic Pasteur pipette and collect the cell lysate for analysis of radioactivity and protein concentration.

9. Transfer 200 μL of cell lysate into a scintillation vial. Add 3 mL of scintillation cocktail and vortex all the samples for an equal period of time. Measure the amount of radioactivity (DPM—disintegrations per minute) using a scintillation counter.

10. Transfer 20 μL of cell lysates into a disposable cuvette and add 980 μL of Bradford protein estimation reagent. Estimate the amount of protein using bovine serum albumin as an internal standard.

11. Normalize the value of DPM in each well to its respective protein concentration.

12. Cellular accumulation can be calculated using Eq. 1:

$$C_{sample} = (DPM_{sample} / DPM_{donor}) \times C_{donor} \qquad (1)$$

DPM_{sample} and DPM_{donor} represent DPM of sample and donor. C_{donor} and C_{sample} represents the concentration of donor and sample.

3.2 Permeability Study

1. Perform transport studies on cells cultured in 12-well transwell inserts (precoat the inserts with collagen if needed—see Subheading 4).

2. Add appropriate number of cells to produce sufficient tight junctions (transepithelial electrical resistance values greater than 300 Ω/cm^2). Post-seeding the cells, change the medium every day until they become confluent. Add 0.5 mL of medium on the apical chamber and 1.5 mL of medium on the basolateral chamber.

3. Prior to experimentation, warm DPBS and drug solutions at 37 °C.

4. Aspirate the culture medium and add 0.5 mL of DPBS to the apical (AP) chamber and 1.5 mL to the basolateral (BL) chamber. Incubate the cells at 37 °C for 10 min. Remove DPBS and add fresh DPBS again. Perform this washing step three times.

5. For AP–BL transport study, add 0.5 mL of radioactive drug solution to the AP side (donor) and 1.5 mL of DPBS to the BL side (receiver). Incubate the cells at 37 °C. For nonradioactive

permeability studies, drugs solubilized in DPBS could be utilized. In that case, the sensitivity of analytical methods is crucial in quantifying the drug amount.

6. Withdraw samples (100 μL) from the BL chamber (receiver) at predetermined time points (e.g., 15, 30, 45, 60, 90, 120, 150, and 180 min) and replace with the same amount (100 μL) of DPBS to maintain sink conditions.

7. Collect each sample into a scintillation vial. Add 3 mL of scintillation cocktail and vortex all the samples for an equal period of time. Measure the amount of radioactivity (DPM) using a scintillation counter.

8. For BL–AP transport study, add 1.5 mL of radioactive drug solution into BL chamber and 0.5 mL of DPBS into AP chamber. Incubate the cells at 37 °C.

9. Withdraw samples (100 μL) from the AP chamber at predetermined time points (e.g., 15, 30, 45, 60, 90, 120, 150, and 180 min) and replace with the same amount (100 μL) of DPBS to maintain sink conditions.

10. Collect each sample into a scintillation vial. Add 3 mL of scintillation cocktail and vortex all the samples for an equal period of time. Measure the amount of radioactivity (DPM) using a scintillation counter.

11. Cumulative amounts of radioactive substance transported across the cell monolayers for the total time period were calculated using Eq. 2:

$$Q = (C_s \times V_t) + \sum_{t=0}^{t=t-1} (C_s \times V_p) \qquad (2)$$

C_s is the amount of radioactive substance transported at a time t, V_p is the volume of the sample withdrawn, and V_t is the total volume in the receiver chamber.

12. Flux value (J) is calculated according to Eq. 3:

$$J = \frac{\Delta Q}{\Delta t \times A} \qquad (3)$$

$\Delta Q / \Delta t$ represents the amount of radioactive substance transported per unit time across the cell monolayers with a cross-sectional area A.

13. Permeability (P) value was then calculated according to Eq. 4:

$$P = \frac{J}{C_{donor}} \qquad (4)$$

14. Thus the efflux ratio was calculated from AP–BL and BL–AP permeability studies as shown in Eq. 5:

$$\text{Efflux ratio} = \frac{\text{BL} - \text{AP permeability}}{\text{AP} - \text{BL permeability}} \tag{5}$$

4   Notes

1. It is best to prepare fresh solutions each time considering the sensitivity of the analyte and the experiment.

2. It is important to note that Triton X-100 is nonselective in nature and may extract proteins along with the lipids [10].

3. If the cells exhibit low TEER values, it is important to precoat the transwells with collagen; this promotes cell attachment and spreading. For collagen treatment, add sufficient amount of collagen into each transwell to spread the entire surface. Remove the collagen immediately and sterilize the coated transwells under UV light overnight.

4. It is suggested to study the permeability of paracellular marker such as mannitol simultaneously with the radioactive material. This would help in predicting the membrane integrity of the cells during the course of the study.

References

1. Siegel R et al (2014) Cancer statistics, 2014. CA Cancer J Clin 64(1):9–29
2. Mellor HR, Callaghan R (2008) Resistance to chemotherapy in cancer: a complex and integrated cellular response. Pharmacology 81(4): 275–300
3. Szakacs G et al (2006) Targeting multidrug resistance in cancer. Nat Rev Drug Discov 5(3):219–234
4. Azzariti A et al (2011) The coordinated role of CYP450 enzymes and P-gp in determining cancer resistance to chemotherapy. Curr Drug Metab 12(8):713–721
5. Wacher VJ, Wu CY, Benet LZ (1995) Overlapping substrate specificities and tissue distribution of cytochrome P450 3A and P-glycoprotein: implications for drug delivery and activity in cancer chemotherapy. Mol Carcinog 13(3):129–134
6. Chen Y et al (2012) Nuclear receptors in the multidrug resistance through the regulation of drug-metabolizing enzymes and drug transporters. Biochem Pharmacol 83(8):1112–1126
7. Tolson AH, Wang H (2010) Regulation of drug-metabolizing enzymes by xenobiotic receptors: PXR and CAR. Adv Drug Deliv Rev 62(13):1238–1249
8. Meijerman I, Beijnen JH, Schellens JH (2008) Combined action and regulation of phase II enzymes and multidrug resistance proteins in multidrug resistance in cancer. Cancer Treat Rev 34(6):505–520
9. Harmsen S et al (2007) The role of nuclear receptors in pharmacokinetic drug-drug interactions in oncology. Cancer Treat Rev 33(4): 369–380
10. Jamur MC, Oliver C (2010) Permeabilization of cell membranes. Methods Mol Biol 588:63–66

Chapter 6

Dynamics of Expression of Drug Transporters: Methods for Appraisal

Marta Gromicho, José Rueff, and António Sebastião Rodrigues

Abstract

Cellular drug resistance remains a major concern in cancer therapy and usually results from increased expression of ABC drug transporters. Imatinib mesylate (IM), a competitive inhibitor of BCR/ABL1 tyrosine kinase activity, is the current standard therapy for chronic myeloid leukaemia (CML) which is caused by the *BCR/ABL1* gene fusion encoding a constitutively active tyrosine kinase. However, up to 33 % of CML patients do not respond to therapy either initially or due to acquired resistance. Usually, IM resistance is due to the presence of *BCR-ABL1* mutations but in many cases resistance is far from being completely understood or from being satisfactorily addressed from a therapeutic standpoint. Although second- and third-generation TKIs (e.g., dasatinib (DA), nilotinib, and bosutinib) were developed to override this phenomenon, resistance remains an unsolved problem. Above all, as more patients are treated with TKIs, more cases of resistance are expected and the discovery of biomarkers of resistance acquires a crucial clinical significance.

We established a valuable in vitro experimental system that mimics the acquired resistance in the absence of mutations. It was developed by the continuous exposure of K562, a human CML-derived cell line expressing BCR-ABL gene, to increasing concentrations of IM and DA (over 36 and 24 weeks, respectively) allowing us to obtain several cell lines with different resistance levels, and therefore to evaluate drug transporters' role in the dynamic cellular responses allied with resistance evolution. The development of such cell models is fundamental to understand the role of drug transporters in resistance since the majority of previous studies were performed on cell lines engineered to over-express a single transporter.

Drug transporters were overexpressed in the majority of resistant cell lines and cell lines from all levels of resistance had increased expression of more than one drug transporter. However, the transporters that attain higher mRNA overexpression (e.g., ABCB1 and ABCG2) did not substantiate a linear relation with the level of resistance. Also, variation in expression of these genes occurs over time of exposure to the same concentration of IM while maintaining resistance, suggesting that resistance mechanisms could vary dynamically in patients as disease progresses. Indeed, we observed that while responding patients demonstrated stable transporters' expression signatures in consecutive samples, in IM-resistant patients they vary significantly over time, advising caution when comparing single-point samples from responsive and resistant patients.

Key words K562 cell line, Drug transporters, Chronic myeloid leukemia, Multidrug resistance, Gene expression

José Rueff and António Sebastião Rodrigues (eds.), *Cancer Drug Resistance: Overviews and Methods*, Methods in Molecular Biology, vol. 1395, DOI 10.1007/978-1-4939-3347-1_6, © Springer Science+Business Media New York 2016

1   Introduction

Cellular drug resistance is a major obstacle in cancer therapy. Cancer cells can acquire resistance to a single drug, to a class of cytotoxic or novel targeted drugs, or to a broad spectrum of unrelated drugs, known as multidrug resistance (MDR). Several mechanisms such as reduced drug uptake, increased drug efflux, activation of DNA repair, and defective apoptotic pathways can contribute to it [1]. Most commonly, however, MDR results from the active, ATP-dependent transport of drugs out of the cell by efflux pumps belonging to the ATP-binding cassette (ABC) family of transporters [1].

The 53 human ABC genes identified in the human genome (www.membranetransport.org) are involved in the efflux of substrates such as phospholipids, sterols, bile salts, peptides, and at least 18 have been described to be associated with drug transport in vitro [2]. It is now well known that different ABC efflux transporters such as P-glycoprotein (ABCB1), multidrug resistance-associated proteins (MRPs, e.g., ABCC1, ABCC3), or the breast cancer resistance protein (ABCG2) actively regulate the traffic of small molecules across the cell membrane, being therefore key determinants of intracellular drug concentrations [3]. Non-ABC proteins such as the major vault protein (MVP) may also be involved, due to their role in the nuclear–cytoplasmic transport of cytotoxic agents [4].

Imatinib mesylate (IM), a competitive inhibitor of BCR/ABL tyrosine kinase activity, is the current standard therapy for chronic myeloid leukaemia (CML) [5, 6]. CML is the most common myeloproliferative disorder accounting for 15–20 % of all leukaemia cases and is characterized by the presence of the Philadelphia chromosome (Ph+) resulting from a balanced translocation that generates a BCR/ABL1 gene fusion encoding a constitutively active tyrosine kinase. Although IM induces complete cytogenetic remission in most patients, about 33 % of newly diagnosed patients exhibit suboptimal response or treatment failure [7–9] and as more patients are treated with IM, more cases of resistance are expected.

CML mutations in the kinase domain of BCR-ABL1 account for about 40 % of all cases of resistance but still many causes of resistance are unclear. Although dasatinib (DA), nilotinib, and bosutinib, the second- and third-generation tyrosine kinase inhibitors (TKIs), have been developed to override the phenomenon resistance is far from being completely understood or from being satisfactorily addressed from a therapeutic standpoint [5, 10]. Therefore the search for early biomarkers of resistance and the elucidation of resistance mechanisms are of crucial clinical significance.

Adequate IM plasma level is essential for a good clinical response [11, 12]. Influx of IM is mediated by SLC22A1, an organic cation transporter (OCT1), which is likely to play an important role in IM resistance [13, 14].

Despite the large number of articles published thus far, it is not known which of the ABC transporters is the most crucial for acquired resistance of CML cells to imatinib. This difficulty is due, in part, to the fact that the majority of previous studies were performed on cell lines engineered to over-express a single transporter [15–18].

To develop an experimental system to perform molecular and dynamic studies of the role of drug transporters on CML drug resistance we continuously exposed K562 cells to increasing concentrations of IM and DA. This methodology allowed us to create a model system that mimics the acquired resistance in the absence of mutations.

Thus, we have been studying expression levels of the genes: ABCB1, ABCC1, ABCC3, ABCG2, SLC22A1, and MVP in K562 cell lines resistant to different concentrations of IM and DA, to evaluate their role in the dynamic cellular responses allied with resistance evolution [19].

We found that several drug transporters (ABCB1, ABCC1, ABCG2, SLC22A1, and MVP) are activated during the development of resistance to increasing doses of IM and DA, in CML K562 cells [19]. Drug transporters were overexpressed in the majority of resistant cell lines and all resistant cell lines had increased expression of more than one drug transporter. However, we found a discrepancy between the level of resistance and the amount of ABC transporter mRNA over-expression as the transporters that attain higher mRNA overexpression (e.g., ABCB1 and ABCG2) were observed in intermediate levels of resistance. Also, variation in expression of these genes occurs over time of exposure to the same concentration of IM while maintaining resistance, suggesting that resistance mechanisms could vary dynamically in patients as disease progresses.

Results excluded both upregulation of BCR-ABL and mutations as relevant resistance mechanisms. And, even if resistant cells had altered DNA repair capacity, it does not explain the observed resistant phenotype [19, 20].

We theorize that drug transporters may mediate the initial stages of resistance but after prolonged exposure and for higher doses of TKI some other mechanisms arise and replace them. Also, ABC transporters other than the three usual suspects (ABCB1, C1, and G2) can also influence treatment outcome. The development of drug-resistant phenotype is usually a multifactorial process, consequence of a complex network of various cellular pathways and molecular mechanisms that are commonly upregulated in tandem in MDR cells.

Hence, to profile the expression signatures of drug transporters through IM therapy and correlate them with resistance, we quantified mRNA expression levels of *SLC22A12*, *ABCB1*, *ABCC1*,

ABCG2, and *MVP* in consecutive samples of CML patients who were either responsive or resistant to IM, followed up from 2 to 73 months with a median observation of 46 months [21]. This amount of follow-up samples allowed us to study drug resistance as a function of time and evaluate the dynamic of drug transporters related with resistance evolution.

The studied genes had higher expression levels in follow-up than in diagnosis samples, indicating a possible induction in expression. IM-sensitive patients presented significantly higher values of *SLC22A1* expression, and highlighted the possible role of the influx transporter SLC22A1 in treatment response, since not only the patients with higher SLC22A1 expression attain better treatment outcome but those with lower expression are more likely to develop resistance. Irrespectively of the response to treatment, gene expression values over time were correlated for most genes suggesting that the equilibrium between influx and efflux of the cell is important to determine drug response. However, while responding patients showed stable expression signatures in consecutive samples, there was considerable variation in IM-resistant patients, indicating that single-point sampling expression signatures are not reliable to predict clinical outcomes or prognostic features in these patients [21].

In light of these results and considering the chronic administration of TKIs, we strongly suggest that any possible interactions with MDR transporters should be studied as a function of time with a detailed characterization, in particular analyzing periods of prolonged drug exposure, as described below.

2 Materials

2.1 Cell Lines

1. The human CML K562 cell line expressing BCR-ABL1 (DSMZ; German National Resource Centre for Biological Material).

2. RPMI 1640 medium (Sigma) supplemented with 10 % fetal bovine serum (Sigma) and 1 % of penicillin–streptomycin (Sigma) under an atmosphere of 5 % CO_2 at 37 °C.

3. Stock dilutions of both imatinib and dasatinib in dimethyl sulfoxide at a concentration of 10 mM and 20 mM, respectively, stored at –20 °C.

4. Laminar flow hood (Bioair Top safe 1.2).

5. Inverted microscope (Nikon Eclipse TE 200-S).

6. Centrifuge (Technica Centric 322A).

7. Thermal Bath (Memmert).

8. Steam autoclave.

9. Freezer –80 °C.

10. Liquid nitrogen container.

2.2 RNA Isolation and cDNA Synthesis

1. All Prep DNA, RNA, Protein Kit (Qiagen).
2. 14.3 M β-mercaptoethanol.
3. Nanodrop spectrophotometer Nd-1000 (Thermo Scientific).
4. High Capacity cDNA Reverse Transcription Kit with RNase Inhibitor (Applied Biosystems).
5. GeneAmp® PCR System 9700 (Applied Biosystems).
6. Centrifuge (Eppendorf 5415D).

2.3 Quantitative Real-Time PCR

1. 7300 qRT-PCR system from Applied Biosystems.
2. MicroAmp® Optical Adhesive Film (Applied Biosystems).
3. MicroAmp® Optical 96-Well Reaction Plate (Applied Biosystems).
4. TaqMan probes and TaqMan Universal PCR Master Mix.
5. Assay-on-Demand Products from Applied Biosystems: ABCB1—Hs00184491_m1; ABCC1—Hs00219905_m1; ABCC3—Hs00358656_m1; ABCG2—Hs00184979_m1; LRP/MVP—Hs00245438_m1; SLC22A1 (SLC22A1)—Hs00427550_m1; BCR-ABL1—Hs03024784_ft; GAPDH—4352934E; and GusB—4333767 F.

3 Methods

3.1 Cell Lines: In Vitro Experimental System

The experimental strategy for establishing a model system that mimics the acquired resistance is described ahead. Then, such a model can be used to perform molecular and dynamic studies of drug resistance at many levels. We used it to study the expression levels of influx (SLC22A1) and efflux (ABCB1, ABCC1, ABCC3, ABCG2) transporter genes and the major vault protein (MVP) in cell lines resistant to different concentrations of IM and DA, to evaluate their role in drug resistance [19].

The human K562 cell line (CML cell line expressing BCR-ABL) was incubated with increasing concentrations of IM, starting with a concentration of 0.05 μM, over a period of 36 weeks, to establish five different IM-resistant cell lines.

After acquiring the ability to grow in the presence of a specific concentration of the drug, a proportion of cells were frozen, and the remaining cells were grown at the next highest drug concentration. The level of resistance was defined by the imatinib concentration at which the growth rate of cells was comparable to that of untreated parental cells. Additionally, we verified that this capacity was maintained several months after cells had been frozen. In this way, subpopulations of cells which were able to grow in the presence of 0.25, 0.5, 1.0, 2.0, and 5.0 μM imatinib were created. The same methodology was applied to develop the cell lines resistant to 0.5, 0.75, and 1.5 nM dasatinib. The latter was achieved after 24 weeks of drug exposure. All resistant cell lines have their respective passage control, meaning K562 wild-type cells (K562_ wt) that were grown in parallel, in the same conditions, except that they were not exposed to the tyrosine kinase inhibitor (TKI).

To determine the stability and/or progression of the resistant phenotype, we selected K562_IM-1.0 and K562_IM-5.0 cell lines, and maintained them in the continuous presence of the correspondent concentration of IM for more than 20 weeks after the acquisition of resistance.

The existence of mutations in kinase domain of BCR-ABL was excluded by direct sequencing in all the resistant and K562_wt cells. We also measured BCR-ABL mRNA transcripts.

3.2 Cell Lines

1. Culture the human CML K562 cell line expressing *BCR-ABL1* in RPMI 1640 medium supplemented with 10 % fetal bovine serum, 100 U/ml penicillin, and 100 μg/ml streptomycin under an atmosphere of 5 % CO_2 at 37 °C.

2. Pass cells twice weekly. At each passage, cells are counted with a hemocytometer to ensure that both resistant and wild-type cells are growing at a similar concentration. This concentration is approximately 1×10^5 cells per mL after each passage (*see* **Note 1**).

3. Dilute both imatinib and dasatinib in dimethyl sulfoxide and prepare stock solutions of 10 mM and 20 mM, respectively. Use a volume of DMSO per culture flask less than 1 % the volume of the culture medium.

3.3 Freezing Cells

1. Transfer cells to a centrifuge tube (15 ml sterile centrifuge tube).

2. Count the cells using a hemocytometer. The viability should be over 90 % to ensure that the cells are healthy enough for freezing.

3. Spin down at $300 \times g$ for 5 min and remove medium.

4. Resuspend cells in enough freezing medium (RPMI supplemented with 20 % FBS and 7.5 % DMSO) to create a cell suspension of 1×10^6 cells per ml. Pipette up and down to ensure even mixture and aliquot about 1 ml into storage vials.

5. Transfer cells immediately to −20 °C for 1 h, followed by −80 °C overnight before permanent storage in liquid nitrogen (*see* **Note 2**).

3.4 RNA Isolation and cDNA Synthesis

1. Use the All Prep DNA, RNA, Protein Kit for RNA extraction.

2. Add 10 μl of β-ME per 1 ml of Lysis Buffer RLT (included in the Kit) (*see* **Note 3**).

3. Determine the number of cells. Pellet the appropriate number of cells by centrifuging at $300 \times g$ for 5 min. Carefully remove all supernatant by aspiration (*see* **Note 4**).

Table 1
Reverse transcription (RT) reaction mix

Component	Volume/reaction (μL)	
	+RT	**−RT**
2× RT buffer	10.0	10.0
20× RT enzyme mix	1.0	–
Nuclease-free H$_2$O	Q.S.[a] to 20 μL	Q.S.[a] to 20 μL
Sample	Up to 9 μL	Up to 9 μL
Total per reaction	20.0	20.0

[a]Quantity sufficient

4. Disrupt the cells by adding Buffer RLT in a proportion of 350 μl per 5×10^6 cells and homogenize the lysate by vortexing for 1 min.

5. Freeze cell lysates at −80 °C until RNA extraction. This was done strictly following the kit manufacturer's protocol.

6. Estimate the concentration and purity of resulting RNA at 260 and 280 nm using the Nanodrop spectrophotometer Nd-1000 and only those samples with A_{260}-to-A_{280} ratios between 1.9 and 2.1 shall be used (*see* **Note 5**).

7. Reverse transcribe 2 μg of total RNA with the High Capacity RNA-to-cDNA Kit in a final volume of 20 μL.

8. Prepare the reverse transcription (RT) reaction mix using the kit components (*see* **Note 6**) according to the table (Table 1).

9. Include no-template controls (NTC) and no-reverse-transcriptase controls (RT negative) for each cDNA synthesis.

10. Perform RT in a thermal cycler according to the following program: 37 °C, 60 min, 95 °C, 5 min, 4 °C ∞.

3.5 Quantitative Real-Time PCR

For quantitative mRNA expression analysis, a real-time RT-PCR protocol is suited. There are different well-established real-time RT-PCR methodologies available. We use an Applied Biosystems instrument (7300) and PCR product detection using TaqMan probes (Table 2) and TaqMan Universal PCR Master Mix.

1. Prepare a reaction mixture of 20 μL as described: 10 μL of TaqMan Universal PCR Master Mix, 1 μL of TaqMan Assay, and 9 μL of diluted cDNA (in each reaction 15 ng of cDNA is quantified). Each assay and sample is run in triplicate.

2. Measure mRNA levels using available Assay-on-Demand Products from Applied Biosystems running the real-time PCR reactions for 40 cycles using universal cycling conditions

Table 2
Information given by the manufacture on assay-on-demand products of the studied target genes

Gene symbol	AB ref	Gene name	Location chromosome	Target exon	Reporter dye	Sequence
ABCB1	Hs00184491_m1	ATP-binding cassette, sub-family B (MDR/TAP), member 1	7	23	FAM-MGB	CAGGTACCATACAGAAACTCTTTGA
ABCC1	Hs00219905_m1	ATP-binding cassette, subfamily C (CFTR/MRP), member 1	16	23	FAM-MGB	GTGGCCAACAGGTGGCTGGCCGTGC
ABCC3	Hs00358656_m1	ATP-binding cassette, subfamily C (CFTR/MRP), member 3	17	8	FAM-MGB	CAGCTGCTCAGCATCCTGATCAGGT
ABCG2	Hs00184979_m1	ATP-binding cassette, subfamily G (WHITE), member 2	4	5	FAM-MGB	AGACTCCAAGGTTGGAACTCAGTTT
MVP	Hs00245438_m1	Major vault protein	16	10	FAM-MGB	TGGCCTACAACTGGCACTTTGAGGT
SLC22A1	Hs00427550_m1	Solute carrier family 22 (organic cation transporter), member 1	6	4	FAM-MGB	TGCTCTACTACTGGTGTGTGCCGGA
BCR-ABL	Hs03024541_ft	Fusion	-	-	FAM-MGB	GTCCACAGCCATTCCGCTGACCATCA
GAPDH	4352934E	Glyceraldehyde-3-phosphate dehydrogenase	12	3	FAM-MGB	
GusB	4333767 F	Beta glucuronidase	7	11	FAM-MGB	

(95 °C for 10 min, followed by 40 cycles of 95 °C for 15 s and 60 °C for 1 min).

3. Exclude all samples from c-DNA synthesis which the NTC and/or the RT negative amplify, showing that primer–dimer formation and genomic DNA contamination occurred.

4. Calculate the relative gene expression levels by the $2^{-\Delta\Delta CT}$ method [22] using a sample of K562_wt from an early passage as calibrator. The threshold cycle (Ct) is defined as the actual PCR cycle when the fluorescence signal increased above the background threshold. Average Ct values from duplicate or triplicate qRT-PCR reactions are normalized to average Ct values for endogenous housekeeping genes *GAPDH* (*glyceraldehyde-3-phosphate dehydrogenase*) for transporters and *GusB* (*beta glucuronidase*) for *BCR-ABL1* (*see* **Note** 7) from the same cDNA preparations and then are compared to the normalized value of the calibrator sample. Report values as average of triplicate analysis.

5. Ensure the validity of the comparison between the mRNA expressions by checking the efficiency of PCR. It should be over 95 % and similar for all qRT-PCR assays. For that, perform a dilution series (10^7–10^0) of any cDNA containing the target gene sequence and the endogenous housekeeping gene sequence. Then, establish standard curves for the target genes and the housekeeping genes and calculate the efficiency by the slope resultant from the linearization of values obtained for each dilution.

3.6 Data Analysis

1. Repeat all qRT-PCR reactions at least three times using cDNA from at least two independent synthesis.

2. Show results as the mean with standard deviation (SD).

3. Calculate the statistical significance of differences between means by Student's *t*-test, at 95 % confidence interval.

3.7 Mutation Analysis

1. Amplify an 863 base pair (bp)–fragment containing the *BCR-ABL1* kinase domain from cDNA in a semi-nested PCR and sequenced in the forward and reverse directions as described by Branford and Hughes [23].

2. A first-stage PCR used forward primer BCRF (5_-TGACCAA CTCGTGTGTGAAACTC) and reverse primer ABL1KinaseR (5_-TCCACTTCGTCTGAGATACTGGATT) and a second-stage PCR used forward primer ABL1kinaseF (5_-CGCAACA AGCCCACTGTCT) and the reverse primer ABL1kinaseR.

3. Compare the direct sequences of the K562 cell line with the GeneBank NM_005157.3 *ABL1* wild-type cDNA reference sequence using the Software of BioEdit Sequence Alignment Editor (Department of Microbiology, North Carolina State University).

4   Notes

1. Morphologically, K562 cells are round large, single cells in suspension. Their doubling time is about 30–40 h. Most cells are undifferentiated mononucleated blasts.

2. This step must be done as soon as the cells are in freezing media. DMSO and some other cryoprotectants are toxic to cells and so should not be exposed to the cells at room temperature for any longer than necessary. Thawing of the vials and placing of the cell suspension back into culture media should also be done very quickly for the same reasons.

3. Buffer RLT containing β-ME can be stored at room temperature (15–25 °C) for up to 1 month.

4. Incomplete removal of cell culture medium will inhibit lysis and dilute the lysate, affecting the conditions for nucleic acid purification. Both effects may reduce nucleic acid yields and purity.

5. The ratio of the readings at 260 and 280 nm (A260/A280) provides an estimate of the purity of RNA with respect to contaminants that absorb in the UV spectrum, such as protein. However, the A260/A280 ratio is influenced considerably by pH. Since water is not buffered, the pH and the resulting A260/A280 ratio can vary greatly. Lower pH results in a lower A260/A280 ratio and reduced sensitivity to protein contamination. For accurate values, we recommend measuring absorbance in 10 mM Tris–Cl, pH 7.5. Pure RNA has an A260/A280 ratio of 1.9–2.1 in 10 mM Tris–Cl, pH 7.5. Always be sure to calibrate the spectrophotometer with the same solution used for dilution.

6. Prepare the RT reaction on ice. Include additional reactions in the calculations to provide excess volume for the loss that occurs during reagent transfers.

7. The ideal endogenous control should have a constant RNA transcription level under different experimental conditions and be sufficiently abundant across different tissues and cell types. It should also be expressed at a similar level as the genes under study. We choose GusB to be the endogenous control of BCR-ABL.

Acknowledgments

This work was supported by grant PEst-OE/SAU/UI0009/2014 from Fundação de Ciência e Tecnologia (FCT).

References

1. Rodrigues AS, Dinis J, Gromicho M et al (2012) Genomics and cancer drug resistance. Curr Pharm Biotechnol 13:651–673

2. Lage H (2008) An overview of cancer multidrug resistance: a still unsolved problem. Cell Mol Life Sci 65:3145–3167

3. Gillet J-P, Efferth T, Remacle J (2007) Chemotherapy-induced resistance by ATP-binding cassette transporter genes. Biochim Biophys Acta 1775:237–262

4. Scheffer GL, Kool M, Heijn M et al (2000) Specific detection of multidrug resistance proteins MRP1, MRP2, MRP3, MRP5, and MDR3 P-glycoprotein with a panel of monoclonal antibodies. Cancer Res 60:5269–5277

5. Bhamidipati PK, Kantarjian H, Cortes J et al (2013) Management of imatinib-resistant patients with chronic myeloid leukemia. Ther Adv Hematol 4:103–117

6. Jabbour E, Cortes J, Kantarjian H (2011) Long-term outcomes in the second-line treatment of chronic myeloid leukemia: a review of tyrosine kinase inhibitors. Cancer 117:897–906

7. Hochhaus A, O'Brien SG, Guilhot F et al (2009) Six-year follow-up of patients receiving imatinib for the first-line treatment of chronic myeloid leukemia. Leukemia 23:1054–1061

8. de Lavallade H, Apperley JF, Khorashad JS et al (2008) Imatinib for newly diagnosed patients with chronic myeloid leukemia: incidence of sustained responses in an intention-to-treat analysis. J Clin Oncol 26:3358–3363

9. Marin D, Milojkovic D, Olavarria E et al (2008) European LeukemiaNet criteria for failure or suboptimal response reliably identify patients with CML in early chronic phase treated with imatinib whose eventual outcome is poor. Blood 112:4437–4444

10. Soverini S, Colarossi S, Gnani A et al (2006) Contribution of ABL kinase domain mutations to imatinib resistance in different subsets of Philadelphia-positive patients: by the GIMEMA Working Party on Chronic Myeloid Leukemia. Clin Cancer Res 12:7374–7379

11. Larson RA, Druker BJ, Guilhot F et al (2008) Imatinib pharmacokinetics and its correlation with response and safety in chronic-phase chronic myeloid leukemia: a subanalysis of the IRIS study. Blood 111:4022–4028

12. Picard S, Titier K, Etienne G et al (2007) Trough imatinib plasma levels are associated with both cytogenetic and molecular responses to standard-dose imatinib in chronic myeloid leukemia. Blood 109:3496–3499

13. White DL, Saunders VA, Dang P et al (2007) Most CML patients who have a suboptimal response to imatinib have low OCT-1 activity: higher doses of imatinib may overcome the negative impact of low OCT-1 activity. Blood 110:4064–4072

14. Clark RE, Davies A, Pirmohamed M et al (2008) Pharmacologic markers and predictors of responses to imatinib therapy in patients with chronic myeloid leukemia. Leuk Lymphoma 49:639–642

15. Brendel C, Scharenberg C, Dohse M et al (2007) Imatinib mesylate and nilotinib (AMN107) exhibit high-affinity interaction with ABCG2 on primitive hematopoietic stem cells. Leukemia 21:1267–1275

16. Burger H, van Tol H, Boersma AWM et al (2004) Imatinib mesylate (STI571) is a substrate for the breast cancer resistance protein (BCRP)/ABCG2 drug pump. Blood 104:2940–2942

17. Houghton PJ (2004) Imatinib mesylate is a potent inhibitor of the ABCG2 (BCRP) transporter and reverses resistance to topotecan and SN-38 in vitro. Cancer Res 64:2333–2337

18. Nakanishi T, Shiozawa K, Hassel BA et al (2006) Complex interaction of BCRP/ABCG2 and imatinib in BCR-ABL-expressing cells: BCRP-mediated resistance to imatinib is attenuated by imatinib-induced reduction of BCRP expression. Blood 108:678–684

19. Gromicho M, Dinis J, Magalhães M et al (2011) Development of imatinib and dasatinib resistance: dynamics of expression of drug transporters ABCB1, ABCC1, ABCG2, MVP, and SLC22A1. Leuk Lymphoma 52:1980–1990

20. Dinis J, Silva V, Gromicho M et al (2012) DNA damage response in imatinib resistant chronic myeloid leukemia K562 cells. Leuk Lymphoma 53:2004–2014

21. Gromicho M, Magalhães M, Torres F et al (2013) Instability of mRNA expression signatures of drug transporters in chronic myeloid leukemia patients resistant to imatinib. Oncol Rep 29:741–750

22. Livak KJ, Schmittgen TD (2001) Analysis of relative gene expression data using real-time quantitative PCR and the 2(-Delta Delta C(T)) method. Methods 25:402–408

23. Branford S, Hughes T (2006) Detection of BCR-ABL mutations and resistance to imatinib mesylate. Methods Mol Med 125:93–106

Fluorimetric Methods for Analysis of Permeability, Drug Transport Kinetics, and Inhibition of the ABCB1 Membrane Transporter

Ana Armada, Célia Martins, Gabriella Spengler, Joseph Molnar, Leonard Amaral, António Sebastião Rodrigues, and Miguel Viveiros

Abstract

The cell membrane P-glycoprotein (P-gp; MDR1, ABCB1) is an energy-dependent efflux pump that belongs to the ATP-binding cassette (ABC) family of transporters, and has been associated with drug resistance in eukaryotic cells. Multidrug resistance (MDR) is related to an increased expression and function of the ABCB1 (P-gp) efflux pump that often causes chemotherapeutic failure in cancer. Modulators of this efflux pump, such as the calcium channel blocker verapamil (VP) and cyclosporine A (CypA), can reverse the MDR phenotype but in vivo studies have revealed disappointing results due to adverse side effects. Currently available methods are unable to visualize and assess in a real-time basis the effectiveness of ABCB1 inhibitors on the uptake and efflux of ABCB1 substrates. However, predicting and testing ABCB1 modulation activity using living cells during drug development are crucial. The use of ABCB1-transfected mouse T-lymphoma cell line to study the uptake/efflux of fluorescent probes like ethidium bromide (EB), rhodamine 123 (Rh-123), and carbocyanine dye $DiOC_2$, in the presence and absence of potential inhibitors, is currently used in our laboratories to evaluate the ability of a drug to inhibit ABCB1-mediated drug accumulation and efflux. Here we describe and compare three in vitro methods, which evaluate the permeability, transport kinetics of fluorescent substrates, and inhibition of the ABCB1 efflux pump by drugs of chemical synthesis or extracted from natural sources, using model cancer cell lines over-expressing this transporter, namely (1) real-time fluorimetry that assesses the accumulation of ethidium bromide, (2) flow cytometry, and (3) fluorescent microscopy using rhodamine 123 and $DiOC_2$.

Key words ABCB1, MDR1, MDR, Real-time fluorimetry, Flow cytometry

1 Introduction

Multidrug resistance (MDR) is a phenomenon in which cancer cells exhibit a cross-resistant phenotype against structurally and functionally unrelated anticancer compounds [1]. Resistance of cancer to therapeutic agents may occur intrinsically (e.g., colorectal and renal cancers) or induced by the agent used in therapy (e.g., leukaemias, lymphomas, myeloma, breast and ovarian carcinomas).

José Rueff and António Sebastião Rodrigues (eds.), *Cancer Drug Resistance: Overviews and Methods*, Methods in Molecular Biology, vol. 1395, DOI 10.1007/978-1-4939-3347-1_7, © Springer Science+Business Media New York 2016

Both forms of resistance can be due to the overexpression of plasma membrane proteins that recognize the agent and extrude it from the cell before it reaches the intended target [2].

The plasma membrane-based proteins involved in the mediation of resistance are members of the superfamily of ATP-binding cassette (ABC) transporters encoded by specific genes which are activated in different cancer cells. These ABC transporters are able to recognize, bind, and extrude a number of chemically and structurally unrelated compounds through biological membranes, including lipids, bile acids, xenobiotics, and peptides for antigen presentation [3].

The human genome consists of 49 known genes for ABC proteins that can be grouped into at least seven family members of transporters (ABCA to ABCG) [4–6]. Among these families, multidrug resistance-associated proteins (MRP1/ABCC1 and MRP2/ABCC2), breast cancer resistance protein (BCRP/MXR/ABCG2), and P-glycoprotein (the product of *ABCB1* gene, also termed multidrug resistance protein 1 or MDR1) function as drug efflux pumps of anticancer drugs, and their overexpression confers an MDR phenotype of cancer cells [7]. Clinically relevant examples of chemoresistance/MDR in a number of cancer types are usually associated with over-expression of ABCB1 and ABCC1 proteins, among others [8].

1.1 P-Glycoprotein (P-gp)/ABCB1 Protein

The permeability-glycoprotein (P-gp), an integral plasma membrane protein of 170 kDa encoded by the human *ABCB1* (*MDR1*) gene, located on chromosome seven (7q21-1) [9], is the most studied member of the ABC transporter superfamily of genes and the first to be discovered [10]. The products of ABC transporter genes are transmembrane-glycosylated proteins [9] involved in energy-dependent transport of a diverse range of molecules and exogenous substances across membranes, and are naturally expressed on intracellular canaliculus of normal human tissues, including liver, kidney, colon, adrenal gland, intestine, placenta, hematopoietic precursor cells, and luminal membrane of brain capillary endothelial cells [11, 12] and everywhere else where physiological barriers exist since its main function is to protect the body from toxic compounds.

The clinical relevance of the transporters is their association with the multidrug resistance phenotype (MDR) which is characterized by cross-resistance to a broad spectrum of chemically distinct chemotherapeutic drugs [13, 14]. In general, patients with ABCB1-positive tumor cells respond less well to chemotherapy, and have a poorer prognosis.

The cytotoxic drugs associated with ABCB1-mediated MDR phenotype are anthracyclines (doxorubicin, daunorubicin), epipodophyllotoxins (etoposide and teniposide), taxanes (paclitaxel), vinca alkaloids (vinblastine, vincristine), anthracenes (bisantrene and mitoxantrone), and camptothecins (topotecan) [9, 15].

In addition, several other exogenous lipophilic compounds can also interact with ABCB1 as substrates or as inhibitors including colchicine-site-binding agents, natural products, calcium channel blockers, calmodulin antagonists, antibiotics, and fluorescent dyes (rhodamine 123, calcein AM, and $DiOC_2$). Among them, the fluorescent dyes are used to evaluate the impact of inhibitors on the transfer function of these transporters, by measuring the capacity of cells to extrude the dye from their compartments. This is the most relevant method to assess clinical MDR in eukaryotic cells [7].

A clinical strategy to overcome resistance and reverse or prevent MDR is the short-term administration of inhibitors that can interfere with overexpressed ABCB1 activity [16]. In vitro studies with inhibitors demonstrate that resistance conferred by the over expression of ABCB1 can be modulated, but in clinical trials the results have been very disappointing because of the wide distribution of efflux pumps in healthy tissues, and their (low) specificity to MDR transporters which requires higher plasma concentrations to reverse multidrug resistance, being associated with unacceptable toxic side effects [16, 17].

So far, four generations of compounds have been identified and developed as ABCB1 modulators for having the capacity to reverse MDR. The first generation of ABCB1 inhibitors were initially developed for having other pharmacological properties. These drugs, which interact with ABCB1 efflux pumps, are not selective, requiring high plasma concentrations to inhibit ABCB1. This can lead to undesirable adverse effects [17]. The best known first-generation ABCB1 modulators are verapamil, tamoxifen, and cyclosporine A. Verapamil is a calcium channel blocker that also competitively inhibits the function of ABCB1 transporter [18].

The approach used to diminish these side effects involves the design of chemical modifications of the drugs. The second generation of ABCB1 inhibitors are a class of molecules with a greater affinity to the ABCB1 pump. They can be divided into two categories, those that are analogues of the first generation, i.e., dexverapamil [18] and those with new chemical structures. In the third generation of compounds, the goal is to overcome the toxicity of the previous generations and to improve the pharmacokinetic interaction of the compounds. These inhibitors have high affinity to ABCB1 at nanomolar concentrations. An example is elacridar [16].

1.2 Natural Compounds as Inhibitors of MDR Transporters

Efforts to identify ABCB1 inhibitors have led to numerous candidates, none of which have passed clinical trials with cancer patients. Human clinical trials with first-, second-, or even third-generation drugs did not show any satisfactory outcomes, evidencing adverse severe side effects, inhibition of cytochrome P-450 monooxygenases, and limited clinical benefits [19, 20].

As an alternative to conventional drug therapy, natural compounds could be less cytotoxic and with fewer side effects, representing the fourth generation of multidrug resistance reversal agents [21]. They exert their action through modulation of transcription factors, growth factors, survival factors, inflammatory pathways, invasion, and angiogenesis [20, 22–25], which leads to inhibition or reversal of early stages of carcinogenesis as shown in experimental models [22]. Some of these molecules of natural origin have reached human clinical trials, such as marine-derived anticancer drugs [26] or compounds of plant origin (phytochemicals).

Most of the studies conducted with phytochemicals investigated polyphenols (e.g., curcumin, genistein, quercetin, resveratrol, epigallocatechin-3-gallate) as described in recent reviews [22, 27–29]. Human clinical trials carried out to test polyphenols reported some positive results in patients with colorectal and prostate cancer [28–30]. The activity of polyphenols was not as potent or as effective as for synthetic drugs, but some polyphenols have been shown to be safe and to increase the efficacy of the drug when used in co-treatments or in chemoprevention through dietary interventions, improving chemotherapy efficacy and decreasing side effects [22, 27, 31]. Natural compounds as ABCB1 modulators have been described in vitro and in vivo [19], alone or in combination with anticancer agents to overcome resistance. They can interact with doxorubicin [32, 33], epirubicin [34], as well as other ABCB1 substrates [29].

Anticancer agents of natural origin have an advantage over all the ABCB1 inhibitors tested because they do not have a "single target" feature, as most drugs used in cancer treatment, but they can modulate multiple pathways as outlined earlier [20, 25]. Genistein, curcumin [31], or quercetin [35] are some of the phytochemicals described to reverse MDR by increasing the sensitivity of cancer cells to classical chemotherapeutics drugs. Therefore, the search for natural ABCB1 modulators that are safe and nontoxic to overcome resistance can be a promising strategy in anticancer chemotherapy and for this purpose the development of robust methods to assess the permeability and drug transport kinetics by the main eukaryotic membrane transporters is essential. A lead compound extracted from a natural source was used in this chapter as an experimental example, assessed against a positive control of chemical origin—verapamil (see Figs. 1, 2, and 3).

1.3 Monitoring ABCB1-Mediated Drug Transport of Fluorescent Compounds

Characterization of ABCB1 transport properties in vivo is challenging because ABCB1 transport properties are difficult to characterize in intact cellular systems. In this chapter we describe three techniques currently used in our laboratories for in vitro evaluation of the activity of putative modulator compounds of ABCB1, using a cell line that overexpresses the ABCB1 transporter (mouse T-lymphoma cells transfected with the human *ABCB1* gene) and its parental cell line. This cell-based model is very useful

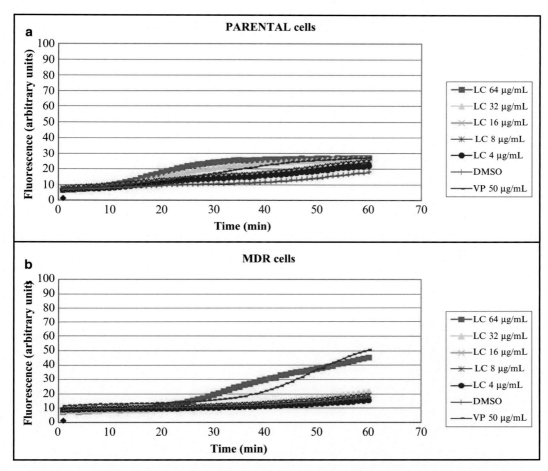

Fig. 1 Comparison of ethidium bromide (EB) accumulation (1 µg/mL) by (**a**) parental mouse lymphoma cells (PAR) and (**b**) human *MDR1* (*ABCB1*) gene-transfected mouse lymphoma cells in the presence of 0.5 % DMSO (control), 50 µM verapamil (VP), and different nontoxic concentrations of "lead compound" (LC) (*see* **Note 1**). In the presence of ABCB1 efflux pumps, retention of EB in resistant cell line in the presence of inhibitors (**a**) was higher than observed in the absence of inhibitors (control)

for the evaluation of agents which can modulate the activity of the ABCB1 transporter because these cells grow easily in vitro and it is possible to carry out quantitative measurements of efflux by following the intra- and extracellular amount of different fluorescent substrates, in the presence and absence of ABCB1 modulators. These eukaryotic cell lines also have low expression of other endogenous drug transporters that could render the interpretation of experimental data more difficult, especially when the reporter fluorescent substrate used for the evaluation of ABCB1 modulators is a common substrate of other efflux pumps. Notably, these techniques can also be optimized for other specific cell lines that overexpress ABC transporters, using specific fluorescent substrates.

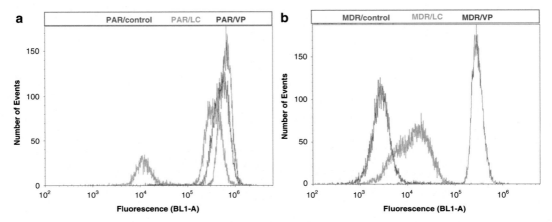

Fig. 2 Flow cytometry analysis of rhodamine 123 (R123) accumulation in L5178y PAR (**a**) and L5178v MDR (**b**) cells (*see* **Note 1**). Representative histogram plots are shown of non-treated cells (control), 10 μM vera-pamil-treated cells (VP), and "lead compound"-treated cells (LC). Measured by flow cytometry using Attune@ Acoustic Focusing Cytometer (Life Technologies)

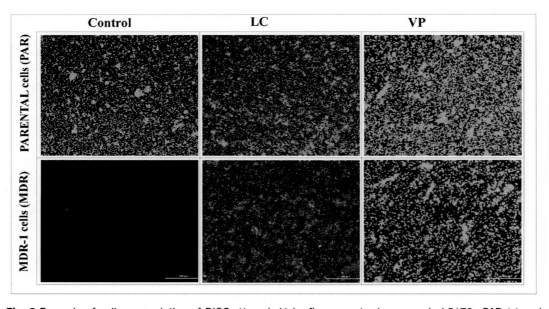

Fig. 3 Example of cell accumulation of DiOC$_2$ (1 μg/mL) by fluorescent microscopy in L5178y PAR (**a**) and L5178v MDR (**b**) cells (*see* **Note 1**). DiOC$_2$ was added for a period of 2 h at 37 °C, and then cells were incubated in the presence of 10 μM verapamil (VP) and "lead compound" (LC) in dye-free medium for a further 45 min at 37 °C and examined by fluorescence microscopy

Several compounds have been used to study drug transport due to their inherent fluorescence, such as anthracyclines, acridines, calcein-AM, rhodamine 123 (Rh-123), and DNA-binding fluoro-chromes such as hoechst 33342 and ethidium bromide [36]. In the three techniques presented in this chapter, rhodamine-123, DiOC$_2$, and ethidium bromide are used as reporter ABCB1 substrates to evaluate the binding activity of new compounds, by monitoring the fluorescent dye transport across the cells (efflux pump activity). Rh-123 and $DiOC_2$ bind specifically to mitochondria of living cells and both can be used to evaluate the ABCB1 pump activity. Ethidium bromide (EB) is a universal substrate of efflux pumps being recognized by bacterial efflux pumps and ABC transporters of eukaryotic cells. EB is highly fluorescent when it goes across the cell membrane and begins to intercalate between nucleic bases of DNA regions especially rich in guanosine/cytosine bases, even in very small concentrations. In aqueous solution, outside the cell, EB loses almost completely its fluorescence, affording the advantage that can be easily quantified by fluorimetry, on a real-time basis [37, 38].

The semiautomated fluorimetric method is an easy and accurate method that evaluates EB efflux by ABC transporters of lymphoma cell lines providing quantitative assessment of the efflux activity of ABCB1 transporter in cell suspensions. The method is carried out by using a new application of the thermocycler Rotor-Gene™ 3000 (Corbett Research) that monitors the accumulation and extrusion of EB on a real-time basis [38, 39]. This assay is conducted for a specific period of time with a nontoxic concentration of EB that does not cause its accumulation inside the cells due to its extrusion by an active efflux pump system of the cell. The slope of EB accumulation over time (number of units of fluorescence per minute) provides an assessment of baseline efflux and possible alterations of this slope induced by efflux inhibitors allow the identification of compounds that act as efflux inhibitors, since more EB is retained inside the cells in suspension and fluorescence increases accordingly. The fluorescence emitted by the accumulation of EB inside all the cells in suspension is followed by real-time fluorimetry at 25 °C (for optimizing conditions of maximum EB accumulation) in the Rotor-Gene™ 3000, using 530 nm band-pass and 585 nm high-pass filters as excitation and emission wavelengths, respectively. As an example *see* Fig. 1.

Another powerful equipment that can also be used to monitor the transport of fluorescent compounds in sensitive and multidrug-resistant tumor cells and provides information about the physical and biological characteristics of individual cells as well as about subpopulations in a heterogeneous cell population is flow cytometry. In flow cytometry, laser light is usually applied to excite the fluorochromes used in the functional assays, monitoring the efflux activity of the cells. These lasers produce light in the UV and/or visible range and fluorochromes are selected based on their ability to fluoresce with the wavelengths of light emitted by the lasers.

Currently, the fluorescence flow cytometers can measure small-angle or forward light scatter, which is related to the size of the cells and large-angle or side scatter, providing information about internal granularity, roughness of the cell surface, and particular characteristics, allowing single-cell identification [40].

Flow cytometry is an ideal technique to measure fluorescent drug retention in tumor cells. Furthermore it can provide information about compounds enhancing the intracellular drug retention and the chemosensitivity of cancer cells by modulating their efflux activity [41]. As an example *see* Fig. 2.

Lastly, fluorescent live-cell imaging assays, using a fluorescence microscope, is also an effective and efficient method to screen ABCB1 efflux inhibitors using the previously described approach of monitoring the transport of fluorescent dyes. The efflux blocking activity of tested compounds is measured by cellular fluorochrome drug retention profiles (e.g., rhodamine 123 or $DiOC_2$) given by the presence and absence of the tested compound. All results are compared with a positive control, namely the fluorescence retention profile of the cells in the presence of verapamil or cyclosporine A, which are well-known inhibitors of the ABCB1 (positive controls). The fluorescence profile given by verapamil or cyclosporin A represents 100 % inhibition in the well-characterized mouse T-lymphoma cell line overexpressing the ABCB1 protein (100 % accumulation). The negative control is given by the untreated cells incubated with the same fluorescent dye (0 % accumulation). As an example *see* Fig. 3.

The semiautomated fluorimetric assay using EB as a fluorescent substrate is a simple, rapid, and sensitive method for screening of large number of samples and the positive samples and/or suitable candidates detected should be further analyzed by single-cell flow cytometry or fluorescent microscopy. These two techniques allow the visualization of intracellular distribution of fluorescent probes in individual living cells (population of cells) and the assessment of cell viability before and after treatment allowing the screening and identification of putative ABCB1 modulators as well as the evaluation of the cytotoxicity of the tested compounds.

2   Materials

2.1   Cell Culture

1. Incubator at 37 °C with 5 % CO_2 and 95 % humidity.

2. Conical sterile 15 mL centrifuge tubes with screw caps.

3. Tissue culture flasks.

4. 24-Well plates for tissue culture.

5. Pasteur pipettes.

6. Colchicine A: Stock solution 1 mg/mL dissolved in DMSO.

7. Phosphate-buffered saline (PBS), pH 7.4. Store at room temperature after autoclaving.

8. Hemocytometer.

9. Inverted microscope.

10. Benchtop centrifuge with swing buckets.

11. Cell lines: L5178 mouse T-cell lymphoma cells transfected with pHa MDR1/A retroviral vector (L5178 MDR) [42] and parental L5178 mouse T-cell lymphoma cells (L5178 PAR) (*see* **Note 1**).

12. Culture medium for L5178 MDR cells and L5178 PAR cells: McCoy's 5A medium supplemented with 10 % heat-inactivated horse serum, L-glutamine (2 mM), and antibiotics (penicillin 100 U/mL, streptomycin 100 µg/mL). The ABCB1-expressing cell line (L5178 MDR) is selected by culturing the infected cells with 60 ng/mL of colchicine to maintain the MDR phenotype.

2.2 Accumulation Assays

1. Modulators of ABCB1 efflux pumps:

    (a) Verapamil: 10 mM stock solution in water, 0.22 µm filter sterilized, stored in aliquots at –20 °C, and recommended to be used as a positive control at a final concentration of 10 µM (cytometry and microscopy) or 50 µM (semiautomated fluorimetric method) per assay.

    (b) Cyclosporin A: 10 mg/mL stock solution in water, 0.22 µm filter sterilized, stored in aliquots at –20 °C. It is a well-described inhibitor of ABCB1, recommended to be also used as a positive control at a concentration of 10 µM per assay (*see* **Note 2**).

    (c) "Lead compound" (LD) used as an example, a natural compound extracted from plants, with known inhibitory properties, was dissolved in DMSO (*see* **Note 3**).

2. Semiautomated fluorimetric method.

    (a) Rotor-Gene™ 3000 thermocycler (Corbett Research, Sydney, Australia), with appropriate excitation and emission wavelengths (530 nm band-pass and 585 nm high-pass, respectively).

    (b) Rotor-Gene Analysis Software 6.1 (Build 93) by Corbett Research (Sydney, Australia).

    (c) Ethidium bromide (EB): Stock solution of 10 mg/mL in sterile bi-distilled water, kept in the dark at 4 °C.

    (d) Phosphate-buffered saline (PBS).

    (e) Eppendorf tubes 0.2 mL

3. Flow cytometry.

    (a) Rhodamine 123: 0.1 mg/mL stock solution in PBS. Sterilize the stock solution by passing through a 0.22 µm filter and store at –20 °C protected from light.

4. Fluorescent microscopy accumulation assays.

    (a) Fluorescent microscope with filters for $DiOC_2$: max. excitation and emission of 488 and 497 nm, respectively.

(b) Carbocyanine dye $DiOC_2$ (3-3′-diethyloxacarbocyanine iodide), 3 mM stock solution prepared in DMSO, filter sterilized, and kept at 4 °C until used.

2.3 Cytotoxicity Assays

1. 3-(4,5-Dimethylthiazol-2-yl)-2,5-diphenyl-2H-tetrazolium bromide (MTT): Sterilize the stock solution (5 mg/mL in PBS) by passing through a 0.22 μm filter and stored at −20 °C protected from light.

2. Coated poly-D-lysin 96-well plates (*see* **Note 4**).

3. Dimethyl sulfoxide (DMSO) (*see* **Note 2**).

4. Microplate reader (excitation filter at 595 nm).

5. PBS.

3 Methods

3.1 Preparation of Cell Cultures

All work must be carried out under sterile conditions.

- Select ABCB1-expressing cell lines by culturing the transfected cells with 60 ng/mL of colchicine to ensure the MDR phenotype. Both cell lines, the parental L5178 (PAR, without ABCB1 overexpression) and multidrug-resistant L5178 human *MDR1*-gene transfected subline (MDR, with ABCB1 overexpression), are cultured in McCoy's 5A medium as described in materials section [43].

- Maintain the cell cultures in a humidified incubator at 37 °C with 5 % CO_2 until use.

3.2 Cytotoxicity Assay

This assay enables one to study the influence of the ABCB1 modulator compounds on cell proliferation. The cytotoxic effects of increasing concentrations of the compounds are studied using the MTT test. The MTT assay is based on a redox potential reaction, where viable cells (with active mitochondria) convert the water-soluble MTT into an insoluble purple formazan (crystals). The formazan crystals formed are then solubilized and the color formed in the wells can be easily measured by optical density [23].

- Seed, using a 96-well plate coated with D-lysine, $1–5 \times 10^4$ of L5178 cells in 100 μL of complete medium and incubate at 37 °C [23].

- After 24 h, add 100 μL of medium containing different concentrations of the compound to be tested (*see* **Note 2**), and incubate for 24–48 h.

- Control wells that have to be present in all plates: blank without cells (only medium), and control with cells plus DMSO. At the end of the treatment period, the medium is removed and 0.5 mg/mL of MTT dissolved in culture medium is added.

Incubate the cells for an additional period of 3 h. Discard media and add 200 μL of DMSO to each well to dissolve the formazan crystals. After a homogenization (careful not to form bubbles), read the absorbance at 595 nm using a microplate reader.

- Inhibition of the cell growth is determined by calculating the ID_{50} according to the formula

$$ID_{50} = 100\left[\left(OD_{sample} - OD_{medium\ control}\right) / \left(OD_{cell\ control} - OD_{medium\ control}\right)\right] \times 100$$

ID_{50} is defined as the inhibitory dose that reduces the growth of the compound-exposed cells by 50 % [44].

3.3 Assays to Monitor ABCB1-Mediated Drug Transport of Fluorescent Compounds

The most common assays to determine ABCB1-mediated drug transport alterations are those that use fluorescent probes. These probes are useful indicators for assessing whether lead compounds are inhibitors of ABCB1 (decreased efflux of fluorescent substrates such as rhodamine 123, $DiOC_2$ and EB—more fluorescent detected inside the cells). Accumulation assays are validated using as positive control known efflux pump inhibitors (EPIs), such as verapamil, that specifically modulate ABCB1 activity [4].

(a) The semiautomated fluorimetric screening method:
This simple method has been developed to assess the real-time accumulation and extrusion of the fluorochrome EB on a real-time basis from the bulk of cells in suspension in an Eppendorf tube, a technique initially developed for monitoring the efflux activity of bacteria [39].

This assay employs a new application of the thermocycler, Rotor-Gene™ 3000, which delivers real-time data on transport kinetics of the fluorochrome substrate, reflecting the balance between accumulation of EB via passive diffusion (through membrane permeability) and extrusion via efflux.

Ethidium bromide is widely used to quantify the transport across the membrane of the cells, because it generates an easily measurable and quantifiable signal between EB inside and outside the cell. This assay is conducted with a nontoxic concentration of EB that does not cause its accumulation inside the cells for a specific period of time. The slope of EB accumulation (number of units of fluorescence/minute) provides an assessment of baseline efflux and alterations to this slope, by the use of compounds that promote EB retention inside the cells (fluorescence increases accordingly), being an easy and high-throughput screening procedure (up to 32 tubes and different experimental conditions can be assessed simultaneously in one assay using this protocol) for the identification of compounds that act as efflux inhibitors [38].

The fluorescence emitted by the accumulation of EB inside cells due to the presence of the efflux inhibitor is monitored by real-time fluorimetry, collecting data each 60 s at 37 °C for 1 h, and the relative final fluorescence activity ratio (RFF) is calculated as the relation of the determined relative fluorescence (RF) at the last point (30′ or 60′) of the EB retention curve of the cells treated with the inhibitor and the RF at the last point of the EB retention curve of the untreated cells (containing the equal volume of the solvent used to solubilize the drug to be tested—solvent control), divided by the RF of the untreated cells. The activity of the compound in test will be quantified according to the following formula: (relative final fluorescence $(RFF) = (RF_{treated} - RF_{untreated})/RF_{untreated}$)—RFFs higher than 1 represent retention of the fluorochrome that is proportional to the inhibitory activity of the tested compound.

- Count L5178 PAR and MDR mouse T-lymphoma cells using a hemocytometer and centrifuge the cells at $2000 \times g$ for 3 min. Remove the supernatant by aspiration and adjust the density of the cell suspension to 2×10^6 cells/mL in PBS.

- The cytotoxicity of EB and the EPI (e.g., verapamil) is determined previously by trypan blue exclusion method or by MTT viability assay to quantify cell viability and guarantee that the concentrations used in the subsequent assays will not in any way affect directly the cell viability. These procedures have previously been described in detail [23].

- Calculation of the lowest ethidium bromide concentration that reflects the natural and intrinsic balance between accumulation and efflux of such a noxious agent: The desired aim is to detect and quantify by fluorimetry the transport of EB through the cell membrane at working concentrations that will not affect cell viability nor perturb cellular functions.

  - Prepare working solutions of EB of 40, 20, 10, and 5 μg/mL in PBS (keep in the dark at 4 °C) (*see* **Note 5**).

  - Distribute the cell suspension in 95 μL aliquots into 0.2 mL tubes and add 5 μL of each concentration of EB in each individual tube. Final concentrations will be 2, 1, 0.5, and 0.25 μg/mL, respectively.

  - Place the tubes into the 32-well rotor of the thermocycler and monitor the fluorescence at 30-s intervals for 30 min or 1 h (*see* **Note 6**).

    From our experiments, we concluded that the optimum concentration of EB for studying accumulation in L5178 PAR and MDR mouse T-lymphoma cells was 1 μg/mL [38].

For testing the inhibitory activity of the compounds distribute the cell suspension in 90 μL aliquots into 0.2 mL tubes, add the compounds at different concentrations (e.g., 0.2–10 μM) in 5 μL volumes of their stock solutions, and incubate the samples for 20 min at room temperature (*see* **Note 7**). After this incubation, add 5 μL of 20 μg/mL of EB to each sample (1 μg/mL final concentration), and place the tubes into the thermocycler as soon as possible; as an example *see* Fig. 1 (*see* **Note 6**).

(b) Rhodamine 123 accumulation test by flow cytometry:
The detection of the rhodamine 123 uptake by the ABCB1 protein is a widely used method to monitor multidrug resistance [45, 46]. Rhodamine is a green fluorescent cationic dye and a specific substrate for ABCB1, often used to detect functional ABCB1-mediated efflux by cells. Since ABCB1 has been intensively studied for more than 35 years, numerous substrates of ABCB1 have been identified such as chemotherapeutic agents, natural products, peptides, amphiphiles, and fluorescent dyes [45]. The rhodamine 123 assay is based on the protocol described previously [44, 47] with minor modifications:

- Adjust the cells to a density of 2×10^6/mL, resuspend them in serum-free McCoy's 5A medium, and distribute in 0.5 mL aliquots into Eppendorf centrifuge tubes. Add the test compounds (10 μL) at different concentrations (0.2–10 μM), and incubate for 10 min at room temperature. Next, add 10 μL (5.2 μM final concentration) of rhodamine 123 to the samples and incubate the cells for a further 20 min at 37 °C. Wash the cells twice with ice-cold PBS, resuspend them in 0.5 mL PBS, and analyze them in the flow cytometer.

- The fluorescence of the cell population will be measured by flow cytometry, as demonstrated in the example in Fig. 2. We recommend the use of verapamil as a positive control in the rhodamine 123 exclusion experiments [48]. The percentage mean fluorescence intensity should be calculated for the treated MDR and parental cell lines as compared to untreated cells. A fluorescence activity ratio (FAR) will be calculated via the following equation, on the basis of the measured fluorescence values:

$$FAR = \frac{MDR_{treated}/MDR_{control}}{parental_{treated}/parental_{control}}$$

- The results provided are from a representative flow cytometric experiment in which 10,000 individual cells are evaluated. Attune flow cytometer research software is used to analyze the recorded data. The data are first presented as histograms and the data will be converted to FAR units that define fluorescence intensity, standard deviation, and peak channel in the total and in the gated populations. FAR higher than 1 represents retention of the fluorochrome in the MDR cells overexpressing the ABCB1 protein that is proportional to the inhibitory activity of the tested compound [49].

(c) DiOC$_2$ or rhodamine 123 accumulation assays by fluorescent microscopy:

This methodology should be used to confirm the preliminary analysis of the data obtained with the semiautomated fluorimetric method and as an alternative method to flow cytometry for evaluation of the EPI activity of putative ABCB1 inhibitors in MDR cancer cells, using DiOC$_2$ as substrate (or rhodamine 123 depending on the available microscope filters). With this method it is not possible to quantify precisely the fluorescence but it is a useful technique to visualize the retention of the fluorochrome on an individual cell basis and confirm the levels of retention/inhibition observed at the end of the semiautomated fluorimetric accumulation assays.

- Count the cells (PAR and MDR mouse T-lymphoma cells), centrifuge, and adjust the cell number to 2×10^6 cell/mL (final concentration in McCoy's 5A culture medium without colchicine) (*see* **Note 8**). Seed 1 mL of the cell suspension (PAR and MDR) into each well of a 24-well tissue culture plates.

  – Add the diluted EPI agent or control solution to each well (*see* **Note 9**). Incubate the plates at 37 °C for 30 min.

  – Add rhodamine 123 (at 5.3 μM final concentration) or DiOC$_2$ (at 1 μg/mL final concentration) dissolved in complete medium without colchicine for another period of 60 min at 37 °C.

  – Harvest the cells by centrifugation at $2000 \times g$ for 5 min. Aspirate the culture medium from the cells; wash the cells twice with ice-cold PBS.

  – Resuspend each pellet in 1 mL of fresh medium without fluorescent probe (with or without inhibitors), transfer to a new plate, and incubate for 1 h (efflux). Harvest the cells by centrifugation and resuspend each pellet in 20 μL ice-cold PBS with inhibitor/without inhibitor (control).

- Cells are examined by fluorescence microscopy at 485 nm excitation laser and 530/30 nm emission filter for green fluorescence (DiOC$_2$), as can be seen in Fig. 3.

4   Notes

1. L5178v MDR and L5178y PAR cell lines used in the representative examples of each method described in this chapter were kindly provided by Professor M.M. Gottesman (National Cancer Institute, Bethesda, MD, USA) [43]. It is also possible to use other cell lines that overexpress the ABCB1 transporter [38].

2. Cyclosporin A gives best results in semiautomated fluorimetric method (Rotor-Gene™ 3000) when using these cell lines.

3. Concentrations of DMSO brought in the final culture medium should be lower than [0.5 %], which does not interfere with the accumulation of EB, rhodamine (Rh-123), or DiOC$_2$.

4. Alternatively, 96-well plates can be coated with poly-D-lysin referring to the standard coating protocols according to the instructions of the manufacturer.

5. The solutions should be kept sterile.

6. This step should not take long, and tubes must be introduced into the thermocycler as quickly as possible for initiating the readings—the transport kinetics is fast.

7. The drug solutions should be prepared immediately before use and should be filter sterilized. The solvent used to solubilize the drug should be included as a separated control at the highest concentration applied in the assay.

8. Colchicine can compete with the fluorescent substrate or inhibitor, so it should be omitted from the medium before the assay.

9. DMSO can interfere with the assay; all the controls should be in water, PBS, or DMSO, depending on the drug solvent and concentration used.

References

1. Saraswathy M, Gong S (2013) Different strategies to overcome multidrug resistance in cancer. Biotechnol Adv 31:1397–1407

2. Dean M (2009) ABC transporters, drug resistance, and cancer stem cells. J Mammary Gland Biol Neoplasia 14:3–9

3. Choudhuri S, Klaassen CD (2006) Structure, function, expression, genomic organization, and single nucleotide polymorphisms of human ABCB1 (MDR1), ABCC (MRP), and ABCG2 (BCRP) efflux transporters. Int J Toxicol 25: 231–259

4. Goodman LS, Hardman JG, Limbird LE, Gilman AG (2001) Goodman & Gilman's the pharmacological basis of therapeutics. McGraw-Hill, New York

5. Kimura Y, Morita SY, Matsuo M, Ueda K (2007) Mechanism of multidrug recognition by MDR1/ABCB1. Cancer Sci 98:1303–1310

6. Higgins CF (2007) Multiple molecular mechanisms for multidrug resistance transporters. Nature 446:749–757

7. Sarkadi B, Homolya L, Szakacs G, Varadi A (2006) Human multidrug resistance ABCB

and ABCG transporters: participation in a chemoimmunity defense system. Physiol Rev 86:1179–1236

8. Taguchi Y, Kino K, Morishima M, Komano T, Kane SE, Ueda K (1997) Alteration of substrate specificity by mutations at the His61 position in predicted transmembrane domain 1 of human MDR1/P-glycoprotein. Biochemistry 36:8883–8889

9. Molnar J, Kars MD, Gunduz U, Engi H, Schumacher U, Van Damme EJ, Peumans WJ, Makovitzky J, Gyemant N, Molnar P (2009) Interaction of tomato lectin with ABC transporter in cancer cells: glycosylation confers functional conformation of P-gp. Acta Histochem 111:329–333

10. Juliano RL, Ling V (1976) A surface glycoprotein modulating drug permeability in Chinese hamster ovary cell mutants. Biochim Biophys Acta 455:152–162

11. Gottesman MM, Pastan I (1993) Biochemistry of multidrug resistance mediated by the multidrug transporter. Annu Rev Biochem 62:385–427

12. Ambudkar SV, Dey S, Hrycyna CA, Ramachandra M, Pastan I, Gottesman MM (1999) Biochemical, cellular, and pharmacological aspects of the multidrug transporter. Annu Rev Pharmacol Toxicol 39:361–398

13. Eckford PD, Sharom FJ (2008) Interaction of the P-glycoprotein multidrug efflux pump with cholesterol: effects on ATPase activity, drug binding and transport. Biochemistry 47:13686–13698

14. Shustik C, Dalton W, Gros P (1995) P-glycoprotein-mediated multidrug resistance in tumor cells: biochemistry, clinical relevance and modulation. Mol Aspects Med 16:1–78

15. Sharom FJ (2014) Complex interplay between the P-glycoprotein multidrug efflux pump and the membrane: its role in modulating protein function. Front Oncol 4:41

16. Yang K, Wu J, Li X (2008) Recent advances in the research of P-glycoprotein inhibitors. Biosci Trends 2:137–146

17. Ding PR, Tiwari AK, Ohnuma S, Lee JW, An X, Dai CL, Lu QS, Singh S, Yang DH, Talele TT, Ambudkar SV, Chen ZS (2011) The phosphodiesterase-5 inhibitor vardenafil is a potent inhibitor of ABCB1/P-glycoprotein transporter. PLoS One 6, e19329

18. Palmeira A, Sousa E, Vasconcelos MH, Pinto MM (2012) Three decades of P-gp inhibitors: skimming through several generations and scaffolds. Curr Med Chem 19:1946–2025

19. Eichhorn T, Efferth T (2012) P-glycoprotein and its inhibition in tumors by phytochemicals derived from Chinese herbs. J Ethnopharmacol 141:557–570

20. Zhu H, Liu Z, Tang L, Liu J, Zhou M, Xie F, Wang Z, Wang Y, Shen S, Hu L, Yu L (2012) Reversal of P-gp and MRP1-mediated multidrug resistance by H6, a gypenoside aglycon from Gynostemma pentaphyllum, in vincristine-resistant human oral cancer (KB/VCR) cells. Eur J Pharmacol 696:43–53

21. Munagala S, Sirasani G, Kokkonda P, Phadke M, Krynetskaia N, Lu P, Sharom FJ, Chaudhury S, Abdulhameed MD, Tawa G, Wallqvist A, Martinez R, Childers W, Abou-Gharbia M, Krynetskiy E, Andrade RB (2014) Synthesis and evaluation of Strychnos alkaloids as MDR reversal agents for cancer cell eradication. Bioorg Med Chem 22:1148–1155

22. Lewandowska U, Gorlach S, Owczarek K, Hrabec E, Szewczyk K (2014) Synergistic interactions between anticancer chemotherapeutics and phenolic compounds and anticancer synergy between polyphenols. Postepy Hig Med Dosw (Online) 68:528–540

23. Martins C, Doran C, Silva IC, Miranda C, Rueff J, Rodrigues AS (2014) Myristicin from nutmeg induces apoptosis via the mitochondrial pathway and down regulates genes of the DNA damage response pathways in human leukaemia K562 cells. Chem Biol Interact 218:1–9

24. Dandawate P, Padhye S, Ahmad A, Sarkar FH (2013) Novel strategies targeting cancer stem cells through phytochemicals and their analogs. Drug Deliv Transl Res 3:165–182

25. Singh BN, Singh HB, Singh A, Naqvi AH, Singh BR (2014) Dietary phytochemicals alter epigenetic events and signaling pathways for inhibition of metastasis cascade: phytoblockers of metastasis cascade. Cancer Metastasis Rev 33(1):41–85

26. Indumathy S, Dass CR (2013) Finding chemo: the search for marine-based pharmaceutical drugs active against cancer. J Pharm Pharmacol 65:1280–1301

27. Khushnud T, Mousa SA (2013) Potential role of naturally derived polyphenols and their nanotechnology delivery in cancer. Mol Biotechnol 55:78–86

28. Aggarwal B, Prasad S, Sung B, Krishnan S, Guha S (2013) Prevention and treatment of colorectal cancer by natural agents from mother nature. Curr Colorectal Cancer Rep 9:37–56

29. Nabekura T (2010) Overcoming multidrug resistance in human cancer cells by natural compounds. Toxins 2:1207–1224

30. Carroll RE, Benya RV, Turgeon DK, Vareed S, Neuman M, Rodriguez L, Kakarala M, Carpenter PM, McLaren C, Meyskens FL Jr, Brenner DE (2011) Phase IIa clinical trial of curcumin for the prevention of colorectal neoplasia. Cancer Prev Res (Phila) 4:354–364

31. Dorai T, Aggarwal BB (2004) Role of chemopreventive agents in cancer therapy. Cancer Lett 215:129–140

32. Martins A, Toth N, Vanyolos A, Beni Z, Zupko I, Molnar J, Bathori M, Hunyadi A (2012) Significant activity of ecdysteroids on the resistance to doxorubicin in mammalian cancer cells expressing the human ABCB1 transporter. J Med Chem 55:5034–5043

33. Kim TH, Shin YJ, Won AJ, Lee BM, Choi WS, Jung JH, Chung HY, Kim HS (2014) Resveratrol enhances chemosensitivity of doxorubicin in multidrug-resistant human breast cancer cells via increased cellular influx of doxorubicin. Biochim Biophys Acta 1840:615–625

34. Gyemant N, Tanaka M, Antus S, Hohmann J, Csuka O, Mandoky L, Molnar J (2005) In vitro search for synergy between flavonoids and epirubicin on multidrug-resistant cancer cells. In Vivo 19:367–374

35. Du G, Lin H, Yang Y, Zhang S, Wu X, Wang M, Ji L, Lu L, Yu L, Han G (2010) Dietary quercetin combining intratumoral doxorubicin injection synergistically induces rejection of established breast cancer in mice. Int Immunopharmacol 10:819–826

36. Krishan A, Fitz CM, Andritsch I (1997) Drug retention, efflux, and resistance in tumor cells. Cytometry 29:279–285

37. Spengler G, Ramalhete C, Martins M, Martins A, Serly J, Viveiros M, Molnar J, Duarte N, Mulhovo S, Ferreira MJ, Amaral L (2009) Evaluation of cucurbitane-type triterpenoids from Momordica balsamina on P-glycoprotein (ABCB1) by flow cytometry and real-time fluorometry. Anticancer Res 29:3989–3993

38. Spengler G, Viveiros M, Martins M, Rodrigues L, Martins A, Molnar J, Couto I, Amaral L (2009) Demonstration of the activity of P-glycoprotein by a semi-automated fluorometric method. Anticancer Res 29:2173–2177

39. Viveiros M, Martins A, Paixao L, Rodrigues L, Martins M, Couto I, Fahnrich E, Kern WV, Amaral L (2008) Demonstration of intrinsic efflux activity of Escherichia coli K-12 AG100 by an automated ethidium bromide method. Int J Antimicrob Agents 31:458–462

40. Shapiro HM (2004) "Cellular astronomy" – a foreseeable future in cytometry. Cytometry A 60:115–124

41. Szakacs G, Paterson JK, Ludwig JA, Booth-Genthe C, Gottesman MM (2006) Targeting multidrug resistance in cancer. Nat Rev Drug Discov 5:219–234

42. Pastan I, Gottesman MM, Ueda K, Lovelace E, Rutherford AV, Willingham MC (1988) A retrovirus carrying an MDR1 cDNA confers multidrug resistance and polarized expression of P-glycoprotein in MDCK cells. Proc Natl Acad Sci U S A 85:4486–4490

43. Weaver JL, Szabo G Jr, Pine PS, Gottesman MM, Goldenberg S, Aszalos A (1993) The effect of ion channel blockers, immunosuppressive agents, and other drugs on the activity of the multi-drug transporter. Int J Cancer 54:456–461

44. Spengler G, Evaristo M, Handzlik J, Serly J, Molnar J, Viveiros M, Kiec-Kononowicz K, Amaral L (2010) Biological activity of hydantoin derivatives on P-glycoprotein (ABCB1) of mouse lymphoma cells. Anticancer Res 30:4867–4871

45. Cornwell MM, Pastan I, Gottesman MM (1987) Certain calcium channel blockers bind specifically to multidrug-resistant human KB carcinoma membrane vesicles and inhibit drug binding to P-glycoprotein. J Biol Chem 262:2166–2170

46. Koizumi S, Konishi M, Ichihara T, Wada H, Matsukawa H, Goi K, Mizutani S (1995) Flow cytometric functional analysis of multidrug resistance by Fluo-3: a comparison with rhodamine-123. Eur J Cancer 31A:1682–1688

47. Molnar J, Szabo D, Mandi Y, Mucsi I, Fischer J, Varga A, Konig S, Motohashi N (1998) Multidrug resistance reversal in mouse lymphoma cells by heterocyclic compounds. Anticancer Res 18:3033–3038

48. Orlowski S, Mir LM, Belehradek J Jr, Garrigos M (1996) Effects of steroids and verapamil on P-glycoprotein ATPase activity: progesterone, desoxycorticosterone, corticosterone and verapamil are mutually non-exclusive modulators. Biochem J 317(Pt 2):515–522

49. Spengler G, Takacs D, Horvath A, Riedl Z, Hajos G, Amaral L, Molnar J (2014) Multidrug resistance reversing activity of newly developed phenothiazines on P-glycoprotein (ABCB1)-related resistance of mouse T-lymphoma cells. Anticancer Res 34:1737–1741

Chapter 8

Resistance to Targeted Therapies in Breast Cancer

Sofia Braga

Abstract

Seventy five percent of all breast cancer (BC) patients express estrogen receptor (ER) but a quarter to half of patients with ER positive BC relapse on ET (endocrine therapy), tamoxifen, aromatase inhibitors (AIs), surgical castration, amongst other treatment strategies. ER positive BC at relapse loses ER expression in 20 % of cases and reduces quantitative ER expression most of the time. ER is not the only survival pathway driving ER positive BC and escape pathways intrinsic or acquired are activated during ET. This overview gives an account of ligand-independent ER activation, namely by receptor networks cross talk, and by the various genomic factors and mechanisms leading to ET response failure. Also the mechanisms of Her1 and Her2 inhibition resistance are dealt within this overview, along with the therapeutic indications and limitations of tyrosine kinase inhibitors, PARP inhibitors, PI3K/AKT/mTOR inhibitors, RAS/RAF/MEK/ERK/MAPK inhibitors, and antiangiogenic drugs. In spite of the many advances in controlling the division of BC cells and the progression of BC tumors these still remain the main cause of death among women in age range of 20–50 years requiring even more efforts in new therapeutic approaches besides the drugs within the scope of the overview.

Key words Breast cancer, Endocrine therapy, Relapse, Resistance, Ligand-independent ER activation

1 Introduction

Early breast cancer (BC) is currently treated with surgery and adjuvant radiotherapy in almost all cases. Chemotherapy (CT) is reserved for tumors larger than 1 cm or with aggressive biological characteristics. Because chemotherapy has toxicity, acute and chronic, we have witnessed, in the last decade, a tendency to safely avoid CT prescription. Well-differentiated tumors, with strong estrogen receptor (ER) expression, low proliferation indexes and progesterone receptor (PgR) expression have been shown to be less responsive to CT and very sensitive to endocrine therapy (ET); therefore CT avoidance has been advocated for patients affected by such BC. The intrinsic BC subtype classification adapted to the clinic is increasingly popular and clinicians classify such BC as luminal A like tumors. Commercially available multigene signatures such as Mammaprint, Oncotype Dx, and others that also aim at

José Rueff and António Sebastião Rodrigues (eds.), *Cancer Drug Resistance: Overviews and Methods*, Methods in Molecular Biology, vol. 1395, DOI 10.1007/978-1-4939-3347-1_8, © Springer Science+Business Media New York 2016

detecting which BC patients can be safely spared adjuvant CT are quite popular. These patients will be spared adjuvant CT and will be treated with ET [1]. HER2 blockade is another targeted therapy in BC used in tumors with c-Erbb2 gene polysomy, but currently these patients cannot be safely spared adjuvant chemotherapy. There is a subtype of BC that expresses no ER, PgR, or HER2 and is therefore called triple negative (TN) BC. These tumors are not treated in the adjuvant setting with targeted therapy.

1.1 Estrogen Receptor Inhibition

ER inhibition is the most important therapeutic target in BC. However, a quarter to half of patients with ER positive BC relapse on ET, tamoxifen, aromatase inhibitors (AIs), surgical castration, LHRH analogs, or fulvestrant. Due to the high volume of ER positive BC patients, 75 % of all BC expresses ER, this group of patients needs further understanding of the cellular pathways at play in endocrine resistance. ER expression in BC increases with patient age, correlates with lower histological tumor grade, lower proliferation, less aneuploidy, less amplification of HER2 oncogenic protein, loss of P53, expression of PgR, bone metastases, and lower recurrence rate. Several tyrosine kinases have been implicated in the phenomenon of endocrine resistance (ligand independent ER activation or non-genomic ER activity) as well as other proliferative or survival pathways (ligand-dependent ER functions or genomic ER activity). ER is not the only survival pathway driving ER positive BC and escape pathways intrinsic or acquired are activated during endocrine therapy. Large scale clinical trials have not looked at the molecular phenotype of ER positive patients relapsing after ET.

1.1.1 Ligand Independent ER Activation

The classic genomic function of ER is to alter gene expression. By binding to estrogen, ER translocates to the nucleus where it alters expression of hundreds of genes, some by upregulation, others by downregulation. There is also a ligand-independent non-genomic activation of ER. ER is phosphorylated by several membrane receptor tyrosine kinases in the cytoplasm that are part of transmembrane growth factors (Fig. 1).

Erbb2, called HER2 by clinicians, is an oncogene that encodes a transmembrane protein that is an important oncogenic driver in BC. Erbb2 is altered through polysomy or mutation. Tumors with HER2 alterations generally have lower or no ER expression, rendering the patients less sensitive to endocrine manipulations. An in vitro model tested this assumption by engineering ER positive cell lines to express HER2 protein, these cells had marked reduction in ER expression which may explain reduced efficacy of ET in BC simultaneously positive for ER and HER2 [2]. The mechanism of ER and PgR downregulation in this setting is not known. Through its genomic activity ER increases expression of several transmembrane receptor ligands like TGFα and IGF. Finally, the activation of tyrosine kinases leads to activation of PI3K/AKT and p42/44 mitogen-activated protein kinases (MAPK) that

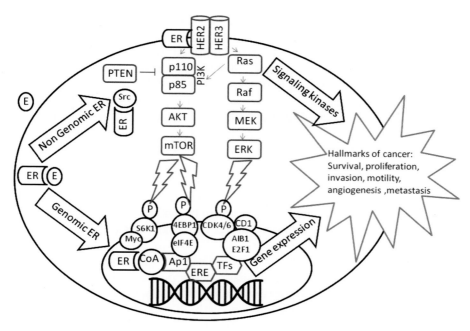

Fig. 1 Schema of a BC cell where some of the growth, differentiation, survival, and apoptosis pathways are depicted. Estrogen receptor signaling which is the most important survival pathway for the BC cell is seen on the *left* with both non-genomic and genomic activity. The epidermal growth factor pathway, the next most relevant growth pathway for the BC cell is depicted. Several other very important cytoplasmic pathways and their interrelations are apparent: *PTEN* PI3K-AKT-mTOR, RAS-RAF-MEK-ERK. Several inhibitors of these pathways are in clinical use. In the nucleus, at the *bottom*, c-myc, cyclins, 4EBP1, AIB1 are shown. The pharmacological targeting of transcription factors and other nuclear molecules has been more elusive, but cyclin inhibitors are being tested in clinical trials presently. *E* estrogen, *ER* estrogen receptor

downregulate expression of ER enabling ET resistance. All these receptor networks that increase phosphorylation of ER, increase the genomic action of ER to produce more ligands for the transmembrane receptors and ultimately decrease ET sensitivity, by decreasing dependence on ER because expression of ER itself is reduced, are termed growth factor cross talk. An important clinical readout of ET resistant BC is the reduced expression of ER. This phenomenon is widely seen in relapsing ER positive BC patients when metastatic sites are re-biopsied (Fig. 1).

IGF1 and insulin are attractive signaling molecules in BC. In 2014 the EBCTCG analyzed data from 80.000 BC patients enrolled in 70 BC adjuvant trials and concluded that obesity is an independent predictor of death in patients under 55 years with ER positive BC. IGF1/IGFR signaling is known to be increased in obese individuals, and therefore the molecular link between IGF1/IGFR signaling and worse prognosis in ER positive BC has an epidemiological confirmation [3]. Regarding BC susceptibility, IGF1 polymorphisms have been associated with higher risk of developing BC [4] and serum IGF1 levels have also been positively associated with ER positive BC risk [5]. Clinical trials with metformin or other IGFR tyrosine kinase inhibitors (TKI) are being run to test these hypotheses.

CDK10 has been identified as a determinant of resistance to ET in BC. A study silenced all kinases in the genome by siRNA, and CDK10 induced tamoxifen resistance when silenced. CDK10 increases ETS2-driven transcription of c-RAF, resulting in MAPK pathway activation and loss of tumor cell reliance upon estrogen signaling. MAPK pathway is a central cytoplasmic growth factor receptor tyrosine kinase pathway upon which several other signaling pathways converge, increasing the ligand independent activation of ER. The lower abundance of CDK10 measured by IHC or qRT-PCR in patient samples effectively correlates with relapse on tamoxifen. The mechanism by which CDK10 expression is reduced seems to be by promoter methylation [6].

1.1.2 Ligand-Dependent/Genomic Action of ER

Nuclear factor kB (NFkB): NFkB is a transcription factor important for mammary gland development and involved in resistance of BC cells to therapy. Endogenous glucocorticoids have been shown to function as BC growth factors once ET resistance appears. Decades ago metastatic (M) BC patients were treated with adrenalectomy and hypophisectomy to control end stage disease. NKkB is the mediator of glucocorticoid action in BC cells. It has been shown to be a pro-survival molecule and to correlate with CT resistance [7]. A study with MCF7, tamoxifen resistant, cell lines showing increased AP-1 and NFkB DNA binding transcriptional activities, were compared to MCF7 tamoxifen sensitive cells. A gene expression signature of upregulated genes due to NFkB and AP-1 transcriptional activity was obtained in the cell line work and then validated in patient material, in a training set and in a validation set. This work showed that tamoxifen resistant BC can have increased NFkB and AP-1 transcriptional responses [8].

AIB1: Expression of AIB1, a nuclear receptor co-activator, impairs tamoxifen-induced cell cycle arrest by potentiating E2F1 activity, a target of pRb-mediated repression [9]. AIB1 or Src3 is an estrogen receptor co-regulator, it is a non-receptor tyrosine kinase. Co-regulators are molecules that bind to the estrogen ER dimer at the DNA level and increase or decrease expression of estrogen target genes. AIB1 also called Sarc3, is amplified in a small percentage but overexpressed in 60 % of BC, the overexpression has been implicated in tamoxifen resistance [10] (Fig. 1).

Myc overexpression: Data regarding oncogenic c-myc in BC is similarly related to ET resistance. C-myc is overexpressed as assessed by IHC and q-RT-PCR in patient samples of ER positive and ER negative BC. In vitro data from MCF7 cell lines show that, upon c-myc expression, marked ET resistance ensued [11]. C-myc is an important oncogene in cancer but has up to now remained untargetable due to absence of the basic helix-loop-helix domain in its structure. In a deep sequencing project of AI resistant human tumor material from two neoadjuvant AI trials with 77 patients, authors inferred pathway-informed gene activities using gene expression and copy number data to identify several "hubs" of activity. The resulting pathways were ESR1, FOXA1 which were expected, and C-MYC, showing in vivo the relevance of c-myc signaling in determining ET resistance [12]. Another study in human material, using a signature derived from BC cell lines that were surviving on long term estrogen deprivation, was able to predict resistance to neoadjuvant anastrozole and letrozole and relapse on adjuvant tamoxifen. This signature was composed of genes involved in proliferation, namely, c-myc [13] (Fig. 1).

Cyclin D1 overexpression and Rb loss are other factors of ET resistance. In MCF-7 cell line models of endocrine resistant breast tumors, cyclin D1 expression and phosphorylation of RB were maintained despite effective ER blockade [14]. PD-0332991, a selective CDK4/6 inhibitor, blocked cell cycle progression of numerous endocrine-resistant breast cancer cell models. The work provides further insight that targeting the downstream mediators of ER action, like cyclin D1 and CDK4/6, may provide a viable means to stabilize endocrine-resistant tumors. Furthermore, recent evidence implicates pRb as a nodal regulator of antiestrogen responsiveness, as pRb inactivation results in antiestrogen resistance [15]. In addition, cyclin E expression by IHC has been known to be a prognostic factor in BC since 2002 [16] (Fig. 1).

Src: Src overexpression also correlates with a more invasive phenotype in cell line models of ER positive BC and inhibition of Src restores ET sensitivity in BC cell lines [17]. The translation of such data to human tumor samples has been elusive due to difficulty in finding a robust biomarker that would also serve as a stratification factor for clinical trials with dasatinib, the molecule that targets this oncogene [18] (Fig. 1).

LRH1: It is not clear if locally produced estrogen influences breast carcinogenesis, but LRH1 is expressed in BC cells and its expression is regulated by ER and their expression coexists in BC cells. LRH1 increases ER actions on proliferation and invasion. Using ChIP-seq and gene expression it was shown that LRH1 and ER share genomic binding sites and cooperate, showing another molecular player in ET resistance [19].

1.1.3 Overcoming ET Resistance

ER positive BC relapses mainly in the bone, and because of this, trials adding zoledronic acid to ET have been performed. In ABCSG12 an Austrian group studied the addition of adjuvant zoledronic acid in ER positive premenopausal BC patients treated with ET (anastrozole or tamoxifen with or without goserelin). Less than 5 % of the patients in this trial received adjuvant CT. The administration of zoledronic acid every 6 months for 3 years reduced the risk of disease recurrence by one third. These data show persistent benefits with zoledronic acid and support its addition to adjuvant endocrine therapy in premenopausal patients with early-stage breast cancer showing that zoledronic acid modulates ET [20]. Further studies focused on all BC patients, of which AZURE was the largest, did not reach similar results. The AZURE study, however, treated more than 95 % of the premenopausal women with CT which renders the premenopausal population very difficult to compare with the population of ABCSG12. In a pre-specified analysis, there was a possible benefit of zoledronic acid in patients who had undergone menopause more than 5 years before BC, these patients were similar to the ABCSG12 patients, in an endocrine perspective, because they had low levels of reproductive hormones [21]. The Tandem trial randomized HER2 and ER positive MBC patients to anastrozole alone or in combination with trastuzumab and aimed to understand if anti-HER2 therapy overcomes ET resistance. Remissions were brief and more common in the combination arm underscoring the need to treat HER2 positive patients with CT. Likewise, in a large cohort of patients with known ER positive and HER2 positive tumors, the addition of lapatinib, an anti-HER2 TKI, to an AI also significantly reduced the risk of progression [22]. Because of the role of EGFR pathway in ET resistance, several trials in MBC have studied the addition of gefitinib to ET. These trials all showed numerical advantage of the combination. However, a large neoadjuvant trial that studied anastrozole alone or in combination with gefitinib, favored the single agent therapy [23]. The other, a smaller, neoadjuvant trial showed activity of the combination only in EGFR positive tumors [24].

1.1.4 The Loss or Change of the Target

ER positive BC at relapse loses ER expression altogether 20 % of the time and reduces quantitative ER expression most of the time [25]. At metastasis, ER positive BC is less ET sensitive or even ET resistant. There are data showing this transformation from immunohistochemical characterization of metastasis and from analysis of circulating tumor cells (CTCs). CTC studies show that in patients who have metastatic ER positive BC up to 70 % of the CTCs do not express ER [26].

1.1.5 Deep Sequencing Studies

In 2009, a deep sequencing study of human material from the primary tumor and the metastatic pleural effusion of a patient with lobular BC, who had a disease free interval of 9 years, was published. Such clinical behavior, with a long disease free interval and metastatic disease with polyserositis is a hallmark of lobular BC, which is sensitive to ET for long periods. In lobular BC, ET resistance can develop after several years of estrogen deprivation. The genes that were shown in this paper to harbor mutations were not related to estrogen biology, which, given the exquisite dependence of lobular BC on ER, was unexpected. However, in the genome to proteome conversion, RNA-editing enzymes that are regulated by estrogen and recode transcripts resulting in a proteome divergent from the genome, were seen to be very active, generating proteins that were quite divergent from their coding sequences. The authors emphasized the need for integrating tumor DNA and RNA sequencing to be able to detect protein variation [27].

In 2013, another deep sequencing study of ER positive BC was published. This study enrolled 11 patients with ER positive metastatic BC. Whole exome and transcriptome sequencing was performed in metastatic material, in the primary tumor and in normal DNA from circulating lymphocytes and epithelial cells from buccal swabs. Nonsynonymous mutations in the ligand binding domain (LBD) of the ESR1 gene were present de novo in six of these patients. All of these patients had been or were still being treated with ET. This study was informative regarding ET resistance, it showed LBD activating mutations are mechanism for acquired resistance [28].

A more recent study published in 2014, sequenced the estrogen response elements (ERE) in the genome of ER positive primary and metastatic breast cancer samples, ER negative tumors were used as controls [29]. Several mutations in ERE that rendered cells insensitive to ET were detected in ER positive metastasis that were not detected in either the corresponding primaries and in the ER negative controls. In vitro data of shRNA silencing of ER, in the MCF7 cell line, shows that ER negative cells acquire a nonepithelial morphology, with mesenchymal characteristics, increased motility with rearrangement and switch from a keratin/ actin to a vimentin based cytoskeleton, and the ability to invade the extracellular matrix [30]. In functional migration essays that compared the migration of parental MCF7 cell line with the shRNA ER depleted MCF7 cell line and a ER negative cell line MDA231 with or without growth factors, it was shown that the two last cell lines were more migratory in serum and in response to growth factors. Additionally, the phosphorylation of Akt and ERK1/2 mediates EGF, IGF-1, and TGFβ signaling [31].

1.2 Her2 Inhibition

1.2.1 Trastuzumab

Several adaptive mechanisms are in place in HER2-positive tumors to resist trastuzumab. In the clinic, unfortunately, there are women who relapse evenduring the year of adjuvant trastuzumab treatment. The tumors might lose HER2 expression as a result of continuous therapy [32], activation of mutations downstream from HER2, of which the PI3K pathway is the best studied example [33], and activation of additional proliferation mechanisms, unrelated to HER2 like cyclin E amplification [34].

In the adjuvant clinical trials performed that validated the use of adjuvant trastuzumab for 1 year combined with standard surgery, CT and RT, retrospective tumor block evaluation of HER2 was performed to assess differential benefit of trastuzumab. In 2009, the HERA trial [35] researchers reported that the benefit of trastuzumab or survival in the observation arm did not vary according to the ratio HER2/CEP17 which corrects for copies of chromosome

17, HER2 copy number or HER2 polysomy [36]. In 2013, the same group reported that neither the differential IHC expression of HER2 protein nor lobular histology influenced response to adjuvant trastuzumab [37, 38].

The North American Adjuvant trials NSABP B31 and NCCTG 9831 reported differing results regarding c-myc amplification and trastuzumab benefit. In the B31 trial, c-myc amplification was found to be predictive of benefit of trastuzumab but there was no differential benefit in the 9831 trial [39, 40]. The authors of the 9831 trial underwent another retrospective biomarker analysis regarding PTEN suppression (Fig. 1). Preclinical and limited clinical evidence suggested than PTEN inactivity would activate PI3K/AKT signaling rendering trastuzumab treatment less active. The evaluation of PTEN status was performed by IHC in tissue microarrays and PTEN negative tumors did not show less benefit from trastuzumab [41]. The authors of the B31 adjuvant trial developed a genomic model where eight genes are assessed and according to their expression differences. Patient samples are classified into one of three groups according to trastuzumab benefit. Of these eight genes, three are associated with HER2 (ERBB2, c17 or f37, GRB7) and five with ER (ESR1, NAT1, GATA3, CA12, IGF1R). These genes were selected for building the predictive model of response to trastuzumab. Tumors with high ESR1 expression and intermediate ERBB2 expression do not benefit from trastuzumab, but tumors without ESR1 expression with intermediate ERBB2 expression that are HER2 positive by IHC benefit from the drug [42].

Another trastuzumab adjuvant trial from Finland tested the addition of 9 weeks of trastuzumab to standard therapy [43], a retrospective analysis of PIK3CA mutations in archived samples could show that PIK3CA mutations correlated with longer disease free survival (DFS) but showed no association with trastuzumab benefit [44]. A French study showed differential survival according to HER2 gene copy number. DFS is shorter in HER2 positive BC patients treated with neoadjuvant trastuzumab if more than ten copies of HER2 gene are present [45]. The correlation between HER2 gene copy number and survival was not tested in the adjuvant trastuzumab trials.

1.2.2 Lapatinib

Lapatinib is an anti-HER2 agent that blocks the TK present in the intracellular domain of HER2.

Preclinical Data

In vitro cell line data and nude mice models with orthotopically implanted tumors, with cells resistant or susceptible to lapatinib, showed that signaling through src kinase mediates resistance to lapatinib via HER2 and EGFR signaling [46]. Another similar experiment showed that a HER2 overexpressing cell line resistant to lapatinib showed evidence of src kinase activity and of AKT and ERK1/2 phosphorylation and that treatment with src or CXCR4 inhibitors abrogated this resistance [47].

Two large scale studies demonstrated the efficacy of lapatinib treatment in HER2 positive MBC: One randomized 324 women with HER2 positive locally advanced or MBC to capecitabine alone or in combination with lapatinib after progression with anthracyclines, taxanes and trastuzumab. The trial was positive for its primary endpoint which was time to progression (TTP), the difference in TTP was 4 months, it increased from 4 to 8 months [48]. Another study randomized 296 women with HER2 positive MBC to lapatinib alone or in combination with trastuzumab after progression with a median of three trastuzumab-containing regimens. The trial was positive for its primary endpoint which was progression free survival (PFS), the difference in PFS was 4 weeks, it increased from 8 to 12 weeks [49]. The very different intervals to progression in these trials reflect the use of no CT in the second trial and a more heavily pretreated study population. A third trial in the metastatic setting, studied the role of lapatinib in overcoming ET resistance. The trial randomized HER2 positive metastatic BC patients receiving first line therapy to letrozole alone or in combination with lapatinib. Letrozole and trastuzumab were allowed as long as they were administered more than a year before randomization. The trial was positive for the primary endpoint; PFS which increased from 3 to 8 months in the combination arm [22].

Neoadjuvant Breast Cancer

The Neo-ALLTO trial randomized 455 patients with HER2 positive primary breast cancer greater than 2 cm in diameter to one of three arms: combination of lapatinib and trastuzumab (L + T), isolated lapatinib (L) or trastuzumab (T). Pathological complete response (pCR), the primary endpoint, was similar in the single agent arms, 24 and 20 %, respectively and the combination arm had a pCR of 51 % [50]. These encouraging results were not translated into differing survival outcomes. The 3-year event-free survival (EFS) was 78, 76, and 84 %, respectively, in the T, the L, or the T + L arm. The 3-year OS was 93, 90 and 95 %, respectively. The landmark result of this trial is the renewed finding that pCR correlates with EFS and OS with a hazard ratio of 0.38 and 0.35 [51]. A retrospective analysis of this trial investigated the influence of PIK3CA mutations on sensitivity to the tested drugs. PIK3CA mutations were found in 23 % of patients, the mutation rate did not vary according to ER status. PIK3CA mutations were associated with lower pCR in all the three regimens. The difference was from 20 to 28 % (T), 14.8–20.4 % (L) and 28.6–55.8 % (T + L), respectively in the mutated vs. WT tumors [52].

Adjuvant

The ALTTO trial randomized 8381 patients with HER2 positive early BC to receive as adjuvant treatment either lapatinib combined with trastuzumab (L + T), trastuzumab followed by lapatinib (TL), trastuzumab (T) or lapatinib (L). The last arm was closed for futility in August 2011. Anti-HER2 therapy was administered after anthracyclines, concomitant with taxanes or concurrent with a non-anthracycline containing platinum regimen. The test for superiority in the L + T vs. T arms was not significant. The DFS at a median follow-up of 4.5 years was 88 vs. 86 %. The test for noninferiority in the TL vs. T arms was not significant. The DFS at a median follow-up of 4.5 years was 87 % vs. 86 % [53]. The DFS of the combination arm was unexpected given the doubling of the pCR rate observed in the NeoALTTO trial. The planned number of DFS events at 4.5 years was 850, but only 555 were actually observed. According to these results, less than 20 % of HER2 positive early BC patients relapse at 5 years, which was less than initially expected. Regarding the lower than expected lapatinib efficacy, it might have been compromised due to toxicity, for only 60–78 % of the patients in the lapatinib arms could receive at least 85 % of the protocol-specified dose.

1.2.3 Pertuzumab

Trastuzumab binds to the subdomain IV of the HER2 extracellular domain, its activity is due to blocking HER2 cleavage and ligand independent mitogenic action as well as antibody-dependent cellular cytotoxicity (ADCC) [54]. HER2 positive BC patients eventually develop resistance to trastuzumab. Pertuzumab binds HER2 in the extracellular subdomain II, prevents dimerization of HER2 with HER3 and also stimulates ADCC [55]. In cell line models trastuzumab and pertuzumab have synergistic activity. The Cleopatra trial compared docetaxel and trastuzumab with docetaxel, trastuzumab, and pertuzumab in HER2 positive MBC patients who had a disease free interval since completion of adjuvant chemotherapy that could have included trastuzumab or not, of at least 1 year. The trial was positive for the primary endpoint which was duration of progression free survival [56]. Consistent with these findings, the overall survival increased nearly 16 months, from 40 to 56 months in the double blockade arm [57].

1.2.4 T-DM1

Trastuzumab emtansine is an antibody–drug conjugate that incorporates the anti-HER2 antitumor properties of trastuzumab with the microtubule inhibitory agent DM1 (T-DM1). The antibody and the cytotoxic are conjugated by a stable linker. The delivery of the cytotoxic to HER2 overexpressing cells improves the therapeutic index minimizing toxicity. Two phase III clinical trials were performed with T-DM1: The Emilia trial in HER2 positive MBC patients resistant to trastuzumab and a taxane. Close to a thousand patients were randomized to receive either lapatinib and capecitabine or T-DM1, the primary endpoint was PFS. The trial was positive for PFS that increased from 6.4 to 9.6 months and for OS that increased from 25.1 to 30.9 months. The TH3RESA trial randomized 602 HER2 positive MBC that had received a median of 4 prior anti-HER2 and chemotherapy regimens, 75 % of the patients had visceral disease. Median PFS increased from 3.3 to 6.2 months [58].

1.2.5   Neratinib

Neratinib is a pan-ErbB (Erb 1, 2, and 4) receptor tyrosine kinase inhibitor that was evaluated in a phase II trial in 124 locally advanced or metastatic HER2 positive BC patients previously treated or not with trastuzumab. The PFS at 16 weeks was 59 % for the first group and 78 % for the second. Objective responses were seen in 24 or 56 % of the patients in the respective groups [58]. Median PFS was 19 weeks. Neratinib was studied in 77 HER2 positive MBC patients, anti-HER2 therapy pretreated with trastuzumab or with trastuzumab and lapatinib. The response rate was 42 and 51 %, respectively [59]. Neratinib is being evaluated in adjuvant trastuzumab-pretreated early stage BC, but not in MBC patients. I-SPY 2 is a phase II neoadjuvant trial in high risk stage II/III BC with more than 2 cm. HER2 negative patients were randomized to neratinib and docetaxel (N+D) followed (→) by doxorubicin and cyclophosphamide (A+C) or D→AC and HER2 positive patients were randomized to N+D→AC or Trastuzumab+D→AC. Results in both groups showed improved pCR rates with the addition of N, from 33 to 56 % and from 17 to 30 %. The TN cohort did not have an improvement in pCR [60].

1.2.6   Afatinib

Afatinib is an irreversible inhibitor of EGFR, HER2 and HER4, it is being studied in BC in phase II and III trials. In a phase II clinical trial 41 patients with HER2 positive MBC, 68.3 % had received for >1 year trastuzumab were treated with afatinib. 46 % of the patients achieved clinical benefit [61].

1.3   Targeted Therapies for Metastatic Breast Cancer

Regarding targeted therapies in MBC a wealth of pathways have, are and will be tested in preclinical models and clinical trials. MBC patients are a population where drug development starts that can be afterwards translated into other carcinomas. Data on growth factor pathways (EGFR and VEGFR) will be detailed, PARP inhibition and other oncogenic survival pathways like Src, PI3K/AKT/mTOR, RAF/ERK/MEK will similarly be reviewed. Finally, less mature data on other survival hallmarks like FGFR pathway, cell cycle regulation and apoptosis inhibition will be reviewed (Fig. 1).

1.3.1   EGFR (Her1) Inhibition

Cetuximab

EGFR has been implicated as a molecular target for treatment of TNBC based on its frequent IHC expression (27–57 %) [62–65]. Cetuximab has demonstrated efficacy in two prospective studies and one retrospective analysis of randomized phase II trials in advanced TNBC. The largest EGFR trial, the international BALI-1, prospectively evaluated the addition of cetuximab to cisplatin for the treatment of first and second-line TNBC patients ($N=173$) [66]. The trial did not meet its primary endpoint of PFS prolongation, although it doubled response rate. The addition of cetuximab to irinotecan and carboplatin in first and second-line MBC patients was tested in the USOR-04-070 trial, a North American endeavor [67]. The trial randomized 72 patients, resulted in improved response rates in TNBC patients (30 % vs. 49 %) with no improvements in PFS or OS. A third cetuximab trial, conducted in North America, added carboplatin to cetuximab in 102 heavily pretreated MTNBC patients and resulted in an response rate of 17 %. Prolonged PFS (2 vs. 8 months) was seen in responders compared with the overall trial population [68].

Tyrosine Kinase inhibitors

Gefitinib: Gefitinib was tested in 2005 in previously treated MBC patients. In 31 assessable patients, 12 had stable disease (SD) including three patients where the disease was stable for over 6 months. Sequential immunohistochemical studies in skin and tumor biopsies showed inhibition of epidermal growth factor receptor (EGFR) phosphorylation in both tissues. The mechanism of action of the drug differs in the two tissues. In the cancer tissue, gefitinib activates apoptosis through activation of p27 that does not happen in the skin. In skin there is reduction in proliferation assessed by ki67 decrease, which did not happen in breast. Both tissues showed inhibition of the mitogen activated protein kinase (MAPK) [67]. In 2007 a neoadjuvant trial tested the concept if increased epidermal growth factor receptor (EGFR) expression promotes breast cancer resistance to endocrine therapy, for this neoadjuvant gefitinib, an EGFR inhibitor, was added to anastrozole. The trial tested Ki67 reductions at 2 and 16 weeks as well as objective response. Contrary to hypothesized the combination did not increase Ki67 reduction nor response rate [23].

Erlotinib: Erlotinib is a tyrosine kinase inhibitor (TKI) that binds competitively at the TK domain of EGFR and inhibits the downstream pathways that control proliferation and survival. EGFR is expressed in up to 91 % of BC cases. A randomized phase II trial assessed the combination of erlotinib with carboplatin and docetaxel in the neoadjuvant treatment of 30 TNBC patients and demonstrated promising activity with pCR of 40 %, nevertheless erlotinib clinical trials in breast cancer are not a priority today [69].

1.3.2 Angiogenesis Inhibition

Bevacizumab

Metastatic Breast Cancer

Bevacizumab was developed on the rational basis of VEGF biology and has activity in numerous tumors with bearable toxicity, it was shown that using the drug, in part, blocks angiogenesis [70]. Bevacizumab created great excitement at the time, in fact it validated that blocking angiogenesis was a therapeutic modality in human cancer, although results were not as promising as initially expected [71]. In Phase II trials with mono-therapy in pretreated MBC patients a 9 % response rate (RR) was observed and less than 20 % stable disease at 6 months [72]. In combination therapy, the addition of bevacizumab to docetaxel or vinorelbine did not result in the expected increased RR [73]. Bevacizumab was combined with trastuzumab in patients progressing on the latter, there were responses and no additive toxicity [74]. This strategy of combination of bevacizumab and trastuzumab has been tested in MBC trials and neoadjuvant trials. The definitive phase III adjuvant trial named BETH, reported results a year ago and will be further discussed in detail [75]. Subsequent phase III trials suggesting improved activity when bevacizumab is administered in conjunction with CT were numerous and randomized thousands of MBC patients. Starting with second line MBC therapy, a phase III trial combined bevacizumab with capecitabine for patients previously treated with anthracycline and taxane therapy. The trial randomized 462 patients to receive capecitabine with or without bevacizumab, 25 % of patients had Her2 positive MBC. There was an increase in overall RR with the addition of bevacizumab to capecitabine (9.1 % vs. 19.8 %), with no significant improvement in PFS or OS [72]. The most important phase III trial combined bevacizumab with taxanes as first line treatment for patients with locally recurrent or MBC. It was an open label trial called ECOG 2100, 673 patients were randomized to paclitaxel plus or minus bevacizumab [73]. In this ECOG trial the ORR more than doubled (22.2 % vs. 48.9 %) as well as PFS (median, 5.8 vs. 11.3 months) with no improvement in OS (median, 25.2 vs. 26.7 months) [76]. Based on these results, the FDA approved bevacizumab for first-line therapy of MBC patients, in the beginning of 2008. This was granted under accelerated approval program, which allowed approval based on data that were not sufficiently complete to permit full approval. One confirmatory trial, the placebo-controlled, AVADO trial, randomized 736 patients to three study arms, two dose levels of bevacizumab combined with docetaxel. At progression, patients received bevacizumab as second-line therapy. At a median follow-up of 25 months, results show that standard dose (15 mg/kg) bevacizumab, in combination with docetaxel, increased ORR modestly (46.4 % vs. 64.1 %), improvement in PFS (median, 8.2 vs. 10.1). The addition of low dose (10 mg/kg) bevacizumab did not significantly improve responses. The final analysis showed no improvement in OS [77]. This trial was larger than the first one and it did not confirm the results of the first trial. These were followed by the RiBBOn 1 and 2 trials which tested chemotherapy with or without bevacizumab in first and second-line setting in 1237 patients [78, 79]. The RIBBON trials were also placebo-controlled. Data were analyzed based on patients receiving: (1) taxane or anthracycline-based CT plus standard dose bevacizumab vs. CT plus placebo and (2) capecitabine plus bevacizumab vs. capecitabine plus placebo. The addition of bevacizumab to CT resulted in improvements in PFS for both the taxane-anthracycline (median, 8.0 vs. 9.2 months) and capecitabine cohorts (median, 5.7 vs. 8.6 months). No significant differences in OS were observed between treatment arms in either the taxane–anthracycline (HR=1.03, $p=0.83$) or capecitabine cohort (HR=0.85, $p=0.27$). In patients with triple-negative breast cancer (TNBC) a significant improvement in PFS (median, 2.7 vs. 6.0 months; HR=0.49, $p=0.0006$) and a trend toward improved OS (median, 12.6 vs. 17.9 months; HR=0.62, $p=0.0534$) was seen [80]. Given these results, that only show small response rates with no survival benefit, the FDA recommended removing the BC indication from the label of bevacizumab in the end of 2010. In June 2011 there was a hearing with Genentech to re-appreciate data but in the end of 2011 the accelerated approval of bevacizumab for BC was withdrawn. The explanation for the different results of ECOG 2100 and the subsequent confirmatory trials will never be clear but the fact that the initial trial was open-label and the other two trials were placebo-controlled has flared up an old discussion in cancer trials.

Neoadjuvant Breast Cancer

Researchers continue to study this drug in other BC settings and BC subtypes, but, outside clinical trials, bevacizumab should not be administered to BC patients. The NSABP B-40 trial added bevacizumab (15 mg/kg) and/or antimetabolites to standard neoadjuvant CT in a randomized phase III trial of Her2-negative BC ($n = 1206$) [81]. Overall, the pCR rate (breast alone, [ypT0/Tis]) improved with the addition of bevacizumab (28.4 % vs. 34.5 %, $p = 0.027$) with the greatest impact observed in the hormone receptor-positive subset (15.2 % vs. 23.3 %, $p = 0.008$). The phase III GeparQuinto trial randomized 1948 Her2-negative BC patients to anthracycline-taxane CT with or without bevacizumab (15 mg/kg). Adding bevacizumab did not significantly increase the pathological complete response rate (pCR in breast and axilla, 15 % vs. 17.5 % or pCR breast alone [ypT0/Tis], 21.3 % vs. 23.9 %) or rates of breast conserving surgery in patients overall, but did improve the pCR rate in a subpopulation of 684 patients with TN disease (breast and axilla [ypT0, ypN0], 27.8 % vs. 36.4 %, $p = 0.021$) [82, 83]. As is expected, adding bevacizumab to CT increases neutropenia, hand–foot syndrome, mucositis, and hypertension. Again, these two large trials, that randomized more than 3000 neoadjuvant BC patients, B40 and GeparQuinto, had discordant results, not only in response but also in BC subtypes.

Adjuvant Setting

The study of this drug in the adjuvant setting started with ECOG 2104, a large randomized phase II pilot trial that incorporated bevacizumab into dose-dense doxorubicin and cyclophosphamide followed by paclitaxel in patients with lymph node positive breast cancer. It was designed to evaluate the safety of two different strategies incorporating bevacizumab; into anthracycline or taxane containing adjuvant therapy, as a precursor to a definitive randomized phase III trial. Patients were treated with dose-dense doxorubicin and cyclophosphamide, followed by paclitaxel, all patients received bevacizumab (10 mg/kg every 2 weeks during 26 weeks), initiated either concurrently with AC or with paclitaxel. The primary end point was incidence of cardiac dysfunction; once the results of the MBC trials had suggested increased cardiac adverse events with the addition of bevacizumab. In 226 enrolled patients, grade 3 hypertension, thrombosis, proteinuria, and hemorrhage were reported for 12, 2, 2, and less than 1 % of patients, respectively. Two patients had grade 3 cerebrovascular ischemia. Three patients in each arm developed congestive heart failure. There was no significant difference between arms in the proportion of patients with an absolute decrease in left ventricular ejection fraction. Bevacizumab, combined with adjuvant therapy, was concluded not to result in prohibitive cardiac toxicity and the definitive, 5000 patient, phase III trial, ECOG 5103, was started in 2007. Fortunately, despite the optimistic safety signals coming from ECOG 2104, systematic and extensive cardiac monitoring was implemented and detected an excess of toxic cardiac events in the bevacizumab containing arm leading to premature termination of the adjuvant trial in 2009 [84]. Beatrice, was an open-label phase III trial, where 2591 patients with triple-negative operable primary invasive breast cancer were randomly assigned to receive a minimum of four cycles of adjuvant chemotherapy alone ($n = 1290$) or with bevacizumab at an equivalent of 5 mg/kg every week for 1 year ($n = 1301$). Beatrice was started in 2008 and has published final results in 2013 at median 5 year follow-up [85]. Similar proportions of patients received anthracycline and taxane therapy (59 and 58 %), non-taxane anthracycline-containing therapy (36 and 37 %), non-anthracycline taxane-containing therapy (5 % of both), and radiation therapy (74 and 73 %). Chemotherapy was completed as planned in 92 % of the chemotherapy group and 93 % of the bevacizumab group, and bevacizumab was completed as planned in 68 %. After median follow-up of 32 months, there was no difference between the bevacizumab group and the chemotherapy group in invasive disease-free survival (IDFS). IDFS events occurred in 14 % of the bevacizumab group vs. 16 % of the chemotherapy group; 3-year IDFS was 83.7 % with bevacizumab and 82.7 % with chemotherapy alone. Subgroup analyses showed no evidence of differences. After 200 deaths, there was no difference in OS. Sites of recurrence were similar in the two treatment groups, with the most common being distant recurrence. The most common sites of distant recurrence were lung (28 and 27 %), liver (20 and 15 %), and bone (17 and 20 %). Distant central nervous system or meningeal recurrence accounted for 7 % of recurrences in the bevacizumab group and 12 % in the chemotherapy group. Exploratory biomarker assessment in approximately 45 % of patients suggested that patients with high pretreatment plasma VEGFR-2 levels might benefit from the addition of bevacizumab and the accompanying editorial suggested longer periods of the drug might give better results [86]. The hazard ratios for invasive disease-free survival for bevacizumab vs. chemotherapy were 0.61 among patients with levels above the median value and 1.24 for those with levels below median ($p = 0.0291$ for interaction). Grade 3 or higher adverse events occurred in 72 % of patients in the bevacizumab group and 57 % of the chemotherapy group. The bevacizumab group had an increased frequency of grade 3 or worse hypertension (12 % vs. 1 %), severe cardiac events occurring at any point during the 18-month safety-reporting period (1 % vs. <0.5 %), and treatment discontinuation (bevacizumab or chemotherapy, 20 % vs. 2 %) [85]. In Beatrice, unlike ECOG 5103, the safety data did not require stopping the trial, and therefore efficacy results of bevacizumab in the adjuvant setting in TNBC patients are available. With respect to Her2 positive breast cancer, as has been alluded to, the Beth trial randomized 3509 women who had either node-positive or high-risk node-negative disease, with the latter group making up 41 % of the population [75]. Patients were enrolled in 1 of 2 chemotherapy regimens: 6 cycles of docetaxel/carboplatin plus trastuzumab (TCH) with or without bevacizumab ($n=3231$) or an anthracycline-based regimen involving three cycles of docetaxel plus trastuzumab given with or without bevacizumab followed by three cycles of fluorouracil, epirubicin, and cyclophosphamide ($n=278$). In both regimens, patients continued trastuzumab with or without bevacizumab after chemotherapy to complete 1 year of targeted therapy. For the primary outcome of the study, which was IDFS, like in the Beatrice trial, there was no statistically significant difference between the patients who received bevacizumab and those who did not. At a median follow-up of 38 months, IDFS rates were 92 % for both groups of the TCH cohort. A secondary endpoint compared IDFS in patients in the anthracycline-based vs. the TCH-based cohorts and also found no significant differences between the regimens, whether with or without bevacizumab. However, the study was not designed to compare these different chemotherapy approaches, and only 278 patients received anthracyclines, less than 5 %. The IDFS of 92 % is in striking contrast with the IDFS of BCIRG 006 Trial of 86 % that had the same investigational arm without bevacizumab. These results will question the use of anthracyclines in the adjuvant treatment of Her2 BC. In grade 3 or 4 adverse events, hypertension was higher in the bevacizumab group (19 % vs. 4 %; $p<0.001$). There was also a trend for more congestive heart failure with bevacizumab (2.1 % vs. <1 %; $p=0.0621$), and a difference in hemorrhage (2 % vs. <1 %; $p<0.0001$). Proteinuria and gastrointestinal perforations were also more common in the bevacizumab group. Contrary to the favorable toxicity results Beatrice trial the Beth trial has again showed important toxicity with adjuvant bevacizumab and ECOG5103 was stopped due to toxicity. In conclusion, bevacizumab in MBC patients does not prolong survival and in used as adjuvant treatment of BC patients increases toxicity and does not prolong IDFS interval.

Tyrosine Kinase Inhibitors

Small molecule oral TKIs target the intracellular catalytic function of the VEGFR family linked tyrosine kinases. The kinases are coupled to the intracellular portion of the transmembrane receptors (e.g., VEGFR-1, 2, and 3). VEGFR-2 is the primary signaling receptor for VEGF-mediated angiogenesis.

Sorafenib

Sorafenib, a small molecule TKI, has both antiangiogenic and antiproliferative effects [87]. Sorafenib has shown single agent activity in pretreated patients [88]. Three randomized, phase IIb trials have shown that sorafenib in combination with standard chemotherapy significantly improved outcomes in first line MBC (PFS; median, 5.6 vs. 6.9 months; HR=0.79, p=0.09 and TTP; median 5.6 vs. 8.1 months; HR=0.67, p=0.017) and second-line (PFS; median, 4.1 vs. 6.4 months; HR=0.58, p=0.0006). Sorafenib showed activity after bevacizumab resistance (PFS; median, 2.7 vs. 3.4 months; HR=0.65, p=0.01) [89–92]. A multivariate analysis of these trials suggested sorafenib to be an interesting drug to pursue development [93]. The toxicity profile includes minimal grade 3/4 hypertension but high rates of grade 3/4 hand-foot syndrome. A different pattern when compared to bevacizumab. A placebo-controlled phase III trial in MBC evaluating capecitabine in combination with sorafenib is currently underway (NCT01234337). Other ongoing randomized trials will evaluate sorafenib in combination with standard chemotherapy (NCT00499525 and NCT01320111), metronomic chemotherapy and/or endocrine therapy (NCT00573755, NCT00954135) in the advanced setting. There are obvious advantages to an oral drug in this setting, that is why these phase III trials are designed with oral regimens.

Sunitinib

Sunitinib is a multi-targeted inhibitor of VEGFR-1, 2, and 3, platelet-derived growth factor receptor, c-Kit, FMS-like tyrosine kinase-3, and RET. A recent phase III trial comparing sunitinib (37.5 mg/day) to capecitabine and a small randomized phase II trial evaluating sunitinib as consolidation therapy following induction chemotherapy [94, 95] have both demonstrated inferior outcomes for single agent sunitinib compared with controls in pretreated MBC. Additionally, two randomized phase III trials in the advanced setting evaluating the addition of sunitinib (37.5 mg/day) to either capecitabine or docetaxel compared with the respective chemotherapies alone demonstrated increased toxicity and comparable PFS with the addition of sunitinib [96, 97]. The definitive first line phase III trial of sunitinib in BC was done comparing docetaxel with or without sunitinib in the first line treatment of MBC. The combination increased response rate but had no effect in PFS and OS [98]. Based on these findings, and the early termination of the phase III sunitinib trial in first line MBC, due to a lack of feasibility due to increased toxicity and weak efficacy results [78] the clinical development of sunitinib in BC was stopped.

Cediranib

Cediranib is being tested in combination with olaparib in recurrent ovarian, fallopian tube, peritoneal, or triple-negative breast cancer. The results are surprising because they show an important prolongation of progression free survival in non-BRCA mutated patients, suggesting that the antiangiogenic drug might be more useful in sporadic diseases [99].

1.3.3 PARP Inhibition

BRCA mutation carriers carry a germ-line mutation in one of the two BRCA genes and have increased susceptibility to develop breast, ovarian, prostate, gastric, and colorectal carcinomas as well as melanomas. BRCA genes are involved in repairing DNA through homologous recombination. The second hit is acquired in the tumor genome, and therefore these tumors are specifically susceptible to DNA damaging agents once they have defects in their DNA repair machinery. TNBC account for 50 % of BC in germ-line BRCA1 gene mutation carriers which is higher than the 15–20 % of TNBC appearing in sporadic BC population. However, BRCA gene products are nonfunctional in a subset of sporadic TNBCs, generally through promoter hyper-methylation or by other means, this property has been termed the BRACness of sporadic TNBC. Additionally, these tumors might have DNA repair defects in homology-directed repair pathways, not BRCA dependent. This is the reason Poly ADP Ribose Polymerase (PARP) inhibitors were investigated in TNBC in general. PARP is an enzyme involved in base excision repair of DNA. If this enzyme is blocked, there will be double strand breaks that will be repaired in case of homologous recombination integrity, which is the case in BRCA competent tumors. In contrast, BRCA1/2-deficient cells would be unable to carry out repair, ultimately leading to persistent double-strand breaks and cell death. In vitro studies confirmed the predicted "synthetic lethality" between PARP and BRCA1/2 pathways [100, 101]. On the other hand, there is overexpression of PARP1 in TNBC [102]. Therefore, the ideal setting to study PARPi in BC was thought to be TNBC, sporadic or genetic.

Iniparib

A phase II trial randomized 123 patients with metastatic TNBC to carboplatin and gemcitabine, alone or in combination with iniparib. Response rate increased from 16 to 48 %, clinical benefit increased from 21 to 62 %, progression free survival increased from 3.6 to 5.9 months and survival increased from 7.7 to 12.3 months. The results of this trial were better than expected. This trial was conducted in unselected TNBC and had an overall survival benefit, uncommon in randomized phase II studies, generally not designed with power to demonstrate survival benefits [103]. These results were not confirmed by the ensuing phase III trial, where 519 patients with metastatic TNBC, in first to third line, were randomized to the same therapeutic regimen. In this trial, neither response nor survival were increased [104]. In a retrospective analysis, it was seen that due to the heterogeneity of TNBC, only 25 % of these patients had basal like TNBC and their BRCA status was unknown. Additionally the PARPi activity of iniparib was put in question [105]. As a consequence of these trials iniparib development in BC has stopped.

Olaparib

Olaparib is an oral PARP inhibitor that has showed activity in BRCA null breast cancer. In a phase II trial where 54 patients with germ-line mutations of BRCA1 or BRCA2 had locally advanced or metastatic BC, the response rate was 41 %. As expected from the profile of genetic BC, 57 % of these patients had TNBC [106]. The activity of olaparib in wild type BRCA TNBC, was shown in another trial. Here, olaparib was used in 23 patients with MBC, where only eight were BRCA mutation carriers, in this trial, there were no objective responses [107]. These results suggest that isolated olaparib is not active in WT BRCA TNBC. The way forward in the development of this agent seems to be BRCA null tumors, TNBC or otherwise, and therefore olaparib is now being planned to be studied in a very challenging adjuvant trial in BRCA mutation carriers called Olympia trial (NCT02032823).

Veliparib

Veliparib has been tested in combination with temozolomide in unselected MBC patients, here, similarly to olaparib, responses were mainly observed in BRCA carriers. The authors created an expansion cohort to the initial trial where only BRCA mutation carriers were enrolled. The overall response rate was 25 % and the clinical benefit rate 50 % these results were not as robust as was expected from the results in BRCA carriers in the initial cohort [108].

1.3.4   Src Inhibition

Src is an important epithelial survival pathway. In BC, Src activation has been shown to correlate with bone metastasis [109]. A Phase I was conducted in MBC patients, the cohort had 60 % of ER positive disease. Dasatinib 100 mg once daily, plus capecitabine 1000 mg/m² twice daily, were associated with clinical benefit, in 56 % of patients. Biomarker changes were consistent with what has been observed with bevacizumab, with significant decreases in plasma VEGF-A and increases in VEGFR-2 and collagen-IV, suggesting dasatinib might have an antiangiogenic effect [110]. Given these findings suggesting antiangiogenic activity, a phase II trial as single agent was run in pretreated metastatic TNBC, where angiogenesis was thought to have a more prominent role, there was a 5 % response rate and a PFS of 2 months [111] (Fig. 1).

1.3.5   PI3K/AKT/mTOR Inhibition

Several pre clinical reports show that genetic alterations in the phosphatidylinositol 3-kinase (PI3K) pathway influence breast cancer prognosis and predict resistance to ET and CT [104]. Mutations in the p110α catalytic subunit of the class IA PI3K encoded by the PIK3CA gene are the most frequent somatic mutations in BC, appearing in a quarter of BC cases [105]. These mutations promote carcinogenesis in mammary gland epithelial models enabling constitutive ligand independent activation of PI3K pathway [106]. These hypothesis have been tested in clinical samples but to date all retrospective analysis of patient samples from endocrine or anti-HER2 therapy trials could not show conclusive evidence of a prognostic role [112] nor a predictive role for response to endocrine [113] or to anti-HER2 therapy [44] for PIK3CA mutations.

The PI3K/AKT/mTOR pathway regulates multiple cellular processes that promote growth, survival, and metastasis. This is the most frequently aberrantly activated pathway in human cancer. This pathway has been implicated in ET and anti-HER2 therapy resistance. Activating mutations of the pathway are detected in 40 % of ER positive and HER2 positive BC, comprising 40 % BC [105]. The mammalian target of rapamycin (mTOR) is a kinase downstream of PI3K pathway. The PI3K/AKT/mTOR pathway is activated in ER positive BC cell lines that have become resistant to ET. It has been shown that rapamycin, an anti-mTOR agent restores BC cell sensitivity to ET [114]. This the rational for the neoadjuvant and MBC trials performed with everolimus. In 2009, results from the neoadjuvant trial showed that everolimus increases letrozole response by clinical and laboratory assessments [115]. Laboratory assessments included Ki67 and cyclin D1 expression by IHC which decreased with both regimens and the phosphorylation of ribosomal S6 unit which was only reduced by the combination therapy. The phosphorylation of serine residues seems to be a specific mitogenic signal of growth factor activation [111]. A number of phase III randomized clinical trials in MBC have been performed with everolimus a rapalogue that inhibits mTORC1, called Breast cancer clinical of OraL EveROlimus (BOLERO) trials: Bolero-1 has been presented and tested the addition of everolimus to standard first line treatment for HER2 positive MBC patients with paclitaxel and trastuzumab. The trial randomized 719 patients. The hypothesis being tested is if mTOR inhibition increases PFS compared with the paclitaxel trastuzumab combination. Bolero-2 randomized 724 patients that had progressed on adjuvant or metastatic nonsteroidal AI to receive examestane alone or in combination with everolimus [116]. Centrally assessed median PFS increased from 4.1 to 10.6 months with the combination arm. These results show that mTOR inhibition can increase the magnitude and duration of response to ET in the clinic in both neoadjuvant and metastatic setting. A massively parallel sequencing effort of the tumor samples from the patients randomized in the Bolero-2 trial [117], showed that 50 % of these tumors harbored PIK3CA mutations, 30 % had amplification of cyclin D1 and over 20 % had mutated or lost P53 and 18 % had amplification of FGFR1. These genomic alterations are not actionable at present and contrary to what was predicted in tumor models, PIK3CA mutated tumors do not respond better to everolimus. Presently it is thought that screening for PIK3CA mutations might be useful for selecting patients for treatment with PIK3CA inhibitors. The Bolero-3 trial randomized 569 HER2 positive MBC patients previously treated with a taxane for MBC, i.e., second line treatment for MBC, to the combination of vinorelbine and trastuzumab vs. the combination plus everolimus. The primary endpoint which was median PFS was significantly increased from 5.8 to 7 months. A retrospective biomarker analysis aimed to find predictive factors of response to everolimus. Phosphorylation of ribosomal protein S6 (pS6) and PTEN levels were assessed by IHC and the PIK3CA gene was sequenced. Patients with low PTEN derived more benefit from everolimus with an increase in PFS from 5.2 to 9.6 months. Patients with high pS6 also derived more benefit from everolimus. As expected, there was marker interaction between these characteristics. Regarding PIK3CA mutations, a trend towards better response to everolimus was seen in this population which was 20 % of the trial patients (126). There are PI3K pan-inhibitors called pan because they inhibit p110 and p85 one of which is buparlisib. In a phase I dose escalation trial those BC that responded most had WT PIK3CA and PTEN. A Phase II trial is ongoing to investigate buparlisib for the treatment of metastatic TNBC (NCT01629615). This drug is being tested in two phase III trials, BELLE 2 and 3, to enhance ET responsiveness with fulvestrant. Buparlisib has been shown to synergize with trastuzumab in BC cell lines, and, in xenografts treatment with buparlisib partially restored sensitivity to trastuzumab [118]. With these preclinical data in mind, a phase I/II trial tested the combination in trastuzumab-resistant HER2 positive MBC patients and found 25 % of clinical benefit. Following this, a neoadjuvant trial is testing the triple therapy: trastuzumab, paclitaxel, and buparlisib (NCT01816594). Still in the metastatic setting, the combination of buparlisib and lapatinib is being tested in trastuzumab-resistant patients harboring PIK3CA or PTEN mutations (NCT01589861). GDC0032 is a p110α-specific, β-sparing inhibitor that targets preferentially PIK3CA mutant tumors and in phase I interesting responses were seen in PIK3CA mutated MBC. This drug is being tested in combination with fulvestrant in a phase Ib trial in MBC patients. It will be similarly tested combined with letrozole and in HER2 positive MBC patients. Similarly, the pan-PI3K inhibitor GDC-0941 and the PI3K/mTOR dual inhibitor GDC-0980 are being investigated in phase II with fulvestrant in patients with advanced disease resistant to aromatase inhibitors (NCT01437566). In addition, a Phase I trial is testing the efficacy of the AKT inhibitor, MK2206, in combination with anastrozole or fulvestrant, or with anastrozole plus fulvestrant in patients with metastatic ER+ disease (NCT01344031). BYL719 is also a p110α inhibitor that was found in preclinical findings to be active in BC with mutations or amplification of PIK3CA and ER expression, and therefore clinical development in phase I has focused on enrichment of patients with tumors with PIK3CA alterations. Perifostine, MK2206, GSK690563. GDC0068 is an ATP-competitive AKT inhibitor and has shown activity in phase I combined with docetaxel in PI3K pathway altered MBC patients (Fig. 1).

Prototypic Ras/MAPK pathway mutations in breast cancer are infrequent. This suggests that breast carcinomas and their normal epithelial cell precursors do not require this pathway to mediate programs of sustained growth and survival. Cells of mammary origin may use alternative redundant signaling cascades to achieve the same results. For instance, alterations in the parallel and highly interconnected phosphatidylinositol-3-kinase (PI3K)/Akt pathway are much more frequent in breast cancer, as has been previously discussed and therefore drug development is well advanced. Clinical trials in BC have not started but in view of the extensive signaling cross talk and shared transcriptional output of these two pathways the combination of PI3K and MEK inhibitors seems to be the best way forward. MEK inhibitors are already being extensively studied in melanoma and the combination of BRAF and MEK inhibition has been shown to be efficacious in metastatic melanoma [72] (Fig. 1).

1.3.7   FGFR Inhibition

FGFR is a transmembrane receptor protein with intracellular tyrosine kinase activity, a genome-wide association study (GWAS) of BC by SNP genotyping in 1145 postmenopausal women of European ancestry with BC and 1142 controls found four SNPs in intron 2 of FGFR2 that were highly associated with BC incidence [119]. Fibroblast growth factor receptor 1 amplification and overexpression has been described in 10 % of ER positive luminal B BC tumor material and correlates with poor survival of patients. In cell lines, overexpression of FGFR1 results in both enhanced ligand-dependent and -independent signaling compromising ET sensitivity [120]. FGFR2 amplification has been described in TNBC [121]. Dovitinib is a TKI that targets FGFR1, VEFGR, and PDFGR, and it was studied in a phase II study in MBC to determine the interaction between FGFR1 amplification and ER positivity and dovitinib activity. The trial showed that dovitinib is more active in MBC patients with FGFR1 amplified, ER positive tumors [122]. A phase III trial overcoming ET resistance with fulvestrant with or without dovitinib is therefore planned. Lucitanib is being evaluated in MBC after one ET line failure. The trial is designed to test the benefit of FGFR-1 blockade in different types of tumor, namely, FGFR1-amplified, FGFR1-non-amplified with neither FGFR1-amplification nor 11q-amplification [122, 123]. FGFR inhibitors with higher specificity have emerged: BGJ398 and AZD4547. The activity of the pan-FGFR inhibitor BGJ398 has been predicted by FGFR amplification in cell line material [72]. BGJ398 is being tested in combination with PI3K inhibitor BYL719 in MBC patients with PIK3CA mutations with or without FGFR1-3 amplification (NCT01928459). AZD4547 has been tested in phase I and is now being studied in FGFR1 or FGFR2 amplified MBC patients (NCT01795768).

1.3.8 Cell Cycle Inhibition via CDK4/6

Palbociclib inhibits 2 cell cycle cyclin-dependent kinases CDK4/6 which facilitate cell cycle entry. CCND1 amplification has been described in 60 % of luminal B BC and in 30 % of luminal A and HER2 positive BC [124], whereas cyclin D1 overexpression has been reported in 50–70 % of BC [125]. Similarly, the luminal B subtype more frequently displays a gain of CDK4 (25 % of luminal B vs. 14 % of luminal A), and a loss of negative regulators including p16INK4A. Basal-like breast cancers do not typically display alterations in cyclin D1 or CDK4/6, but 20 % of cases harbor mutations that lead to the homozygous loss of RB1, indicating these patients might respond less to inhibitors of CDK4/6. In 2014, a phase II study that combined palbociclib with letrozole in the first line treatment of ER positive HER2 negative MBC patients, reported an increase in PFS from 10 to 20 months [126]. In attempts to define biomarkers predictive of a response to palbociclib, authors analyzed the amplification of CCND1 and/or the loss of CDKN2A (p16INK4A) by fluorescence in situ hybridization which was found not to be predictive of palbociclib benefit. The FDA designated palbociclib in April 2013 as breakthrough therapy. Two phase III trials are ongoing in MBC, PALOMA 2 and 3, the first one with same design the other in combination with fulvestrant, and an adjuvant trial called PENELOPE is also underway. LY2835219 is a selective inhibitor of CDK4 and CDK6 it has been tested in phase I. PFS in ER positive MBC patients with median of seven prior systemic regimens were 9 months [127]. LEE011 is a selective inhibitor of CDK4/6 kinases. In solid tumors, Rb protein is frequently inactivated by increased CDK4/6 activity which can in turn originate from inactivation of CDKN2A, amplification of d-cyclins, mutations of BRAF/PIK3CA or PTEN deletion. LEE011 more frequently stops growth than promotes lesion regression. LEE011 is being tested in combinations with the PI3K inhibitor BYL719 and letrozole in MBC patients and in combination with everolimus and examestane in MBC. LEE011 is also being tested with letrozole in the neoadjuvant setting and the phase III trials MONALEESA 1 and 2 have started (Fig. 1).

1.3.9 Apoptosis Induction

LCL161 is an antagonist of the inhibitor of apoptosis protein (IAP). IAPs protect cells from apoptosis by inhibiting caspases. This drug is being tested in BC in the neoadjuvant setting in combination with weekly paclitaxel. Tumor samples are being tested for the activation of caspase 3 by immunohistochemistry (NCT01617668).

2   Conclusion

BC has served as a model for drug development due to its high incidence and prevalence, systemic therapy sensitivity, and the relative fitness of MBC patients when compared to patients with other metastatic carcinomas. Additionally, BC is the leading cancer killer among women aged 20–59 years worldwide, and therefore prevention, early diagnosis, and better treatments are needed. MBC patients have a median survival of 2 years, including all BC subtypes. ER positive MBC patients have a median survival of 5 years. HER2 positive MBC patients in the era of new anti-HER2 agents pertuzumab and TDM1 will increase survival to a median of 5 years [57], whereas TN MBC patients have a lower median survival. BC cells in vivo eventually become resistant to targeted agents as they become resistant to classic chemotherapy. For several targeted drugs resistance mechanisms are not yet clear and further experiments are needed. The understanding of target definition, surrogate markers of efficacy with easily accessible readouts and resistance mechanisms are very useful because these might help position and recover various drugs. The most important examples in the last few years are the antiangiogenic drugs and the PARP inhibitors. Another recent example of selecting targets is the discovery that palbociclib and PI3K inhibitors will pursue development in ER positive BC patients in an effort overcome ET resistance.

References

1. Senkus E, Kyriakides S, Penault-Llorca F, Poortmans P, Thompson A, Zackrisson S, Cardoso F, Group EGW (2013) Primary breast cancer: ESMO clinical practice guidelines for diagnosis, treatment and follow-up. Ann Oncol 24(Suppl 6):7–23

2. Konecny G, Pauletti G, Pegram M, Untch M, Dandekar S, Aguilar Z, Wilson C, Rong HM, Bauerfeind I, Felber M, Wang HJ, Beryt M, Seshadri R, Hepp H, Slamon DJ (2003) Quantitative association between HER-2/neu and steroid hormone receptors in hormone receptor-positive primary breast cancer. J Natl Cancer Inst 95:142–153

3. Pan H, Gray R, EBCTCG (2014) Effect of obesity in premenopausal ER+ early breast cancer: EBCTCG data on 80,000 patients in 70 trials. J Clin Oncol 22:503

4. Al-Delaimy WK, Flatt SW, Natarajan L, Laughlin GA, Rock CL, Gold EB, Caan BJ, Parker BA, Pierce JP (2011) IGF1 and risk of additional breast cancer in the WHEL study. Endocr Relat Cancer 18:235–244

5. Endogenous H, Breast Cancer Collaborative G, Key TJ, Appleby PN, Reeves GK, Roddam AW (2010) Insulin-like growth factor 1 (IGF1), IGF binding protein 3 (IGFBP3), and breast cancer risk: pooled individual data analysis of 17 prospective studies. Lancet Oncol 11:530–542

6. Iorns E, Turner NC, Elliott R, Syed N, Garrone O, Gasco M, Tutt AN, Crook T, Lord CJ, Ashworth A (2008) Identification of CDK10 as an important determinant of resistance to endocrine therapy for breast cancer. Cancer Cell 13:91–104

7. Khan S, Lopez-Dee Z, Kumar R, Ling J (2013) Activation of NFkB is a novel mechanism of pro-survival activity of glucocorticoids in breast cancer cells. Cancer Lett 337:90–95

8. Zhou Y, Yau C, Gray JW, Chew K, Dairkee SH, Moore DH, Eppenberger U, Eppenberger-Castori S, Benz CC (2007) Enhanced NF kappa B and AP-1 transcriptional activity associated with antiestrogen resistant breast cancer. BMC Cancer 7:59

9. Louie MC, Zou JX, Rabinovich A, Chen HW (2004) ACTR/AIB1 functions as an E2F1 coactivator to promote breast cancer cell proliferation and antiestrogen resistance. Mol Cell Biol 24:5157–5171

10. Osborne CK, Bardou V, Hopp TA, Chamness GC, Hilsenbeck SG, Fuqua SA, Wong J, Allred DC, Clark GM, Schiff R (2003) Role of the estrogen receptor coactivator AIB1 (SRC-3) and HER-2/neu in tamoxifen resistance in breast cancer. J Natl Cancer Inst 95:353–361

11. McNeil CM, Sergio CM, Anderson LR, Inman CK, Eggleton SA, Murphy NC, Millar EK, Crea P, Kench JG, Alles MC, Gardiner-Garden M, Ormandy CJ, Butt AJ, Henshall SM, Musgrove EA, Sutherland RL (2006)

c-Myc overexpression and endocrine resistance in breast cancer. J Steroid Biochem Mol Biol 102:147–155

12. Ellis MJ, Ding L, Shen D, Luo J, Suman VJ, Wallis JW, Van Tine BA, Hoog J, Goiffon RJ, Goldstein TC, Ng S, Lin L, Crowder R, Snider J, Ballman K, Weber J, Chen K, Koboldt DC, Kandoth C, Schierding WS, McMichael JF, Miller CA, Lu C, Harris CC, McLellan MD, Wendl MC, DeSchryver K, Allred DC, Esserman L, Unzeitig G, Margenthaler J, Babiera GV, Marcom PK, Guenther JM, Leitch M, Hunt K, Olson J, Tao Y, Maher CA, Fulton LL, Fulton RS, Harrison M, Oberkfell B, Du F, Demeter R, Vickery TL, Elhammali A, Piwnica-Worms H, McDonald S, Watson M, Dooling DJ, Ota D, Chang LW, Bose R, Ley TJ, Piwnica-Worms D, Stuart JM, Wilson RK, Mardis ER (2012) Whole-genome analysis informs breast cancer response to aromatase inhibition. Nature 486:353–360

13. Miller TW, Balko JM, Ghazoui Z, Dunbier A, Anderson H, Dowsett M, Gonzalez-Angulo AM, Mills GB, Miller WR, Wu H, Shyr Y, Arteaga CL (2011) A gene expression signature from human breast cancer cells with acquired hormone independence identifies MYC as a mediator of antiestrogen resistance. Clin Cancer Res 17:2024–2034

14. Thangavel C, Dean JL, Ertel A, Knudsen KE, Aldaz CM, Witkiewicz AK, Clarke R, Knudsen ES (2011) Therapeutically activating RB: reestablishing cell cycle control in endocrine therapy-resistant breast cancer. Endocr Relat Cancer 18:333–345

15. Varma H, Skildum AJ, Conrad SE (2007) Functional ablation of pRb activates Cdk2 and causes antiestrogen resistance in human breast cancer cells. PLoS One 2, e1256

16. Keyomarsi K, Tucker SL, Buchholz TA, Callister M, Ding Y, Hortobagyi GN, Bedrosian I, Knickerbocker C, Toyofuku W, Lowe M, Herliczek TW, Bacus SS (2002) Cyclin E and survival in patients with breast cancer. N Engl J Med 347:1566–1575

17. Fan P, Agboke FA, McDaniel RE, Sweeney EE, Zou X, Creswell K, Jordan VC (2014) Inhibition of c-Src blocks oestrogen-induced apoptosis and restores oestrogen-stimulated growth in long-term oestrogen-deprived breast cancer cells. Eur J Cancer 50:457–468

18. Elsberger B (2014) Translational evidence on the role of Src kinase and activated Src kinase in invasive breast cancer. Crit Rev Oncol Hematol 89:343–351

19. Lai CF, Flach KD, Alexi X, Fox SP, Ottaviani S, Thiruchelvam PT, Kyle FJ, Thomas RS, Launchbury R, Hua H, Callaghan HB, Carroll JS, Charles Coombes R, Zwart W, Buluwela L, Ali S (2013) Co-regulated gene expression by oestrogen receptor alpha and liver receptor homolog-1 is a feature of the oestrogen response in breast cancer cells. Nucleic Acids Res 41:10228–10240

20. Gnant M., Mlineritsch B., Stoeger H., Luschin-Ebengreuth G., Heck D., Menzel C., Jakesz R., Seifert M., Hubalek M., Pristauz G., Bauernhofer T., Eidtmann H., Eiermann W., Steger G., Kwasny W., Dubsky P., Hochreiner G., Forsthuber E. P., Fesl C., Greil R., Austrian B., Colorectal Cancer Study Group VA (2011) Adjuvant endocrine therapy plus zoledronic acid in premenopausal women with early-stage breast cancer: 62-month follow-up from the ABCSG-12 randomised trial. Lancet Oncol 12:631–641

21. Coleman RE, Marshall H, Cameron D, Dodwell D, Burkinshaw R, Keane M, Gil M, Houston SJ, Grieve RJ, Barrett-Lee PJ, Ritchie D, Pugh J, Gaunt C, Rea U, Peterson J, Davies C, Hiley V, Gregory W, Bell R, Investigators A (2011) Breast-cancer adjuvant therapy with zoledronic acid. N Engl J Med 365:1396–1405

22. Johnston S, Pippen J Jr, Pivot X, Lichinitser M, Sadeghi S, Dieras V, Gomez HL, Romieu G, Manikhas A, Kennedy MJ, Press MF, Maltzman J, Florance A, O'Rourke L, Oliva C, Stein S, Pegram M (2009) Lapatinib combined with letrozole versus letrozole and placebo as first-line therapy for postmenopausal hormone receptor-positive metastatic breast cancer. J Clin Oncol 27:5538–5546

23. Smith IE, Walsh G, Skene A, Llombart A, Mayordomo JI, Detre S, Salter J, Clark E, Magill P, Dowsett M (2007) A phase II placebo-controlled trial of neoadjuvant anastrozole alone or with gefitinib in early breast cancer. J Clin Oncol 25:3816–3822

24. Polychronis A, Sinnett HD, Hadjiminas D, Singhal H, Mansi JL, Shivapatham D, Shousha S, Jiang J, Peston D, Barrett N, Vigushin D, Morrison K, Beresford E, Ali S, Slade MJ, Coombes RC (2005) Preoperative gefitinib versus gefitinib and anastrozole in postmenopausal patients with oestrogen-receptor positive and epidermal-growth-factor-receptor-positive primary breast cancer: a double-blind placebo-controlled phase II randomised trial. Lancet Oncol 6:383–391

25. Lindstrom LS, Karlsson E, Wilking UM, Johansson U, Hartman J, Lidbrink EK, Hatschek T, Skoog L, Bergh J (2012) Clinically used breast cancer markers such as estrogen receptor, progesterone receptor, and human epidermal growth factor receptor 2 are unstable throughout tumor progression. J Clin Oncol 30:2601–2608

26. Babayan A, Hannemann J, Spotter J, Muller V, Pantel K, Joosse SA (2013) Heterogeneity of estrogen receptor expression in circulating tumor cells from metastatic breast cancer patients. PLoS One 8, e75038

27. Shah S, Morin R, Khattra J, Prentice L, Pugh T, Burleigh A, Delaney A, Gelmon K, Guliany R, Senz J, Steidl C, Holt R, Jones S, Sun M, Leung G, Moore R, Severson T, Taylor G, Teschendorff A, Tse K, Turashvili G, Varhol R, Warren R, Watson P, Zhao Y, Caldas C, Huntsman D, Hirst M, Marra M, Aparicio S (2009) Mutational evolution in a lobular breast tumour profiled at single nucleotide resolution. Nature 461:809–813

28. Robinson DR, Wu YM, Vats P, Su F, Lonigro RJ, Cao X, Kalyana-Sundaram S, Wang R, Ning Y, Hodges L, Gursky A, Siddiqui J, Tomlins SA, Roychowdhury S, Pienta KJ, Kim SY, Roberts JS, Rae JM, Van Poznak CH, Hayes DF, Chugh R, Kunju LP, Talpaz M, Schott AF, Chinnaiyan AM (2013) Activating ESR1 mutations in hormone-resistant metastatic breast cancer. Nat Genet 45:1446–1451

29. Jeselsohn R, Yelensky R, Buchwalter G, Frampton G, Meric-Bernstam F, Gonzalez-Angulo AM, Ferrer-Lozano J, Perez-Fidalgo JA, Cristofanilli M, Gomez H, Arteaga CL, Giltnane J, Balko JM, Cronin MT, Jarosz M, Sun J, Hawryluk M, Lipson D, Otto G, Ross JS, Dvir A, Soussan-Gutman L, Wolf I, Rubinek T, Gilmore L, Schnitt S, Come SE, Pusztai L, Stephens P, Brown M, Miller VA (2014) Emergence of constitutively active estrogen receptor-alpha mutations in pretreated advanced estrogen receptor-positive breast cancer. Clin Cancer Res 20:1757–1767

30. Al Saleh S, Al Mulla F, Luqmani YA (2011) Estrogen receptor silencing induces epithelial to mesenchymal transition in human breast cancer cells. PLoS One 6, e20610

31. Khajah MA, Al Saleh S, Mathew PM, Luqmani YA (2012) Differential effect of growth factors on invasion and proliferation of endocrine resistant breast cancer cells. PLoS One 7, e41847

32. Mittendorf EA, Wu Y, Scaltriti M, Meric-Bernstam F, Hunt KK, Dawood S, Esteva FJ, Buzdar AU, Chen H, Eksambi S, Hortobagyi GN, Baselga J, Gonzalez-Angulo AM (2009) Loss of HER2 amplification following trastuzumab-based neoadjuvant systemic therapy and survival outcomes. Clin Cancer Res 15:7381–7388

33. Berns K, Horlings HM, Hennessy BT, Madiredjo M, Hijmans EM, Beelen K, Linn SC, Gonzalez-Angulo AM, Stemke-Hale K, Hauptmann M, Beijersbergen RL, Mills GB, van de Vijver MJ, Bernards R (2007) A functional genetic approach identifies the PI3K pathway as a major determinant of trastuzumab resistance in breast cancer. Cancer Cell 12:395–402

34. Scaltriti M, Eichhorn PJ, Cortes J, Prudkin L, Aura C, Jimenez J, Chandarlapaty S, Serra V, Prat A, Ibrahim YH, Guzman M, Gili M, Rodriguez O, Rodriguez S, Perez J, Green SR, Mai S, Rosen N, Hudis C, Baselga J (2011) Cyclin E amplification/overexpression is a mechanism of trastuzumab resistance in HER2+ breast cancer patients. Proc Natl Acad Sci U S A 108:3761–3766

35. Piccart-Gebhart MJ, Procter M, Leyland-Jones B, Goldhirsch A, Untch M, Smith I, Gianni L, Baselga J, Bell R, Jackisch C, Cameron D, Dowsett M, Barrios CH, Steger G, Huang CS, Andersson M, Inbar M, Lichinitser M, Lang I, Nitz U, Iwata H, Thomssen C, Lohrisch C, Suter TM, Ruschoff J, Suto T, Greatorex V, Ward C, Straehle C, McFadden E, Dolci MS, Gelber RD, Herceptin Adjuvant Trial Study T (2005) Trastuzumab after adjuvant chemotherapy in HER2-positive breast cancer. N Engl J Med 353:1659–1672

36. Dowsett M, Procter M, McCaskill-Stevens W, de Azambuja E, Dafni U, Rueschoff J, Jordan B, Dolci S, Abramovitz M, Stoss O, Viale G, Gelber RD, Piccart-Gebhart M, Leyland-Jones B (2009) Disease-free survival according to degree of HER2 amplification for patients treated with adjuvant chemotherapy with or without 1 year of trastuzumab: the HERA trial. J Clin Oncol 27:2962–2969

37. Zabaglo L, Stoss O, Ruschoff J, Zielinski D, Salter J, Arfi M, Bradbury I, Dafni U, Piccart-Gebhart M, Procter M, Dowsett M, Team HTS (2013) HER2 staining intensity in HER2-positive disease: relationship with FISH amplification and clinical outcome in the HERA trial of adjuvant trastuzumab. Ann Oncol 24:2761–2766

38. Metzger-Filho O, Procter M, de Azambuja E, Leyland-Jones B, Gelber RD, Dowsett M, Loi S, Saini KS, Cameron D, Untch M, Smith I, Gianni L, Baselga J, Jackisch C, Bell R, Sotiriou C, Viale G, Piccart-Gebhart M (2013) Magnitude of trastuzumab benefit in patients with HER2-positive, invasive lobular breast carcinoma: results from the HERA trial. J Clin Oncol 31:1954–1960

39. Kim CBJ, Horne Z (2004) Trastuzumab sensitivity of breast cancer with coamplification

of HER2 and C-MYC suggests proapoptotic function of dysregulated c-MYC in-vivo. Breast Cancer Res Treat 88:S6

40. Perez EA, Jenkins RB, Dueck AC, Wiktor AE, Bedroske PP, Anderson SK, Ketterling RP, Sukov WR, Kanehira K, Chen B, Geiger XJ, Andorfer CA, McCullough AE, Davidson NE, Martino S, Sledge GW, Kaufman PA, Kutteh LA, Gralow JR, Harris LN, Ingle JN, Lingle WL, Reinholz MM (2011) C-MYC alterations and association with patient outcome in early-stage HER2-positive breast cancer from the north central cancer treatment group N9831 adjuvant trastuzumab trial. J Clin Oncol 29:651–659

41. Perez EA, Dueck AC, McCullough AE, Chen B, Geiger XJ, Jenkins RB, Lingle WL, Davidson NE, Martino S, Kaufman PA, Kutteh LA, Sledge GW, Harris LN, Gralow JR, Reinholz MM (2013) Impact of PTEN protein expression on benefit from adjuvant trastuzumab in early-stage human epidermal growth factor receptor 2-positive breast cancer in the North Central Cancer Treatment Group N9831 trial. J Clin Oncol 31:2115–2122

42. Pogue-Geile KL, Kim C, Jeong JH, Tanaka N, Bandos H, Gavin PG, Fumagalli D, Goldstein LC, Sneige N, Burandt E, Taniyama Y, Bohn OL, Lee A, Kim SI, Reilly ML, Remillard MY, Blackmon NL, Kim SR, Horne ZD, Rastogi P, Fehrenbacher L, Romond EH, Swain SM, Mamounas EP, Wickerham DL, Geyer CE Jr, Costantino JP, Wolmark N, Paik S (2013) Predicting degree of benefit from adjuvant trastuzumab in NSABP trial B-31. J Natl Cancer Inst 105:1782–1788

43. Joensuu H., Kellokumpu-Lehtinen P. L., Bono P., Alanko T., Kataja V., Asola R., Utriainen T., Kokko R., Hemminki A., Tarkkanen M., Turpeenniemi-Hujanen T., Jyrkkio S., Flander M., Helle L., Ingalsuo S., Johansson K., Jaaskelainen A. S., Pajunen M., Rauhala M., Kaleva-Kerola J., Salminen T., Leinonen M., Elomaa I., Isola J., FinHer Study I (2006) Adjuvant docetaxel or vinorelbine with or without trastuzumab for breast cancer. N Engl J Med 354:809–820

44. Loi S, Michiels S, Lambrechts D, Fumagalli D, Claes B, Kellokumpu-Lehtinen PL, Bono P, Kataja V, Piccart MJ, Joensuu H, Sotiriou C (2013) Somatic mutation profiling and associations with prognosis and trastuzumab benefit in early breast cancer. J Natl Cancer Inst 105:960–967

45. Guiu S, Gauthier M, Coudert B, Bonnetain F, Favier L, Ladoire S, Tixier H, Guiu B, Penault-Llorca F, Ettore F, Fumoleau P, Arnould L (2010) Pathological complete response and survival according to the level of HER-2 amplification after trastuzumab-based neoadjuvant therapy for breast cancer. Br J Cancer 103:1335–1342

46. Formisano L, Nappi L, Rosa R, Marciano R, D'Amato C, D'Amato V, Damiano V, Raimondo L, Iommelli F, Scorziello A, Troncone G, Veneziani B, Parsons SJ, De Placido S, Bianco R (2014) Epidermal growth factor-receptor activation modulates Src-dependent resistance to lapatinib in breast cancer models. Breast Cancer Res 16:R45

47. De Luca A, D'Alessio A, Gallo M, Maiello MR, Bode AM, Normanno N (2014) Src and CXCR4 are involved in the invasiveness of breast cancer cells with acquired resistance to lapatinib. Cell Cycle 13:148–156

48. Geyer CE, Forster J, Lindquist D, Chan S, Romieu CG, Pienkowski T, Jagiello-Gruszfeld A, Crown J, Chan A, Kaufman B, Skarlos D, Campone M, Davidson N, Berger M, Oliva C, Rubin SD, Stein S, Cameron D (2006) Lapatinib plus capecitabine for HER2-positive advanced breast cancer. N Engl J Med 355:2733–2743

49. Blackwell KL, Burstein HJ, Storniolo AM, Rugo H, Sledge G, Koehler M, Ellis C, Casey M, Vukelja S, Bischoff J, Baselga J, O'Shaughnessy J (2010) Randomized study of Lapatinib alone or in combination with trastuzumab in women with ErbB2-positive, trastuzumab-refractory metastatic breast cancer. J Clin Oncol 28:1124–1130

50. Baselga J, Bradbury I, Eidtmann H, Di Cosimo S, de Azambuja E, Aura C, Gomez H, Dinh P, Fauria K, Van Dooren V, Aktan G, Goldhirsch A, Chang TW, Horvath Z, Coccia-Portugal M, Domont J, Tseng LM, Kunz G, Sohn JH, Semiglazov V, Lerzo G, Palacova M, Probachai V, Pusztai L, Untch M, Gelber RD, Piccart-Gebhart M, Neo AST (2012) Lapatinib with trastuzumab for HER2-positive early breast cancer (NeoALTTO): a randomised, open-label, multicentre, phase 3 trial. Lancet 379:633–640

51. de Azambuja E, Holmes AP, Piccart-Gebhart M, Holmes E, Di Cosimo S, Swaby RF, Untch M, Jackisch C, Lang I, Smith I, Boyle F, Xu B, Barrios CH, Perez EA, Azim HA Jr, Kim SB, Kuemmel S, Huang CS, Vuylsteke P, Hsieh RK, Gorbunova V, Eniu A, Dreosti L, Tavartkiladze N, Gelber RD, Eidtmann H, Baselga J (2014) Lapatinib with trastuzumab for HER2-positive early breast cancer (NeoALTTO): survival outcomes of a randomised, open-label, multicentre, phase 3 trial and their association with pathological

complete response. Lancet Oncol 15:1137–1146

52. Baselga J, Majewski I, Nuciforo PG, Eidtmann H, Holmes E, Sotiriou C, Fumagalli D, Delgado D, Piccart-Gebhart M, Bernards R (2013) PI3KCA mutations and correlation with pCR in the NeoALTTO trial (BIG 01-06). Eur J Cancer 49:S402

53. Piccart-Gebhart MJ, Holmes AP, Baselga J, De Azambuja E, Dueck AC, Viale G, Zujewski JA, Goldhirsch A, Santillana S, Pritchard KIACW, Christian Jackisch, Istvan Lang, Michael Untch, Ian E. Smith, Frances Boyle, Binghe Xu, Henry Leonidas Gomez, Richard D. Gelber, Edith A. Perez (2014) First results from the phase III ALTTO trial (BIG 2-06; NCCTG [Alliance] N063D) comparing one year of anti-HER2 therapy with lapatinib alone (L), trastuzumab alone (T), their sequence (T→L), or their combination (T+L) in the adjuvant treatment of HER2-positive early breast cancer (EBC). J Clin Oncol 32:LBA4

54. Molina MA, Codony-Servat J, Albanell J, Rojo F, Arribas J, Baselga J (2001) Trastuzumab (herceptin), a humanized anti-Her2 receptor monoclonal antibody, inhibits basal and activated Her2 ectodomain cleavage in breast cancer cells. Cancer Res 61:4744–4749

55. Scheuer W, Friess T, Burtscher H, Bossenmaier B, Endl J, Hasmann M (2009) Strongly enhanced antitumor activity of trastuzumab and pertuzumab combination treatment on HER2-positive human xenograft tumor models. Cancer Res 69:9330–9336

56. Baselga J., Cortes J., Kim S. B., Im S. A., Hegg R., Im Y. H., Roman L., Pedrini J. L., Pienkowski T., Knott A., Clark E., Benyunes M. C., Ross G., Swain S. M., Group C. S. (2012) Pertuzumab plus trastuzumab plus docetaxel for metastatic breast cancer. N Engl J Med 366:109–119

57. Swain S, Kim S, Cortes J, Ro J, Semiglazov V, Campone M, Ciruelos E, Ferrero J, Schneeweiss A, Heeson S, Clark E, Ross G, Benyunes MC, Baselga J (2014) Final overall survival (OS) analysis from the CLEOPATRA study of first-line (1L) pertuzumab (PTZ), trastuzumab (T), and docetaxel (D) in patients (PTS) with HER2-positive metastatic breast cancer (MBC). Ann Oncol 25

58. Burstein HJ, Sun Y, Dirix LY, Jiang Z, Paridaens R, Tan AR, Awada A, Ranade A, Jiao S, Schwartz G, Abbas R, Powell C, Turnbull K, Vermette J, Zacharchuk C, Badwe R (2010) Neratinib, an irreversible ErbB receptor tyrosine kinase inhibitor, in

patients with advanced ErbB2-positive breast cancer. J Clin Oncol 28:1301–1307

59. Awada A, Dirix L, Manso Sanchez L, Xu B, Luu T, Dieras V, Hershman DL, Agrapart V, Ananthakrishnan R, Staroslawska E (2013) Safety and efficacy of neratinib (HKI-272) plus vinorelbine in the treatment of patients with ErbB2-positive metastatic breast cancer pretreated with anti-HER2 therapy. Ann Oncol 24:109–116

60. Park JW, Liu MC, Yee D, DeMichele A, Veer L, Hylton N, Symmans F, Buxton MB, Chien AJ, Wallace A, Melisko M, Schwab R, Boughey J, Tripathy D, Kaplan H, Nanda R, Chui S, Albain KS, Moulder S, Elias A, Lang JE, Edminston K, Northfelt D, Euhus D, Khan Q, Lyandres J, Davis SE, Yau C, Sanil A, Esserman LJ, Berry DA (2014) Abstract CT227: neratinib plus standard neoadjuvant therapy for high-risk breast cancer: efficacy results from the I-SPY 2 TRIAL. Cancer Res 74:227

61. Lin C, Buxton MB, Moore D, Krontiras H, Carey L, DeMichele A, Montgomery L, Tripathy D, Lehman C, Liu M, Olapade O, Yau C, Berry D, Esserman LJ, Investigators IST (2012) Locally advanced breast cancers are more likely to present as Interval Cancers: results from the I-SPY 1 TRIAL (CALGB 150007/150012, ACRIN 6657, InterSPORE trial). Breast Cancer Res Treat 132:871–879

62. Kreike B, van Kouwenhove M, Horlings H, Weigelt B, Peterse H, Bartelink H, van de Vijver MJ (2007) Gene expression profiling and histopathological characterization of triple-negative/basal-like breast carcinomas. Breast Cancer Res 9:R65

63. Tan DS, Marchio C, Jones RL, Savage K, Smith IE, Dowsett M, Reis-Filho JS (2008) Triple negative breast cancer: molecular profiling and prognostic impact in adjuvant anthracycline-treated patients. Breast Cancer Res Treat 111:27–44

64. Viale G, Rotmensz N, Maisonneuve P, Bottiglieri L, Montagna E, Luini A, Veronesi P, Intra M, Torrisi R, Cardillo A, Campagnoli E, Goldhirsch A, Colleoni M (2009) Invasive ductal carcinoma of the breast with the "triple-negative" phenotype: prognostic implications of EGFR immunoreactivity. Breast Cancer Res Treat 116:317–328

65. Rakha EA, El-Sayed ME, Green AR, Lee AH, Robertson JF, Ellis IO (2007) Prognostic markers in triple-negative breast cancer. Cancer 109:25–32

66. Baselga J, Gomez P, Greil R, Braga S, Climent MA, Wardley AM, Kaufman B, Stemmer SM, Pego A, Chan A, Goeminne JC, Graas MP,

Kennedy MJ, Ciruelos Gil EM, Schneeweiss A, Zubel A, Groos J, Melezinkova H, Awada A (2013) Randomized phase II study of the anti-epidermal growth factor receptor monoclonal antibody cetuximab with cisplatin versus cisplatin alone in patients with metastatic triple-negative breast cancer. J Clin Oncol 31:2586–2592

67. Baselga J, Albanell J, Ruiz A, Lluch A, Gascon P, Guillem V, Gonzalez S, Sauleda S, Marimon I, Tabernero JM, Koehler MT, Rojo F (2005) Phase II and tumor pharmacodynamic study of gefitinib in patients with advanced breast cancer. J Clin Oncol 23:5323–5333

68. Carey LA, Rugo HS, Marcom PK, Mayer EL, Esteva FJ, Ma CX, Liu MC, Storniolo AM, Rimawi MF, Forero-Torres A, Wolff AC, Hobday TJ, Ivanova A, Chiu WK, Ferraro M, Burrows E, Bernard PS, Hoadley KA, Perou CM, Winer EP (2012) TBCRC 001: randomized phase II study of cetuximab in combination with carboplatin in stage IV triple-negative breast cancer. J Clin Oncol 30:2615–2623

69. Sharma P, Khan Q, Kimler B, Klemp J, Connor C, McGinness M, Mammen J, Tawfik O, Fan F, Fabian C (2010) Abstract P1-11-07: results of a phase II study of neoadjuvant platinum/taxane based chemotherapy and erlotinib for triple negative breast cancer. Cancer Res 70:1

70. Kerbel RS, Viloria-Petit A, Klement G, Rak J (2000) 'Accidental' anti-angiogenic drugs. Anti-oncogene directed signal transduction inhibitors and conventional chemotherapeutic agents as examples. Eur J Cancer 36:1248–1257

71. Kerbel RS (1997) A cancer therapy resistant to resistance. Nature 390:335–336

72. Miller LD, Smeds J, George J, Vega VB, Vergara L, Ploner A, Pawitan Y, Hall P, Klaar S, Liu ET, Bergh J (2005) An expression signature for p53 status in human breast cancer predicts mutation status, transcriptional effects, and patient survival. Proc Natl Acad Sci U S A 102:13550–13555

73. Miller K, Wang M, Gralow J, Dickler M, Cobleigh M, Perez EA, Shenkier T, Cella D, Davidson NE (2007) Paclitaxel plus bevacizumab versus paclitaxel alone for metastatic breast cancer. N Engl J Med 357:2666–2676

74. Pegram MD, Reese DM (2002) Combined biological therapy of breast cancer using monoclonal antibodies directed against HER2/neu protein and vascular endothelial growth factor. Semin Oncol 29:29–37

75. Susman E (2014) Bevacizumab fails to improve outcomes in HER2 positive breast cancer. Oncol Times 36:13–14

76. Gray R, Bhattacharya S, Bowden C, Miller K, Comis RL (2009) Independent review of E2100: a phase III trial of bevacizumab plus paclitaxel versus paclitaxel in women with metastatic breast cancer. J Clin Oncol 27:4966–4972

77. Miles D, Bridgewater J, Ellis P, Harrison M, Nathan P, Nicolson M, Raouf S, Wheatley D, Plummer C (2010) Using bevacizumab to treat metastatic cancer: UK consensus guidelines. Br J Hosp Med 71:670–677

78. Robert NJ, Dieras V, Glaspy J, Brufsky AM, Bondarenko I, Lipatov ON, Perez EA, Yardley DA, Chan SY, Zhou X, Phan SC, O'Shaughnessy J (2011) RIBBON-1: randomized, double-blind, placebo-controlled, phase III trial of chemotherapy with or without bevacizumab for first-line treatment of human epidermal growth factor receptor 2-negative, locally recurrent or metastatic breast cancer. J Clin Oncol 29:1252–1260

79. Brufsky AM, Hurvitz S, Perez E, Swamy R, Valero V, O'Neill V, Rugo HS (2011) RIBBON-2: a randomized, double-blind, placebo-controlled, phase III trial evaluating the efficacy and safety of bevacizumab in combination with chemotherapy for second-line treatment of human epidermal growth factor receptor 2-negative metastatic breast cancer. J Clin Oncol 29:4286–4293

80. Brufsky A, Valero V, Tiangco B, Dakhil S, Brize A, Rugo HS, Rivera R, Duenne A, Bousfoul N, Yardley DA (2012) Second-line bevacizumab-containing therapy in patients with triple-negative breast cancer: subgroup analysis of the RIBBON-2 trial. Breast Cancer Res Treat 133:1067–1075

81. Bear HD, Tang G, Rastogi P, Geyer CE Jr, Robidoux A, Atkins JN, Baez-Diaz L, Brufsky AM, Mehta RS, Fehrenbacher L, Young JA, Senecal FM, Gaur R, Margolese RG, Adams PT, Gross HM, Costantino JP, Swain SM, Mamounas EP, Wolmark N (2012) Bevacizumab added to neoadjuvant chemotherapy for breast cancer. N Engl J Med 366:310–320

82. Gerber B, Loibl S, Eidtmann H, Rezai M, Fasching PA, Tesch H, Eggemann H, Schrader I, Kittel K, Hanusch C, Kreienberg R, Solbach C, Jackisch C, Kunz G, Blohmer JU, Huober J, Hauschild M, Nekljudova V, Untch M, von Minckwitz G, German Breast Group I (2013) Neoadjuvant bevacizumab and anthracycline-taxane-based chemotherapy in 678 triple-negative primary breast cancers; results from the GeparQuinto study (GBG 44). Ann Oncol 24:2978–2984

83. Gerber B, von Minckwitz G, Eidtmann H, Rezai M, Fasching P, Tesch H, Eggemann H, Schrader I, Kittel K, Hanusch C, Solbach C, Jackisch C, Kunz G, Blohmer JU, Huober J, Hauschild M, Nekljudova V, Loibl S, Untch M (2014) Surgical outcome after neoadjuvant chemotherapy and bevacizumab: results from the GeparQuinto study (GBG 44). Ann Surg Oncol 21:2517–2524

84. Miller KD, O'Neill A, Perez EA, Seidman AD, Sledge GW (2012) A phase II pilot trial incorporating bevacizumab into dose-dense doxorubicin and cyclophosphamide followed by paclitaxel in patients with lymph node positive breast cancer: a trial coordinated by the Eastern Cooperative Oncology Group. Ann Oncol 23:331–337

85. Cameron D, Brown J, Dent R, Jackisch C, Mackey J, Pivot X, Steger GG, Suter TM, Toi M, Parmar M, Laeufle R, Im YH, Romieu G, Harvey V, Lipatov O, Pienkowski T, Cottu P, Chan A, Im SA, Hall PS, Bubuteishvili-Pacaud L, Henschel V, Deurloo RJ, Pallaud C, Bell R (2013) Adjuvant bevacizumab-containing therapy in triple-negative breast cancer (BEATRICE): primary results of a randomised, phase 3 trial. Lancet Oncol 14:933–942

86. Hutchinson L (2013) Breast cancer: BEATRICE bevacizumab trial – every cloud has a silver lining. Nat Rev Clin Oncol 10:548

87. Wilhelm SM, Adnane L, Newell P, Villanueva A, Llovet JM, Lynch M (2008) Preclinical overview of sorafenib, a multikinase inhibitor that targets both Raf and VEGF and PDGF receptor tyrosine kinase signaling. Mol Cancer Ther 7:3129–3140

88. Moreno-Aspitia A, Morton RF, Hillman DW, Lingle WL, Rowland KM Jr, Wiesenfeld M, Flynn PJ, Fitch TR, Perez EA (2009) Phase II trial of sorafenib in patients with metastatic breast cancer previously exposed to anthracyclines or taxanes: North Central Cancer Treatment Group and Mayo Clinic Trial N0336. J Clin Oncol 27:11–15

89. Bianchi G, Loibl S, Zamagni C, Salvagni S, Raab G, Siena S, Laferriere N, Pena C, Lathia C, Bergamini L, Gianni L (2009) Phase II multicenter, uncontrolled trial of sorafenib in patients with metastatic breast cancer. Anticancer Drugs 20:616–624

90. Gomez P, Roché H, Costa F, Segalla J, Pinczowski H, Ciruelos E, Cabral Filho S, Van Eyll B, Baselga J (2010) Abstract P2–16-01: overall survival data from SOLTI-0701: a multinational, double-blind, placebo-controlled, randomized phase 2b study evaluating the oral combination of sorafenib and capecitabine in patients with locally advanced or metastatic HER2-negative breast cancer. Cancer Res 702:P2–16-01

91. Hudis C, Tauer KW, Hermann RC, Makari-Judson G, Isaacs C, Beck JT, Kaklamani VG, Stepanski EJ, Rugo HS, Wang W, Bell-McGuinn KM, Chera H, Zaugg B, Ro SK, Li S, Schwartzberg LS (2011) Sorafenib (SOR) plus chemotherapy (CRx) for patients (pts) with advanced (adv) breast cancer (BC) previously treated with bevacizumab (BEV). J Clin Oncol 29

92. Bondarde S, Kaklamani V, Prasad Sahoo T, Lokanatha D, Raina V, Jain M, Schwartzberg L, Gradishar W (2010) Abstract P2–16-03: sorafenib in combination with paclitaxel as a first-line therapy in patients with locally recurrent or metastatic breast cancer: overall survival results from a double-blind, randomized, placebo-controlled, phase 2b trial. Cancer Res 70:P2-16-03

93. Gradishar WJ (2005) The future of breast cancer: the role of prognostic factors. Breast Cancer Res Treat 89(Suppl 1):S17–26

94. Wildiers H, Fontaine C, Vuylsteke P, Martens M, Canon JL, Wynendaele W, Focan C, De Greve J, Squifflet P, Paridaens R (2010) Multicenter phase II randomized trial evaluating antiangiogenic therapy with sunitinib as consolidation after objective response to taxane chemotherapy in women with HER2-negative metastatic breast cancer. Breast Cancer Res Treat 123:463–469

95. Barrios CH, Liu MC, Lee SC, Vanlemmens L, Ferrero JM, Tabei T, Pivot X, Iwata H, Aogi K, Lugo-Quintana R, Harbeck N, Brickman MJ, Zhang K, Kern KA, Martin M (2010) Phase III randomized trial of sunitinib versus capecitabine in patients with previously treated HER2-negative advanced breast cancer. Breast Cancer Res Treat 121:121–131

96. Mayer EL, Dhakil S, Patel T, Sundaram S, Fabian C, Kozloff M, Qamar R, Volterra F, Parmar H, Samant M, Burstein HJ (2010) SABRE-B: an evaluation of paclitaxel and bevacizumab with or without sunitinib as first-line treatment of metastatic breast cancer. Ann Oncol 21:2370–2376

97. Crown J, Dieras V, Staroslawska E, Yardley DA, Davidson N, Bachelot TD, Tassell VR, Huang X, Kern KA, Romieu G (2010) Phase III trial of sunitinib (SU) in combination with capecitabine (C) versus C in previously treated advanced breast cancer (ABC). J Clin Oncol 28, LBA1011

98. Bergh J, Bondarenko IM, Lichinitser MR, Liljegren A, Greil R, Voytko NL, Makhson AN, Cortes J, Lortholary A, Bischoff J, Chan

A, Delaloge S, Huang X, Kern KA, Giorgetti C (2012) First-line treatment of advanced breast cancer with sunitinib in combination with docetaxel versus docetaxel alone: results of a prospective, randomized phase III study. J Clin Oncol 30:921–929

99. Liu J, Barry WT, Birrer MJ, Lee J-M, Buckanovich RJ, Fleming GF, Rimel B, Buss MK, Nattam SR, Hurteau J, Luo W, Quy P, Obermayer E, Whalen C, Lee H, Winer EP, Kohn EC, Ivy SP, Matulonis U (2014) A randomized phase 2 trial comparing efficacy of the combination of the PARP inhibitor olaparib and the antiangiogenic cediranib against olaparib alone in recurrent platinum-sensitive ovarian cancer. J Clin Oncol 32:LBA5500

100. Farmer H, McCabe N, Lord CJ, Tutt AN, Johnson DA, Richardson TB, Santarosa M, Dillon KJ, Hickson I, Knights C, Martin NM, Jackson SP, Smith GC, Ashworth A (2005) Targeting the DNA repair defect in BRCA mutant cells as a therapeutic strategy. Nature 434:917–921

101. Bryant HE, Schultz N, Thomas HD, Parker KM, Flower D, Lopez E, Kyle S, Meuth M, Curtin NJ, Helleday T (2005) Specific killing of BRCA2-deficient tumours with inhibitors of poly(ADP-ribose) polymerase. Nature 434:913–917

102. Fong PC, Boss DS, Yap TA, Tutt A, Wu P, Mergui-Roelvink M, Mortimer P, Swaisland H, Lau A, O'Connor MJ, Ashworth A, Carmichael J, Kaye SB, Schellens JH, de Bono JS (2009) Inhibition of poly(ADP-ribose) polymerase in tumors from BRCA mutation carriers. N Engl J Med 361:123–134

103. O'Shaughnessy J, Osborne C, Pippen JE, Yoffe M, Patt D, Rocha C, Koo IC, Sherman BM, Bradley C (2011) Iniparib plus chemotherapy in metastatic triple-negative breast cancer. N Engl J Med 364:205–214

104. Engelman JA (2009) Targeting PI3K signalling in cancer: opportunities, challenges and limitations. Nat Rev Cancer 9:550–562

105. Andre F, Bachelot T, Commo F, Campone M, Arnedos M, Dieras V, Lacroix-Triki M, Lacroix L, Cohen P, Gentien D, Adelaide J, Dalenc F, Goncalves A, Levy C, Ferrero JM, Bonneterre J, Lefeuvre C, Jimenez M, Filleron T, Bonnefoi H (2014) Comparative genomic hybridisation array and DNA sequencing to direct treatment of metastatic breast cancer: a multicentre, prospective trial (SAFIR01/UNICANCER). Lancet Oncol 15:267–274

106. Isakoff SJ, Engelman JA, Irie HY, Luo J, Brachmann SM, Pearline RV, Cantley LC, Brugge JS (2005) Breast cancer-associated PIK3CA mutations are oncogenic in mammary epithelial cells. Cancer Res 65:10992–11000

107. Gelmon KA, Tischkowitz M, Mackay H, Swenerton K, Robidoux A, Tonkin K, Hirte H, Huntsman D, Clemons M, Gilks B, Yerushalmi R, Macpherson E, Carmichael J, Oza A (2011) Olaparib in patients with recurrent high-grade serous or poorly differentiated ovarian carcinoma or triple-negative breast cancer: a phase 2, multicentre, open-label, non-randomised study. Lancet Oncol 12:852–861

108. Isakoff SJ, Overmoyer B, Tung NM, Gelman RS, Giranda VL, Bernhard KM, Habin KR, Ellisen LW, Winer EP, Goss PE (2010) A phase II trial of the PARP inhibitor veliparib (ABT888) and temozolomide for metastatic breast cancer. J Clin Oncol 28:Abstr. 1019

109. Zhang XH, Wang Q, Gerald W, Hudis CA, Norton L, Smid M, Foekens JA, Massague J (2009) Latent bone metastasis in breast cancer tied to Src-dependent survival signals. Cancer Cell 16:67–78

110. Somlo G, Atzori F, Strauss LC, Geese WJ, Specht JM, Gradishar WJ, Rybicki A, Sy O, Vahdat LT, Cortes J (2013) Dasatinib plus capecitabine for advanced breast cancer: safety and efficacy in phase I study CA180004. Clin Cancer Res 19:1884–1893

111. Sturgill TW, Ray LB, Erikson E, Maller JL (1988) Insulin-stimulated MAP-2 kinase phosphorylates and activates ribosomal protein S6 kinase II. Nature 334:715–718

112. Cizkova M, Susini A, Vacher S, Cizeron-Clairac G, Andrieu C, Driouch K, Fourme E, Lidereau R, Bieche I (2012) PIK3CA mutation impact on survival in breast cancer patients and in ERalpha, PR and ERBB2-based subgroups. Breast Cancer Res 14:R28

113. Ellis MJ, Lin L, Crowder R, Tao Y, Hoog J, Snider J, Davies S, DeSchryver K, Evans DB, Steinseifer J, Bandaru R, Liu W, Gardner H, Semiglazov V, Watson M, Hunt K, Olson J, Baselga J (2010) Phosphatidyl-inositol-3-kinase alpha catalytic subunit mutation and response to neoadjuvant endocrine therapy for estrogen receptor positive breast cancer. Breast Cancer Res Treat 119:379–390

114. Ghayad SE, Bieche I, Vendrell JA, Keime C, Lidereau R, Dumontet C, Cohen PA (2008) mTOR inhibition reverses acquired endocrine therapy resistance of breast cancer cells at the cell proliferation and gene-expression levels. Cancer Sci 99:1992–2003

115. Baselga J, Semiglazov V, van Dam P, Manikhas A, Bellet M, Mayordomo J, Campone M, Kubista E, Greil R, Bianchi G, Steinseifer J, Molloy B, Tokaji E, Gardner H, Phillips P, Stumm M, Lane HA, Dixon JM, Jonat W,

Rugo HS (2009) Phase II randomized study of neoadjuvant everolimus plus letrozole compared with placebo plus letrozole in patients with estrogen receptor-positive breast cancer. J Clin Oncol 27:2630–2637

116. Baselga J, Campone M, Piccart M, Burris HA 3rd, Rugo HS, Sahmoud T, Noguchi S, Gnant M, Pritchard KI, Lebrun F, Beck JT, Ito Y, Yardley D, Deleu I, Perez A, Bachelot T, Vittori L, Xu Z, Mukhopadhyay P, Lebwohl D, Hortobagyi GN (2012) Everolimus in postmenopausal hormone-receptor-positive advanced breast cancer. N Engl J Med 366:520–529

117. Hortobagyi GN, Piccart-Gebhart MJ, Rugo HS, Burris HA, Campone M, Noguchi S, Alejandra TP, Deleu I, Shtivelband M, Provencher L, Masuda N, Dakhil SR, Anderson I, Chen D, Damask A, Huang A, McDonald R, Taran T, Sahmoud T, Baselga J (2013) Correlation of molecular alterations with efficacy of everolimus in hormone receptor–positive, HER2-negative advanced breast cancer: results from BOLERO-2. J Clin Oncol 31:509

118. O'Brien NA, Browne BC, Chow L, Wang Y, Ginther C, Arboleda J, Duffy MJ, Crown J, O'Donovan N, Slamon DJ (2010) Activated phosphoinositide 3-kinase/AKT signaling confers resistance to trastuzumab but not lapatinib. Mol Cancer Ther 9:1489–1502

119. Easton DF, Pooley KA, Dunning AM, Pharoah PD, Thompson D, Ballinger DG, Struewing JD, Morrison J, Field H, Luben R, Wareham N, Ahmed S, Healey CS, Bowman R, Collaborators S, Meyer KB, Haiman CA, Kolonel LK, Henderson BE, Le Marchand L, Brennan P, Sangrajrang S, Gaborieau V, Odefrey F, Shen CY, Wu PE, Wang HC, Eccles D, Evans DG, Peto J, Fletcher O, Johnson N, Seal S, Stratton MR, Rahman N, Chenevix-Trench G, Bojesen SE, Nordestgaard BG, Axelsson CK, Garcia-Closas M, Brinton L, Chanock S, Lissowska J, Peplonska B, Nevanlinna H, Fagerholm R, Eerola H, Kang D, Yoo KY, Noh DY, Ahn SH, Hunter DJ, Hankinson SE, Cox DG, Hall P, Wedren S, Liu J, Low YL, Bogdanova N, Schurmann P, Dork T, Tollenaar RA, Jacobi CE, Devilee P, Klijn JG, Sigurdson AJ, Doody MM, Alexander BH, Zhang J, Cox A, Brock IW, MacPherson G, Reed MW, Couch FJ, Goode EL, Olson JE, Meijers-Heijboer H, van den Ouweland A, Uitterlinden A, Rivadeneira F, Milne RL, Ribas G, Gonzalez-Neira A, Benitez J, Hopper JL, McCredie M, Southey M, Giles GG, Schroen C, Justenhoven C, Brauch H, Hamann U, Ko YD, Spurdle AB, Beesley J, Chen X, kConFab, Group AM, Mannermaa A, Kosma VM, Kataja V,

Hartikainen J, Day NE, Cox DR, Ponder BA (2007) Genome-wide association study identifies novel breast cancer susceptibility loci. Nature 447:1087–1093

120. Turner N, Pearson A, Sharpe R, Lambros M, Geyer F, Lopez-Garcia MA, Natrajan R, Marchio C, Iorns E, Mackay A, Gillett C, Grigoriadis A, Tutt A, Reis-Filho JS, Ashworth A (2010) FGFR1 amplification drives endocrine therapy resistance and is a therapeutic target in breast cancer. Cancer Res 70:2085–2094

121. Turner N, Lambros MB, Horlings HM, Pearson A, Sharpe R, Natrajan R, Geyer FC, van Kouwenhove M, Kreike B, Mackay A, Ashworth A, van de Vijver MJ, Reis-Filho JS (2010) Integrative molecular profiling of triple negative breast cancers identifies amplicon drivers and potential therapeutic targets. Oncogene 29:2013–2023

122. Andre F, Bachelot T, Campone M, Dalenc F, Perez-Garcia JM, Hurvitz SA, Turner N, Rugo H, Smith JW, Deudon S, Shi M, Zhang Y, Kay A, Porta DG, Yovine A, Baselga J (2013) Targeting FGFR with dovitinib (TKI258): preclinical and clinical data in breast cancer. Clin Cancer Res 19:3693–3702

123. Zamora E, Muñoz-Couselo E, Cortes J, Perez-Garcia J (2014) The fibroblast growth factor receptor: a new potential target for the treatment of breast cancer. Curr Breast Cancer Rep 6:51–58

124. Cga N (2012) Comprehensive molecular portraits of human breast tumours. Nature 490:61–70

125. Chung J, Noh H, Park KH, Choi E, Han A (2014) Longer survival in patients with breast cancer with cyclin d1 over-expression after tumor recurrence: longer, but occupied with disease. J Breast Cancer 17:47–53

126. Finn RS, Crown JP, Lang I, Boer K, Bondarenko IM, Kulyk SO, Ettl J, Patel R, Pinter T, Schmidt M, Shparyk Y, Thummala AR, Voytko NL, Fowst C, Huang X, Kim ST, Randolph S, Slamon DJ (2015) The cyclin-dependent kinase 4/6 inhibitor palbociclib in combination with letrozole versus letrozole alone as first-line treatment of oestrogen receptor-positive, HER2-negative, advanced breast cancer (PALOMA-1/TRIO-18): a randomised phase 2 study. Lancet Oncol 16(1):25–35

127. Patnaik A, Rosen LS, Tolaney SM, et al (2014) Clinical activity of LY2835219, a novel cell cycle inhibitor selective for CDK4 and CDK6, in patients with metastatic breast cancer. AACR Annual Meeting. Abstract CT232. Presented April 7, 2014.

Chapter 9

MicroRNAs and Cancer Drug Resistance

Bruno Costa Gomes, José Rueff, and António Sebastião Rodrigues

Abstract

The discovery of small regulatory noncoding RNAs revolutionized our thinking on gene regulation. The class of microRNAs (miRs), a group of small noncoding RNAs (20–22 nt in length) that bind imperfectly to the 3′-untranslated region of target mRNA, has been insistently implicated in several pathological conditions including cancer. Indeed, major hallmarks of cancer, such as cell differentiation, cell proliferation, cell cycle, cell survival, and cell invasion, has been described as being regulated by miRs. Recent studies have also implicated miRs in cancer drug resistance. Regardless of the several studies done until now, drug resistance still is a burden for cancer therapy and patients' outcome, often resulting in more aggressive tumors that tend to metastasize to distant organs. Hence, with this review, we aim to summarize the miRs that influence molecular pathways that are involved in cancer drug resistance, such as drug metabolism, drug influx/efflux, DNA damage response (DDR), epithelial-to-mesenchymal transition (EMT), and cancer stem cells.

Key words MicroRNA, Drug resistance, Noncoding RNAs, Cancer

1 Introduction

MicroRNAs (miRs) were discovered by Victor Ambros and colleagues [1] in 1993, who observed that the *C. elegans lin-4* gene coded for a pair of small RNAs with antisense complementary to multiple sites on the 3′-UTR of *lin-14* gene. This small RNA substantially reduced the amount of LIN-14 protein without noticeably changing the level of *lin-14* mRNA. This landmark study showed that small RNAs possessed regulatory functions and soon the presence of other regulatory RNAs (e.g., *let-7*) was observed in other species namely humans [2]. This group of regulatory RNAs was called microRNAs (miRs) [3], an evolutionary conserved class of small RNAs that was found to control many developmental and cellular processes in eukaryotic organisms. The latest version (June 2014) of the miRBase database (miRbase 21) listed 24 521 miRs loci from 206 species, processed to produce 30,424 mature miR products. Of these, 1881 sequences belonged to the human genome [4].

José Rueff and António Sebastião Rodrigues (eds.), *Cancer Drug Resistance: Overviews and Methods*, Methods in Molecular Biology, vol. 1395, DOI 10.1007/978-1-4939-3347-1_9, © Springer Science+Business Media New York 2016

MiRs posttranscriptionally modulate gene expression by binding to their target mRNAs. miRs can be intergenic or intragenic and are produced from endogenous hairpin transcripts named pri-miR. Then, the nuclear Drosha/DGCR8 heterodimer cleaves pri-miR hairpin stem, producing the pre-miR (60–100 nucleotides) which is exported to the cytoplasm by Exportin5 and RAN-GTP. The pre-miR is then processed by the RNAse III endonuclease Dicer and its TRBP (HIV transactivating response RNA-binding protein) partner, releasing a duplex with 22–25 nucleotides. This duplex associates with the Argonaute protein forming a RNA-induced silencing complex (RISC). The mature miR stays in the complex and the passenger strand is degraded. The RISC complex is the functional complex that will interact with mRNA and trigger the regulatory effect [5]. Due to their small size, miRs are capable of binding to several regions in the 3′-UTR region of several mRNAs and in turn mRNAs can be targeted by several miRs. Consequently there is a biological redundancy in gene regulation executed by miRs. Thus, their action is extremely broad and their involvement in gene expression and cellular phenotype is well established. Although miR binding sites have also been found in 5′-UTR and in the coding sequences of mRNAs [6], they preferentially interact with seed-matching sequences in the 3′-UTR of mRNA. Several studies have shown that miRs could regulate cell differentiation [7–9], cell proliferation [10, 11], cell cycle [12, 13], cell survival [14, 15], and cell invasion [16–18]. Therefore, any misexpression of miRs can lead to altered cell phenotypes and consequently cancer initiation and progression [19]. Many miRs are located at fragile sites on chromosomes known for having common alterations (i.e., amplification, deletion, and rearrangements) in cancer [20]. MiRs that inhibit translation of proto-oncogenes are considered tumor suppressor miRs, and are usually downregulated in cancer. Other miRs are upregulated in cancer and may act as oncogenic miRs by downregulating tumor suppressor genes [21]. Recent studies have highlighted the intratumoral heterogeneity in expression of miRs [22]. This might explain the different miR expression profiles described by several groups for the same types of cancer and underlines the importance in analyzing numerous sample locations of the primary tumor in order to obtain an accurate profile of miR expression.

As stated in previous chapters, drug resistance is frequently classified into two broad types: intrinsic and acquired. Intrinsic drug resistance is not essentially a genetic attribute of the cancer cells, but can be defined as preexisting to the therapeutical challenge endowing the cancer cell with competence to survive treatment, thus rendering therapy potentially ineffective from the beginning. More often than not, intrinsic resistance could be conceived as the result of the pharmacogenetic/pharmacogenomic configuration of the host of the tumor. On the other hand, acquired drug resistance is developed during therapy and usually due to adaptive processes, such as compensatory signaling pathways, drug inactivation, increased expression of drug target, alterations in drug targets, increased expression of drug efflux pumps, cell death inhibition, epigenetic phenomena, tumor microenvironment, and DNA damage response and repair augmentation [23–26]. Drug resistance usually results in a more aggressive tumor and cancer cells often tend to metastasize to distant organs.

Within the molecular complexity of the cancer cells and their readily capacity to change the circuitry of molecular regulation, the discovery of miRs and their roles in gene expression quickly led to studies that assessed the influence of miRs in drug resistance. As a consequence, many groups have focused on the role of these small regulatory RNAs in the development of cancer drug resistance. Several studies have shown that drug resistance can also be influenced by miRs, since they can regulate drug resistance-related genes, alter drug targets, change drug concentrations, influence therapeutic-induced cell death, regulate angiogenesis, and be involved in the development of tumor stem cells.

2   MicroRNAs in Cancer Drug Resistance

As stated above, miRs have been linked to several hallmarks of cancer in tumor cells. Differential expression of miRs in tumor cells before treatment has been associated with response to chemotherapy, while changes in miR expression have been observed in cancer cells following treatment. Table 1 summarizes the studies that showed a regulation of drug resistance by miRs. The table is divided into the main categories of drug resistance pathways and the respective regulator miR. Thus, we elaborate on miRs influencing on drug metabolism, drug transporters, DNA repair, epithelial to mesenchymal transition (EMT), and cancer stem cells. Recent studies have attempted to identify single nucleotide polymorphisms either in miR loci or target loci and correlate their presence with altered therapeutic response [27, 28].

2.1   Drug Metabolism

Drug metabolism is a complex pathway of xenobiotic detoxification that involves multiple proteins, and can be divided in three main phases: modification, conjugation, and excretion. Xenobiotics are foreign compounds (such as drugs) that are not normally produced or expected to be present in an organism. Concerted actions of drug-metabolizing enzymes (DME) and drug transporters lead primarily to an increase in the polarity of xenobiotics, called Phase I reactions, followed by conjugation reactions (Phase II reactions) that increase their polarity but block the reactivity of polar groups introduced in Phase I reactions. Thereafter the transmembrane transport of the resulting metabolites is performed by membrane transporter proteins, essentially ABC transporters (Phase III reactions).

Table 1
Pathways of drug resistance regulated by miRs (*NS not specified*)

Target gene	microRNA	Type of cancer/established cell line	Drug	Reference
Drug metabolism				
CYP1B1	miR-27b	Human uterine cervix adenocarcinoma cell line HeLa; Human breast adenocarcinoma cell line MCF-7; Human embryonic kidney cell line HEK293; Human leukemic T-cell line Jurkat; Breast cancerous and adjacent noncancerous tissue	NS	Tsuchiya et al. [30]
CYP2E1	miR-378	Human embryonic kidney cell line HEK293	NS	Mohri et al. [37]
CYP3A4	miR-27b	Human pancreas cancer PANC1; Human colon carcinoma LS-180; Human embryonic kidney cell line HEK293	Cyclophosphamide	Pan et al. [38]
SULT1A1	miR-631	Human breast cancer cell lines ZR75-1 and MCF7; Human mammary epithelial cell line MCF10A; US Cooperative Human Tissue Network under an Institutional Review Board (IRB)-approved protocol	Actinomycin D	Yu et al. [41]
GSTP1	miR-133a	Human head and neck Squamous Cell Carcinoma (SCC); Human esophageal SCC and bladder cell lines	cisplatin and carboplatin	Moriya et al. [42]
Drug transport				
ABCB1	miR-451	Human breast adenocarcinoma cell line MCF-7	Doxorubicin	Kovalchuk et al. [52]; Zhu et al. [56]
ABCB1	miR-200c	Breast cancerous tissue; Human breast adenocarcinoma cell line MCF-7	Doxorubicin	Chen et al. [53]
ABCB1	miR-298	Human breast adenocarcinoma cell lines MCF-7 and MDA-MB-231	Doxorubicin	Bao et al. [54]
ABCB1	miR-27a	Human breast adenocarcinoma cell line MCF-7	Doxorubicin	Zhu et al. [56]
ABCB1	miR-145	Human colon carcinoma cell line Caco-2; Human embryonic kidney cell line HEK293	NS	Ikemura et al. [57]
ABCB1	miR-381	Human chronic myelogenous leukemia cell line K562	Adriamycin	Xu et al. [58]
ABCB1	miR-495	Human chronic myelogenous leukemia cell line K562	Adriamycin	Xu et al. [58]
ABCG2	miR-181a	Human breast adenocarcinoma cell line MCF-7	Mitoxantrone	Jiao et al. [63]
ABCG2	miR-328	Human breast adenocarcinoma cell line MCF-7	Mitoxantrone	Pan et al. [64]
ABCG2	miR-487a	Human breast adenocarcinoma cell line MCF-7	Mitoxantrone	Ma et al. [65]
ABCG2	miR-519c	Human embryonic kidney HEK293; Human breast adenocarcinoma cell line MCF-7	Mitoxantrone	Li et al. [66]
ABCG2	miR-328	Human embryonic kidney HEK293; Human breast adenocarcinoma cell line MCF-7	Mitoxantrone	Li et al. [66]

ABCC1	miR-326	Normal breast and breast tumor tissues; Human breast adenocarcinoma cell line MCF-7	VP-16 and doxorubicin	Liang et al. [69]
ABCC1	miR-345	Human breast adenocarcinoma cell line MCF-7	Cisplatin	Pogribny et al. [68]
ABCC1	miR-7	Human breast adenocarcinoma cell line MCF-7	Cisplatin	Pogribny et al. [68]
ABCC1	miR-1291	Human pancreatic carcinoma cell line PANC-1; Human small lung cancer cell line H69; Human embryonic kidney cell line HEK293	Doxorubicin	Pan et al. [70]
ABCC2	miR-297	Human ileocecal colorectal adenocarcinoma cell line HCT-8 and HCT-116; Colorectal cancerous and adjacent noncancerous tissue	oxaliplatin and vincristine	Xu et al. [72]
SLC15A1	miR-92b	Human colon carcinoma cell line Caco-2-BBE	NS	Dalmasso et al. [74]
SLC16A1	miR-29a	Hepatoma cell line mhAT3F; Pancreatic beta cell line MIN6; Human embryonic kidney cell line HEK293	NS	Pullen et al. [75]
SLC16A1	miR-29b	Hepatoma cell line mhAT3F; Pancreatic beta cell line MIN6; Human embryonic kidney cell line HEK293	NS	Pullen et al. [75]
SLC16A1	miR-124	Hepatoma cell line mhAT3F; Pancreatic beta cell line MIN6; Human embryonic kidney cell line HEK293	NS	Pullen et al. [75]
DNA repair				
RAS	let-7 family	Human non-small-cell lung cancer cells cell line A549		Weidhaas et al. [93]
ERCC1	miR-138	Human non-small-cell lung cancer cells cell line A549	Cisplatin	Wang et al. [95]
MSH2	miR-21	Human Dukes' type C, colorectal adenocarcinoma cell lines Colo-320 DM and SW620; Human colorectal adenocarcinoma cell line HCT-116; Human Dukes' type B, colorectal adenocarcinoma cell line SW480, Human colon carcinoma cell line RKO	5-fluorouracil	Valeri et al. [96]
MSH6	miR-21	Human Dukes' type C, colorectal adenocarcinoma cell lines Colo-320 DM and SW620; Human colorectal adenocarcinoma cell line HCT-116; Human Dukes' type B, colorectal adenocarcinoma cell line SW480, Human colon carcinoma cell line RKO	5-fluorouracil	Valeri et al. [96]
REV1	miR-96	Human Bone Osteosarcoma Epithelial Cell line U2OS; Human uterine cervix adenocarcinoma cell line HeLa; Human breast cancer cell line HCC1937; Human breast adenocarcinoma cell line MDA-MB-231	Cisplatin; PARP inhibitor AZD2281	Wang et al. [97]
RAD51	miR-96	Human Bone Osteosarcoma Epithelial Cell line U2OS; Human uterine cervix adenocarcinoma cell line HeLa; Human breast cancer cell line HCC1937; Human breast adenocarcinoma cell line MDA-MB-231	Cisplatin; PARP inhibitor AZD2281	Wang et al. [97]
RAD51	miR-155	Human breast adenocarcinoma cell line MCF-7; triple-negative breast cancer tissue	NS	Gasparini et al. [92]

(continued)

Table 1
(continued)

Target gene	microRNA	Type of cancer/established cell line	Drug	Reference
BRCA1	miR-182	Human acute promyelocytic leukemia cell line HL60; Human chronic myelogenous leukemia cell line K562; Human breast adenocarcinoma cell line MCF-7	PARP inhibitor	Moskwa et al. [100]
BRCA1	miR-146a miRNA-146-5p	Breast cancer tissue Breast cancer tissue	NS NS	Garcia et al. [101] Garcia et al. [101]
BRCA1	miR-193a-5p	Human Mammary Epithelial progenitor Cell line HMEpC ; Human Small Airway Epithelial progenitor Cell line HSAEpC; H226; H460; Human breast cancer cell lines MDA-MB-231, MDA-MB-157, and SK-BR-3	Cisplatin	van Jaarsveld et al. [88]
BRCA1	miR-296-5p	Human Mammary Epithelial progenitor Cell line HMEpC; Human Small Airway Epithelial progenitor Cell line HSAEpC; H226; H460; Human breast cancer cell lines MDA-MB-231, MDA-MB-157, and SK-BR-3	Cisplatin; doxorubicin and paclitaxel	van Jaarsveld et al. [88]
BRCA1	miR-183	Human Mammary Epithelial progenitor Cell line HMEpC; Human Small Airway Epithelial progenitor Cell line HSAEpC; H226; H460; Human breast cancer cell lines MDA-MB-231, MDA-MB-157, and SK-BR-3		van Jaarsveld et al. [88]
BRCA1	miR-16	HSAEpCs	Cisplatin and doxorubicin	van Jaarsveld et al. [88]
EMT				
	miR-200c miR-200b	Human breast adenocarcinoma cell line MCF-7 Human breast adenocarcinoma cell line MCF-7 and resistant derivates	Doxorubicin 4-hydroxytamoxifen, fulvestrant	Chen et al. [105] Manavalan et al. [106]
	miR-200c	Human breast adenocarcinoma cell line MCF-7 and resistant derivates	4-hydroxytamoxifen, fulvestrant	Manavalan et al. [106]
MIG6	miR-200c miR-200c	Human breast cancer cell line SKBr-3 Several human cancer cell lines	trastuzumab	Bai et al. [107] Izumchenko et al. [108]

MAGI2	miR-134/miR-487b/miR-655 cluster	Human lung adenocarcinoma cell lines A549, LC2/ad, PC3, PC9, RERF-LCKJ, RERF-LCMS, PC14, and ABC-1	Gefitinib	Kitamura et al. [109]
	miR-147	Human colon cancer cell line HCT116 and SW480; Human lung cancer cell line A549	Gefitinib	Lee et al. [110]
SMAD3	miR-489	Human breast adenocarcinoma cell line MCF-7	Doxorubicin	Jiang et al. [111]
Fbw7	miR-223	The human pancreatic cancer cells AsPC-1 and PANC-1	Gemcitabine	Ma et al. [112]
Stem cells				
Nanog/Oct4	let-7a	Human head and neck cancer tissues	Cisplatin	Yu et al. [125]
Oct4 and Sox2	miR-145	Glioblastoma	Temozolomide	Yang et al. [126]
TP53INP1	miR-130b	Human liver tumor and adjacent non-tumor tissue	Doxorubicin	Ma et al. [127]
p53–Nanog	miR-214	Human ovarian cancer A2780, OV2008, OV8, and SKOV3	Cisplatin and doxorubicin	Xu et al. [128]
ABCB1	miR-451	Human colon carcinoma cell lines DLD1, HT29, LS513, SW620, LoVo, and RKO; Colorectal cancer tissue	Irinotecan	Bitarte et al. [129]
	miR-302	Human head and neck squamous cell carcinoma (HNSCC) cell line HSC-3	Cisplatin	Bourguignon et al. [130]

Although extensive studies have been performed on transcriptional regulation of the DMEs, there is a lack of understanding of their posttranscriptional regulation [29]. Recent studies have shown that miRs also control the expression of some DME [30–32]. However few studies have shown a direct involvement of miRs and DME with drug resistance. One of the key players of the Phase I (modification) are cytochrome P450 (CYP) enzymes that catalyze oxidation reactions of the xenobiotics and occasionally reduction reactions [33]. More than 90 % of the reactions involved in the metabolism of all chemicals, whether general chemicals, natural, physiological compounds, and drugs, are catalyzed by P450s [34]. Three-fourths of the human CYP reactions can be accounted for by a set of five CYPs: 1A2, 2C9, 2C19, 2D6, and 3A4, with the largest fraction of the CYP reactions being catalyzed by CYP 3A enzymes. The importance of CYP 3A4 in metabolic reactions of drugs varies from 13 % for general chemicals to 27 % for drugs [34]. Therefore the regulation of DMEs is crucial to drug efficacy and may be related to drug failure or drug resistance.

Tsuchiya et al. [30] showed a direct association of miR-27b and CYP1B1 in breast cancer. The authors not only validated CYP1B1 as a mirR-27b target in cell lines but also showed that in tissue samples there is an inverse correlation between miR-27b expression and CYP1B1 protein expression. Indeed, the authors showed that miR-27b decreased in expression along the group staining of CYP1B1 by immunohistochemistry, being more expressed in the weak staining group and less expressed in the strong staining group. CYP1B1 is highly expressed in estrogen target tissues, and catalyzes the metabolic activation of various pro-carcinogens and the 4-hydroxylation of 17β-estradiol, and is also abundant in cancerous tissues. However, the authors did not show an association with drug resistance. Nevertheless, since deactivation of 4-hydroxy-tamoxifen, a biotransformation product of tamoxifen that has 100-fold increased affinity to estrogen receptors then tamoxifen itself, occurs via CYP1B1 [35], the increased expression of CYP1B1 in breast cancer cells could augment the resistance to tamoxifen, a widely used drug in breast cancer treatment.

CYP2E1 is the fourth most abundant isoform (approximately 7 % of total P450 protein) after CYP3A4 (30 % of total P450), CYP2C (20 % of total P450), and CYP1A2 (approximately 13 % of total P450). CYP2E1 catalyzes the metabolism of numerous low-molecular-weight xenobiotics, including organic solvents (e.g., ethanol, acetone, carbon tetrachloride, chloroform, vinyl chloride, glycerol, hexane, and toluene), and several procarcinogens, such as N-nitrosodimethylamine and N-nitrosomethylethylamine. Interestingly, the ectopic expression of CYP2E1 induced ROS generation, affected autophagy, and inhibited migration in breast cancer cells, thus potentially being involved in breast cancer metastasis [36]. Mohri et al. identified a possible miR-responsive element (MRE378) in the 3′-UTR of human CYP2E1 mRNA, and luciferase assays using HEK293 cells confirmed that miR-378 functionally recognized this region [37]. The overexpression of miR-378 significantly decreased the CYP2E1 protein level and enzyme activity in cells expressing CYP2E1 including 3′-UTR, but not in the cells expressing CYP2E1 excluding 3′-UTR, indicating that the 3′-UTR plays a role in the miR-378-dependent repression. However, the presence of miR-378 did not facilitate the degradation of the CYP2E1 mRNA. Therefore, according to the authors, the downregulation of CYP2E1 by miR-378 would mainly be due to the translational repression, not mRNA degradation. Additionally the relationship between the expression levels of miR-378, CYP2E1 mRNA and protein as well as enzyme activity was assessed using a panel of 25 human livers. CYP2E1 protein levels were significantly correlated with the enzymatic activities but were inversely correlated with CYP2E1 mRNA levels, while miR-378 levels showed a significant inverse correlation with the CYP2E1 protein levels [37]. In another study, Pan et al. [38] showed that miR-27b interacts with the 3′-UTR of CYP3A4, thus regulating its expression. Moreover an overexpression of miR-27b in PANC1 Human pancreas cancer cells led to a lower sensitivity to cyclophosphamide, indicating that miR-27b can alter CYP3A4-catalyzed drug activation, and consequently impact on drug response and resistance.

Regarding Phase II reactions even fewer studies have linked miR-mediated regulation and drug resistance. One example is the sulfotransferase isoform 1A1 (SULT1A1), a member of the sulfotransferase (SULT) family of phase II detoxification enzymes that catalyze the transfer of the sulfonyl group from 3′-phosphoadenosine 5′-phosphosulfate (PAPS) to nucleophilic groups of a variety of xenobiotic and endogenous compounds, thus increasing their solubility and excretion [39]. SULT1A1 is the most highly expressed SULT in the liver. Several therapeutic agents, including 4-hydroxytamoxifen, are substrates for SULT1A1, and variability in the activity levels of the enzyme can markedly influence the efficacy of these drugs and consequently drug resistance [40]. Interestingly, a common single nucleotide polymorphism (SNP) in the coding region of SULT1A1, several proximal promoter SNPs, and copy number variation (CNV) are associated with altered enzymatic activity, but these variants do not fully account for the observed variation of SULT1A1 activity in human populations. Thus, Yu et al. [41] looked for SNPs in the 3′-UTR region of this gene. In silico analyses predicted that the 973C→T SNP would influence the binding of miR-631 to the SULT1A1 3′-UTR. Accordingly, in vitro luciferase reporter assays and overexpression of miR inhibitors in ZR75-1, MCF7, and MCF10A breast cell lines confirmed that SULT1A1 is a direct target of miR-631 [41].

Finally, Moriya et al. [42] found that miR-133a was a potential regulator of GSTP1. Transfection of miR-133a repressed GSTP1 expression at both mRNA and protein levels in several different cell lines. The functional significance of miR-133a was investigated using head and neck Squamous Cell Carcinoma (SCC), esophageal SCC, and bladder cell lines, and the authors showed that restoration of miR-133a expression inhibited cancer cell proliferation, invasion, and migration, suggesting that miR-133a may function as a tumor suppressor. GSTP1 is a member of the GST enzyme superfamily, and catalyzes the conjugation of electrophiles to glutathione in phase II detoxification reactions, including platinum drugs such as cisplatin and carboplatin [43]. GSTP1 has several critical roles in both normal and neoplastic cells, including phase II xenobiotic metabolism, stress responses, signaling, and apoptosis. Overexpression of GSTP1 has been observed in many types of cancer and in human tumor cell lines either inherently or made resistant to chemotherapy drugs, including cisplatin and various alkylating agents [44]. For example, GSTP1 knockdown selectively influenced cisplatin and carboplatin chemosensitivity; cell cycle progression was unaffected, but cell invasion and migration was significantly reduced [45]. The reduced expression of miR-133a may thus lead to an increased expression of GSTP1, contributing to drug resistance.

In spite of these results, miR-dependent regulation of expression in DMEs does not seem to be the most important mode of regulation as few miR-binding regions are found in the 3′-UTR of DME genes. Furthermore, the miR binding sites described for most of the DMEs are poorly conserved, leading one to speculate that other forms of regulation are more important.

2.2 Drug Transport

Drug transport through cell membranes is a critical step in allowing access of pharmacologic agents to intracellular targets. The involvement of drug transport is probably amongst the most studied mechanisms in cancer drug resistance [46]. Multidrug resistance (MDR) is frequently linked to overexpression of one or more of drug transport proteins present in the cytoplasmic membrane. The ABC transporters have an important cellular role in the efflux and influx of several substrates necessary to the cell and also in the efflux of toxic endogenous molecules and xenobiotics (See chapters by Mitra, Viverios, and Gromicho, in this book). Up to now, 49 different ABC transporters were identified and classified in seven families from ABCA through ABCG [47, 48]. The relevance of miRs in regulating the expression of ABC transporters has been recently reviewed [31, 49].

One of the most well-known ABC transporters is ABCB1, also known as MDR1 or P-gp transporter. In chemotherapeutic-resistant cancer cell lines, ABCB1 is often observed to be upregulated. The increased expression of ABCB1 leads to an increased resistance of several chemotherapeutics, such as taxanes (e.g., paclitaxel and docetaxel), epipodophyllotoxins derivates (e.g., etoposide and teniposide), anthracyclines (e.g., doxorubicin), antibiotics (e.g., actinomycin D), vinca alkaloids (e.g., vinblastine and vincristine), and tyrosine kinase inhibitors (e.g., imatinib and erlotinib) [47, 50]. To date, several authors have published data about misexpression of miRs and ABCB1 [51–54]. Kovalchuck and colleagues [51] showed that the *ABCB1* gene is highly expressed in the MCF-7/DOX breast tumor cell lines resistant to doxorubicin when compared with wild type MCF-7. Conversely, miR-451 expression is undetected, showing a negative correlation between ABCB1 and miR-451 expression. These authors then showed that miR-451 targets the *ABCB1* 3′-UTR regulatory region which consequently leads to a depletion of the drug transporter and increased sensitivity to doxorubicin. Transfection of miR-451 reestablished the sensitivity of the MCF-7/DOX cells to doxorubicin. Similarly, Chen and colleagues [52] showed the same pattern but with miR-200c. The authors also showed a correlation of miR-200c with poor response to neoadjuvant chemotherapeutics using breast cancer tissues. Low expression of miR-200c leads to poor neoadjuvant therapeutic outcomes. However, they did not follow *ABCB1* gene and protein expression in the patients. Although published studies suggest a decreased expression of miR-451 correlated with higher expression of ABCB1 in drug resistant cells [51, 55], in a human ovarian cancer cell line, and its multidrug resistant counterpart, as well as in a human cervix carcinoma cell line and its multidrug resistant variant, expressions of miR-27a and miR-451 were upregulated in multidrug resistant cells as compared with their parental lines, downregulating expression of the *ABCB1* gene [56]. These results seem to point that the involvement of specific miRs in drug resistance should be cautiously taken at the moment, since the results could depend on various factors, including the cell lines under study. Bao et al. [53] used a different breast tumor cell line, MDA-MB-231, to show that miR-298 regulates *ABCB1* gene expression and increases resistance to doxorubicin. Remarkably, the authors also showed that the miR processing is altered in the resistant cell lines, due to the fact that DICER is weakly expressed and higher levels of miR-298 precursor was detected instead of mature form. Other authors also demonstrated a regulation of ABCB1 by miR-145 [57] in intestinal epithelial cells, and mir-381 and miR-495 in leukemia K562 cells resistant to adriamycin (K562/ADM cells) [58]. In this last study, functional analysis indicated that restoring expression of miR-381 or miR-495 in K562/ADM cells was correlated with reduced expression of the *ABCB1* gene and its protein product and increased drug uptake by the cells [58].

ABCG2 is another ABC transporter that, in normal tissues, functions as a defense mechanism against toxins and xenobiotics, with expression in the gut, bile canaliculi, placenta, blood–testis and blood–brain barriers. ABCG2 recognizes and transports a variety of chemotherapeutic drugs out of cancer cells, thereby resulting in reduced drug concentration, and subsequent drug resistance. Consequently ABCG2 plays a critical role in the development of MDR in breast cancer [59]. Increased ABCG2 expression has been found in breast cancer cells that exhibit resistance to mitoxantrone (MX), topotecan, and 7-ethyl-10-hydroxycamptothecin (SN-38) [60]. Upregulation of ABCG2 also confers resistance to tamoxifen in breast cancer cells [61]. In addition, ABCG2 expression correlates with chemotherapeutic response to anthracycline in patients with breast cancer [62]. Jiao et al. [63], performed microarray analysis to determine the differential expression patterns of miRs that target ABCG2 between the MX resistant breast cancer cell line MCF-7/MX and its parental MX sensitive cell line MCF-7. MiR-181a was found to be the most significantly downregulated miRNA in MCF-7/MX cells. Overexpression of miR-181a downregulated ABCG2 expression, and sensitized MX-resistant MCF-7/MX cells to MX. Moreover, in a nude mouse xenograft model, intratumoral injection of miR-181a mimics inhibited ABCG2 expression, and enhanced the antitumor activity of MX. Other authors have shown that ABCG2 is regulated by other miRs, including miR-328 [64] and 487a [65], and can influence MX resistance. miR-519c and miR-328 were also described as ABCG2 regulators and Li et al. [66] showed intracellular accumulation of MX in cells lacking ABCG2 expression. Interestingly, the authors also showed differences in expression of this miRs in stem-like ABCG2$^+$ cells and their ABCG2$^-$ counterparts. Thus, further investigation of miR regulation in stem cells may provide new insights into chemoresistance.

Another well-known ABC transporter is ABCC1, also known as MRP1. The main subtracts of ABCC1 are vincristine and etoposide and ABCC1 also confers resistance to anthracyclines (doxorubicin, daunorubicin, epirubicin), mitoxantrone, flutamide, and methotrexate. Curiously, many drugs are only transported in the presence of glutathione [67]. Regarding ABCC1, three reports were published showing a regulation by miRs [68–70]. Pogribny and colleagues [68] revealed that miR-345 and miR-7 increases sensitivity to cisplatin through a negative correlation with ABCC1. For that, the authors used a MCF-7 cell line resistant to cisplatin which expresses high levels of ABCC1 and lower levels of miR-345 and miR-7. Liang et al. [69] showed that miR-326 represses ABCC1 expression and sensitizes VP-16 resistant MCF-7 cells to VP-16 and doxorubicin. Pan et al. [70] reported that miR-1291 targets the 3'UTR of ABCC1 and consequently regulates its expression. This has impact in drug disposition and consequently in drug resistance. Interestingly, miR-1291 was described by these authors as being originated from a small nucleolar RNA, SNORA34.

ABCC2, also known as MRP2, and ABCC1 share a 49 % amino acid identity. As ABCC1, this efflux pump needs the presence of glutathione and can transport methotrexate, cisplatin, irinotecan, paclitaxel, and vincristine. ABCC2 is expressed in some solid tumors from the kidney, colon, breast, lung, ovary, and as well as in cells from patients with acute myelogenous leukemia [71]. Regarding ABCC2, to our knowledge, only one article has been published associating miR misexpression and ABCC2. Xu et al. [72] showed that miR-297 targets the 3′ UTR region of ABCC2 transcripts and consequently downregulates its expression. They also showed an inverse correlation between both molecules in colorectal carcinoma cell lines. Moreover, cell lines resistant to oxaliplatin and vincristine were sensitized when miR-297-mimics were transfected into these cells, *in vitro* and *in vivo*.

Intestinal epithelial cells are responsible for the absorption of most cancer drugs, and they express a variety of influx transporters specific for drugs, amino acids, peptides, organic anions, organic cations, and other nutrients. Peptide transporter 1 (PEPT1/ SLC15A1), organic cation/carnitine transporter 2 (SLC22A5), organic anion transporting polypeptide 2B1 (SLCO2B1), and monocarboxylate transporter 1 (MCT1/SLC16A1) are expressed at the brush-border membrane, whereas organic cation transporter 1 (SLC22A1) is mainly expressed at the basolateral membrane in the small intestine [31]. Recent studies have indicated that the regional differences in the expression of these transporters are dependent on the differentiation of intestinal epithelial cells [73]. Hence, misexpression of miRs could have a marked impact on absorption of cancer drugs. There are a limited number of reports on the SLC transporters regulated by miRNAs (Table 1). Dalmasso et al. [74] showed for the first time that SLC15A1 is regulated by a miR, namely miR-92b, causing diminished influx activity. Moreover, it suppresses bacterial peptide-induced proinflammatory responses in intestinal epithelial cells by inhibiting SLC15A. Pullen et al. [75] showed that miR-29a, miR-29b, and miR-124 can target SLC16A1, resulting in decreased expression at the protein level. The authors also refer that this regulation mechanism is not the main regulator but complements other transcriptional mechanisms and mutations that alter SLC16A1 expression.

2.3 DNA Repair

DNA damage by endogenous or exogenous agents elicits a powerful cellular response called the DNA Damage Response (DDR), which call up concerted molecular pathways to detect, repair, induce cell cycle arrest to allow repair, or in cases of high numbers of DNA lesions or irreparable damage, apoptosis, or cellular senescence (permanent cell cycle arrest) [76–79]. In the past few years evidence has accumulated that drug resistance is also linked to alterations in these pathways [26, 80–85]. The DDR pathways include DNA tolerance mechanisms by error-prone polymerases, the direct reversal of lesions, essentially de-alkylation of alkylated bases by O^6-methyl-guanine-DNA methyltransferase (*MGMT*), alkylation repair homolog 2 (*ALKBH2*) and alkylation repair homolog 3 (*ALKBH3*); nucleotide excision repair (NER); base excision repair (BER); mismatch repair (MMR); and the double strand break repair by homologous recombination (HR) and non-homologous end joining (NHEJ) [86, 87]. Besides these signaling cascades, the DDR also elicits the induction of several noncoding RNAs, including miRs. A large number of miRs are transcriptionally induced upon DNA damage and the level of induction is variable depending on cell type and the nature and the intensity of DNA damage and time after DNA damage [88–93]. Conversely many miRs target DDR genes, thus controlling feed-back and feed-forward loops to fine-tune the response (for a review see refs. 88, 94, 95). Wouters et al. found that 74 (52 %) mammalian DNA repair and DNA damage checkpoint genes contain conserved microRNA target sites predicted in their 3′-UTR by the algorithms Targetscan, Miranda, or both [95].

One of the first indications that implicated miR-mediated regulation of the DDR was knockdown of the miR biogenesis pathway (Dicer and Ago2), which resulted in increased sensitivity to UV and altered cell cycle after UV damage [90]. Following this study many reports have shown that different DNA damaging agents induce different patterns of miR expression [95]. Thus it is conceivable that alterations in miRs are involved in tumor response to anticancer agents.

A few examples indicate indeed that misexpression of miR is associated with drug responsiveness [96, 97]: members of the let-7 family of miRs are rapidly downregulated upon ionizing radiation in A549 lung cancer cells. Interestingly, the let-7 family of miRs regulates expression of oncogenes, such as RAS, and is specifically downregulated in many cancer subtypes. Low levels of let-7 predict a poor outcome in lung cancer. Overexpression of the let-7 family leads to radiosensitization in vitro of lung cancer cells and in vivo in a *Caenorhabditis elegans* model of radiation-induced cell death, whereas decreasing their levels causes radioresistance. In *C. elegans*, this was shown to occur partly through control of the proto-oncogene homologue let-60/RAS and genes in the DNA damage response pathway [96].

In another example, miR-138 was shown to target the ERCC1 gene, involved in NER, and to increase the sensitivity of A549/DDP cells to cisplatin in vitro and augmented apoptosis, suggesting that miR-138 could play an important role in the development of cisplatin resistance [98].

Valeri et al. [99] showed that MMR proteins MSH2 and MSH6 are inhibited by miR-21 overexpression causing a reduction in 5-fluorouracil (5-FU) induced G2/M damage arrest and apoptosis, in vitro. Moreover, xenograft studies demonstrate that miR-21 overexpression reduced the therapeutic efficacy of 5-FU.

REV1, an error-prone Y-family DNA polymerase required for translesion synthesis across interstrand crosslinks, was validated as a target of miR-96. Overexpression of miR-96 promoted cellular hypersensitivity to cisplatin in vitro and in vivo and enhanced sensitivity to the PARP inhibitor AZD2281. This miR also targets RAD51, a recombinase that promotes HR repair of double strand breaks (DSBs) and interstrand DNA crosslink (ICLs) [100]. RAD51 is also targeted by miR-155 in human breast cancer cells and affects the cellular response to ionizing radiation (IR). Due to this interaction, the efficiency of HR repair is reduced and sensitivity to IR augmented in vitro and in vivo. Indeed, overexpression of miR-155 was related with low levels of RAD51 and with better overall survival of patients with triple-negative breast cancers (TNBC) [101]. This emphasizes the possibility of how personalized therapy in TNBC patients could be used, knowing the miR-155 levels.

BRCA1 is an important component of the DDR pathway. BRCA1 encodes a nuclear phosphoprotein and primarily functions to maintain genomic stability via critical roles in DNA repair, cell cycle checkpoint control, transcriptional regulation, apoptosis, and mRNA splicing [102]. Mutations in BRCA1 are associated with an increased risk of developing breast and ovarian cancer. BRCA1 is also a target of miRNA-182 [103], indeed, the authors showed that high expression of this miR in multiple breast tumor cell lines influences BRCA1 levels and sensitivity to PARP1 inhibition. MiRNA-146a and miRNA-146-5p also bind to the same site in the 3′-UTR of BRCA1 and downregulate its expression. In breast tumors, levels of these miRs are inversely correlated with that of the BRCA1 protein and these miRs are overexpressed in triple negative breast cancers, a common type of breast cancer in women with BRCA1 mutations [104].

In another study, although the authors did not show specific targets, miR-296-5p and miR-193a-3p overexpression induced resistance to cisplatin, whereas miR-183 overexpression induced sensitivity. This study was done in breast cancer cells and also showed that miR-296-5p overexpression led to doxorubicin and paclitaxel resistance. These authors also examined whether overexpression of miR-16, miR-21, and miR-382 in Human Small Airway Epithelial progenitor (HSAEpCs) cells could modulate chemotherapy sensitivity. Thus, they found that miR-382 and miR-21 had no effect in resistance, while miR-16 promoted sensitivity to cisplatin and doxorubicin [91].

2.4 Epithelial to Mesenchymal Transition

Metastasis is the ultimate cause of death in most cancer patients. The growth of cancer cells at distant organs of a different tissue requires complex processes of detaching from the original tissue; invasion through the basement membrane; movement in the bloodstream or lymphatic system; and anchorage in other organs. The initial process is called epithelial-to-mesenchymal transition

(EMT) and is characterized by a phenotypic change of the tumor cells from cell–cell adhesion and polarity to motility, invasiveness, and some of the features of stem cells. This process not only enable the spread of the tumor cells but also their anchorage in distant organs, since tumor cells that undergo EMT can reverse this characteristic acquiring the epithelial phenotype again, in a process called mesenchymal-to-epithelial transition (MET). In EMT, cells lose the expression of E-cadherin and gain the expression of vimentin, N-cadherin, and fibronectin, markers of mesenchymal phenotype. Presumably, EMT is sustained by transient molecular changes and not by permanent genetic alterations. Indeed, the reversible nature of EMT must be associated with reversible epigenetic mechanisms, which allows stable but reversible modifications that do not directly affect the DNA primary sequence [105–107].

MiRs, as posttranscriptional regulators, are good candidates as EMT regulators and, as with epigenetic mechanisms, do not affect the DNA primary sequence and can press tumor cells to acquire an EMT phenotype in the tumor microenvironment. The most studied case is the miR-200 family that targets at least two transcriptional repressors of E-cadherin, ZEB1, and ZEB2.

It is known that the sensitivity to some cancer drugs like etoposide, taxol, and epidermal growth factor receptor inhibitors is increased with restoration of E-cadherin expression. Chen et al. [108] showed that miR-200c increases drug sensitivity of breast cancer cells to doxorubicin through the E-cadherin-mediated upregulation of PTEN. Similarly, Manavalan et al. [109] showed that an increased expression of miR-200b and miR-200c enhances the sensitivity to growth inhibition by 4-hydroxytamoxifen (4-OHT) and fulvestrant in breast cancer cells. Although it is known that miR-200 family regulates EMT through ZEB1 and E-cadherin, the real mechanism through which the miR-200 family regulates drug resistance is not known, and thus further studies are necessary to understand these phenomena. In order to answer this question, Bai et al. [110] published interesting data about miR-200c and feedback circuits of miR-200c/ZEB1 and miR-200c/ZNF217/TGF-β/ZEB1. The authors showed that these circuits contribute to trastuzumab resistance and metastasis of breast cancers. Interestingly, this feedback circuits might be related with reverse EMT in metastasis formation, since ZEB1 can inhibit miR-200c expression. The authors also showed that low levels of miR-200c activate the TGF-β signaling pathway and consequently trastuzumab resistance in breast cancer cells. Indeed, restoring miR-200c was sufficient to resensitize cells to trastuzumab and reverse the mesenchymal phenotype by inhibiting TGF-β signaling and ZEB1 expression. Similarly, Izumchenko et al. [111] reported that a high MIG6 expression and a suppression of miR-200c expression is a consequence of TGF-β-induced EMT and a signature for resistance to erlotinib.

Kitamura and colleagues [112] also showed, in lung adenocarcinoma, the importance of TGF-β signaling in drug resistance and EMT, namely, they showed that miR-134/miR-487b/miR-655 cluster promotes the EMT through TGF-β signaling and induces resistance to gefitinib by directly targeting MAGI2, whose suppression is encompassed by loss of PTEN stability [112].

Another example is the overexpression of miR-147, which alone induced reversal of EMT and consequently reversal of the native drug resistance of the colon cancer cell line HCT116 to gefitinib. Although the specific mechanism of action of miR-147 is still unknown, the authors found that miR-147 significantly upregulates CDH1 and represses ZEB1, known EMT markers, and inhibited TGF-β1 expression and also repressed Akt phosphorylation, leading to gefitinib sensitivity [113]. Jiang et al. [114] reported that miR-489 is underexpressed in a MCF7 breast cancer cell line resistant to doxorubicin, a cell line that shows mesenchymal phenotype. On the contrary, SMAD3, involved in TGF-β-induced EMT, is overexpressed in the same cell line. Ectopic expression of mir-489 not only reversed mesenchymal features, as well as sensitized the breast cell line to doxorubicin, through inhibition of SMAD3. No matter what miR and the respective target might be deregulated, all these studies show a point in common that is TGF-β signaling. This enhances the importance of TGF-β signaling in EMT and the regulation of EMT influenced drug resistance by miRs. miR-223 was also associated with drug resistance and EMT in pancreatic cancer. miR-223 is upregulated in gemcitabine resistant pancreatic cancer cells, thus acting as an oncogene, most probably, through inhibition of Fbw7 which consequently overexpresses Notch-1. The authors also showed that by inhibiting miR-223, pancreatic cancer cells were sensitized to gemcitabine [115].

2.5 Cancer Stem Cells and Drug Resistance

Somatic stem cells are typically slowly cycling cells capable of self-renewing mitotic divisions in which one or both of the daughter cells are faithful reproductions of the parent stem cell. The experimental observation that certain minority subpopulations of primary human acute myeloid leukemias (AMLs) could propagate the disease in immunodeficient mouse hosts at higher frequencies than the bulk populations of leukemic cells, led to the basis of what was later called the stem cell hypothesis. These cells made up the so-called side population (SP) cells, described as a subset of cells highly expressing ABC transporters and exhibiting cancer stem cell (CSC)-like phenotypes. Initially they were isolated by fluorescence-activated cell sorting (FACS) techniques based on Hoechst 33342 efflux. The SP cells were first isolated from the hematopoietic system but were then identified in normal tissues and several solid tumors.

Although it is accepted that most tumors arise from a single mutated cell, i.e., their origin is monoclonal, the tumor itself is a sum of several types of cells, due to the heterogeneity derived from a continuous evolution of the primitive cancer cell. Not all of these cells will display characteristics of cancer cells, such as metastization or unlimited replication potential. Operationally, (CSC) make up subpopulations of neoplastic cells within a tumor that have an elevated ability to seed new tumors upon experimental implantation in appropriate animal hosts [116]. They share many of the features of normal stem cells, including the capacity for self-renewal and differentiation, although their ability to differentiate into more than a few cell types has not been unequivocally proven, besides leukemias [117]. Although CSCs have been well characterized in hematological malignancies, their existence in other tissues has been much debated (for a review see Ref. [118]). Over the past few years CSC have been identified using stem cell specific markers in several solid tumors including breast, brain, colon, prostate, and pancreatic cancer [119–122]. It is often difficult to strictly define CSCs by associating them with traits beyond their tumor-initiating capability [118, 123]. Moreover, the possible existence of CSCs within tumors is intimately linked to tumor heterogeneity and tumor dedifferentiation. Nevertheless, several miRs have been shown to regulate stemness, or what we consider as properties of tumor-initiating and maintaining cancer cells, of different cancer types.

Recent studies showed differential expression of certain miRs between CSC and their differentiated counterparts [6, 124, 125], suggesting that miRs could also be involved in the regulation of CSC. For example, miR-200c and miR-34 have been shown to regulate CSC properties by targeting Bmi1 and downregulating Bcl2 and Notch, respectively [125, 126]. Additionally, miR-134, miR-296, and miR-470 modulate embryonic stem cell differentiation by suppressing the expression of the stem cell transcription factors Nanog, Oct4, and Sox2 [6]. Therefore miRs may impact on cancer drug resistance and several miRs have been reported to regulate stem cell properties and drug resistance concomitantly [127].

Yu et al. showed that let-7a expression was significantly decreased and Nanog/Oct4 expression was increased in head and neck cancer (HNC) tissues as compared to adjacent normal cells [128]. HNC–ALDH1+ cells displayed a decreased level of let-7a than HNC–ALDH1– cells. The overexpression of let-7a in vitro and in vivo showed that the self-renewal, resistance to cisplatin, and tumor initiation properties were significantly suppressed in let7a-overexpressing HNC–ALDH1+ cells, suggesting that the resistance of HNC–ALDH1+ cells to chemotherapy is partially due to the preferential activation of let-7a miRNA gene expression.

In another study, expression of miR-145, a tumor-suppressive miR, was shown to be inversely correlated with the levels of Oct4 and Sox2 in glioblastoma-CD133+ (GBM-CD133+) cells and malignant glioma specimens [129]. CD133 is a putative CSC marker in glioblastomas. The authors subsequently showed that miR-145 negatively regulates GBM tumorigenesis by targeting Oct4 and Sox2 in GBM-CD133$^+$ cells. miR-145 delivery to GBM-CD133$^+$ cells using polyurethane-short branch polyethylenimine (PU-PEI) significantly inhibited their tumorigenic and CSC-like abilities and facilitated their differentiation into CD133$^-$-non-CSCs. Moreover, PU-PEI-miR145-treated GBM-CD133$^+$ cells suppressed the expression of stemness (Nanog, c-Myc, and Bmi-1), drug-resistance (ABCG2, ABCC5, ABCB1), and anti-apoptotic genes (Bcl-2, Bcl-xL) and increased the sensitivity of the cells to radiation and temozolomide. The in vivo delivery of PU-PEI-miR145 alone significantly suppressed tumorigenesis with stemness, and synergistically improved the survival rate when used with radiotherapy and temozolomide in orthotopic GBM-CD133$^+$-transplanted immunocompromised mice [129].

Some miRs possess the ability to promote the generation of CSC by downregulating tumor suppressors. In hepatocellular carcinoma, miR-130b was shown to be associated with CSC growth that leads to worse overall survival and more frequent recurrence of cancer in patients. The increased miR-130b occurs in parallel with the reduction of tumor protein 53-induced nuclear protein 1, a known miR-130b target. Moreover, cells transfected with miR-130b presented a higher resistance to doxorubicin [130].

Similarly, other studies have revealed a regulation of stem cell properties through stem cell factors, including the p53–Nanog axis. For example, Xu et al. [131] showed that miR-214 regulates ovarian cancer cell stemness and chemoresistance towards cisplatin and doxorubicin treatment by targeting p53–Nanog, and expression of p53 abrogated miR-214-induced ovarian CSC properties.

Bitarte et al. [132] prepared colonospheres with CSCs properties from different colon carcinoma cells, and after performing miR profiling observed that miR-451 was downregulated in colonospheres versus parental cells. Expression of miR-451 caused a decrease in self-renewal, tumorigenicity, and chemoresistance to irinotecan, through a downregulation of the ABCB1 transporter.

Bourguignon et al. [133] observed that human head and neck squamous cell carcinoma (HNSCC) derived HSC-3 cells contain a subpopulation of (CSCs) characterized by high levels of CD44v3 and aldehyde dehydrogenase-1 (ALDH1) expression. These tumor cells also expressed stem cell markers (Oct4, Sox2, and Nanog) and displayed the hallmark CSC properties of self-renewal/clonal formation and the ability to generate heterogeneous cell populations. Hyaluronan (HA) activation of CD44v3 (an HA receptor) lead to nuclear accumulation of oncogenic transcription factors (Nanog, Oct4, Sox2), and CSCs in HNSCC display upregulated miR-302 expression which, in turn, upregulates several survival proteins responsible for clonal formation, self-renewal, and cisplatin resistance. MiR-302 is controlled by an upstream promoter containing

Oct4-Sox2-Nanog binding sites, while stimulation of miR-302 expression by HA-CD44 is Oct4-Sox2-Nanog-dependent in HNSCC-specific CSCs. This process results in suppression of several epigenetic regulators (AOF1/AOF2 and DNMT1) and the upregulation of several survival proteins (cIAP-1, cIAP-2, and XIAP) leading to self-renewal, clonal formation, and cisplatin resistance [133].

Several of these studies have used cell lines in vitro that express stem cell markers; however, one must keep in mind that these cell lines have vastly altered karyotypes (e.g., several translocations, insertions, and deletions) that will obviously alter their biological behavior. Therefore, caution must be exercised in interpreting the results described.

Acknowledgments

This work was supported by grant PEst-OE/SAU/UI0009/2014 from Fundação de Ciência e Tecnologia (FCT). B.C.G. was supported by SFRH/BD/64131/2009 from FCT.

References

1. Lee RC, Feinbaum RL, Ambros V (1993) The C. elegans heterochronic gene lin-4 encodes small RNAs with antisense complementarity to lin-14. Cell 75:843–854

2. Pasquinelli AE, Reinhart BJ, Slack F, Martindale MQ, Kuroda MI, Maller B, Hayward DC, Ball EE, Degnan B, Muller P, Spring J, Srinivasan A, Fishman M, Finnerty J, Corbo J, Levine M, Leahy P, Davidson E, Ruvkun G (2000) Conservation of the sequence and temporal expression of let-7 heterochronic regulatory RNA. Nature 408:86–89

3. Lee RC, Ambros V (2001) An extensive class of small RNAs in Caenorhabditis elegans. Science 294:862–864

4. Kozomara A, Griffiths-Jones S (2014) miRBase: annotating high confidence microRNAs using deep sequencing data. Nucleic Acids Res 42:D68–D73

5. Di Leva G, Garofalo M, Croce CM (2014) MicroRNAs in cancer. Annu Rev Pathol 9:287–314

6. Tay Y, Zhang J, Thomson AM, Lim B, Rigoutsos I (2008) MicroRNAs to Nanog, Oct4 and Sox2 coding regions modulate embryonic stem cell differentiation. Nature 455:1124–1128

7. Deng L, Shang L, Bai S, Chen J, He X, Trevino RM, Chen S, Li X, Meng X, Yu B, Wang X, Liu Y, McDermott SP, Ariazi AE, Ginestier C, Ibarra I, Ke J, Luther TK, Clouthier SG, Xu L, Shan G, Song E, Yao H, Hannon GJ, Weiss SJ, Wicha MS, Liu S (2014) MicroRNA100 inhibits self-renewal of breast cancer stem-like cells and breast tumor development. Cancer Res 74(22):6648–6660

8. Kang IH, Jeong BC, Hur SW, Choi H, Choi SH, Ryu JH, Hwang YC, Koh JT (2014) MicroRNA-302a stimulates osteoblastic differentiation by repressing COUP-TFII expression. J Cell Physiol 230:911–921

9. Lazare SS, Wojtowicz EE, Bystrykh LV, de Haan G (2014) microRNAs in hematopoiesis. Exp Cell Res 329(2):234–238

10. Janaki Ramaiah M, Lavanya A, Honarpisheh M, Zarea M, Bhadra U, Bhadra MP (2014) miR-15/16 complex targets p70S6 kinase1 and controls cell proliferation in MDA-MB-231 breast cancer cells. Gene 552: 255–264

11. Zhong K, Chen K, Han L, Li B (2014) microRNA-30b/c inhibits non-small cell lung cancer cell proliferation by targeting Rab18. BMC Cancer 14:703

12. Lerner M, Lundgren J, Akhoondi S, Jahn A, Ng HF, Akbari Moqadam F, Oude Vrielink JA, Agami R, Den Boer ML, Grander D, Sangfelt O (2011) MiRNA-27a controls FBW7/hCDC4-dependent cyclin E degradation and cell cycle progression. Cell Cycle 10:2172–2183

13. Liang LH, He XH (2011) Macro-management of microRNAs in cell cycle progression of tumor cells and its implications in anti-cancer therapy. Acta Pharmacol Sin 32:1311–1320

14. Zhou L, Zhang WG, Wang DS, Tao KS, Song WJ, Dou KF (2014) MicroRNA-183 is involved in cell proliferation, survival and poor prognosis in pancreatic ductal adenocarcinoma by regulating Bmi-1. Oncol Rep 32:1734–1740

15. Floyd DH, Zhang Y, Dey BK, Kefas B, Breit H, Marks K, Dutta A, Herold-Mende C, Synowitz M, Glass R, Abounader R, Purow BW (2014) Novel anti-apoptotic microRNAs 582-5p and 363 promote human glioblastoma stem cell survival via direct inhibition of caspase 3, caspase 9, and Bim. PLoS One 9, e96239

16. Li R, Yuan W, Mei W, Yang K, Chen Z (2014) MicroRNA 520d-3p inhibits gastric cancer cell proliferation, migration, and invasion by downregulating EphA2 expression. Mol Cell Biochem 396:295–305

17. Li W, Zang W, Liu P, Wang Y, Du Y, Chen X, Deng M, Sun W, Wang L, Zhao G, Zhai B (2014) MicroRNA-124 inhibits cellular proliferation and invasion by targeting Ets-1 in breast cancer. Tumour Biol 35(11): 10897–10904

18. Zhang R, Luo H, Wang S, Chen Z, Hua L, Wang HW, Chen W, Yuan Y, Zhou X, Li D, Shen S, Jiang T, You Y, Liu N, Wang H (2014) miR-622 suppresses proliferation, invasion and migration by directly targeting activating transcription factor 2 in glioma cells. J Neurooncol 121(1):63–72

19. Melo SA, Esteller M (2011) Dysregulation of microRNAs in cancer: playing with fire. FEBS Lett 585:2087–2099

20. Calin G, Sevignani C, Dumitru C, Hyslop T, Noch E, Yendamuri S, Shimizu M, Rattan S, Bullrich F, Negrini M (2004) Human microRNA genes are frequently located at fragile sites and genomic regions involved in cancers. Proc Natl Acad Sci U S A 101: 2999–3004

21. Shenouda SK, Alahari SK (2009) MicroRNA function in cancer: oncogene or a tumor suppressor? Cancer Metastasis Rev 28:369–378

22. Raychaudhuri M, Schuster T, Buchner T, Malinowsky K, Bronger H, Schwarz-Boeger U, Hofler H, Avril S (2012) Intratumoral heterogeneity of microRNA expression in breast cancer. J Mol Diagn 14:376–384

23. Housman G, Byler S, Heerboth S, Lapinska K, Longacre M, Snyder N, Sarkar S (2014) Drug resistance in cancer: an overview. Cancers 6:1769–1792

24. Holohan C, Van Schaeybroeck S, Longley DB, Johnston PG (2013) Cancer drug resistance: an evolving paradigm. Nat Rev Cancer 13:714–726

25. Longley DB, Johnston PG (2005) Molecular mechanisms of drug resistance. J Pathol 205: 275–292

26. Rodrigues AS, Dinis J, Gromicho M, Martins C, Laires A, Rueff J (2012) Genomics and cancer drug resistance. Curr Pharm Biotechnol 13:651–673

27. Rukov JL, Shomron N (2011) MicroRNA pharmacogenomics: post-transcriptional regulation of drug response. Trends Mol Med 17:412–423

28. Manikandan M, Munirajan AK (2014) Single nucleotide polymorphisms in microRNA binding sites of oncogenes: implications in cancer and pharmacogenomics. Omics 18:142–154

29. Urquhart BL, Tirona RG, Kim RB (2007) Nuclear receptors and the regulation of drug-metabolizing enzymes and drug transporters: implications for interindividual variability in response to drugs. J Clin Pharmacol 47:566–578

30. Tsuchiya Y, Nakajima M, Takagi S, Taniya T, Yokoi T (2006) MicroRNA regulates the expression of human cytochrome P450 1B1. Cancer Res 66:9090–9098

31. Ikemura K, Iwamoto T, Okuda M (2014) MicroRNAs as regulators of drug transporters, drug-metabolizing enzymes, and tight junctions: implication for intestinal barrier function. Pharmacol Ther 143:217–224

32. Koturbash I, Beland FA, Pogribny IP (2012) Role of microRNAs in the regulation of drug metabolizing and transporting genes and the response to environmental toxicants. Expert Opin Drug Metab Toxicol 8:597–606

33. Rodriguez-Antona C, Ingelman-Sundberg M (2006) Cytochrome P450 pharmacogenetics and cancer. Oncogene 25:1679–1691

34. Rendic SP, Guengerich FP (2015) Survey of human oxidoreductases and cytochrome P450 enzymes involved in the metabolism of chemicals. Chem Res Toxicol 28(1):38–42

35. Crewe HK, Notley LM, Wunsch RM, Lennard MS, Gillam EM (2002) Metabolism of tamoxifen by recombinant human cytochrome P450 enzymes: formation of the 4-hydroxy, 4′-hydroxy and N-desmethyl metabolites and isomerization of trans-4-hydroxytamoxifen. Drug Metab Dispos 30: 869–874

36. Leung T, Rajendran R, Singh S, Garva R, Krstic-Demonacos M, Demonacos C (2013) Cytochrome P450 2E1 (CYP2E1) regulates the response to oxidative stress and migration of breast cancer cells. Breast Cancer Res 15:R107

37. Mohri T, Nakajima M, Fukami T, Takamiya M, Aoki Y, Yokoi T (2010) Human CYP2E1 is regulated by miR-378. Biochem Pharmacol 79:1045–1052

38. Pan YZ, Gao W, Yu AM (2009) MicroRNAs regulate CYP3A4 expression via direct and indirect targeting. Drug Metab Dispos 37:2112–2117

39. Duffel MW, Marshal AD, McPhie P, Sharma V, Jakoby WB (2001) Enzymatic aspects of the phenol (aryl) sulfotransferases. Drug Metab Rev 33:369–395

40. Mercer KE, Apostolov EO, da Costa GG, Yu X, Lang P, Roberts DW, Davis W, Basnakian AG, Kadlubar FF, Kadlubar SA (2010) Expression of sulfotransferase isoform 1A1 (SULT1A1) in breast cancer cells significantly increases 4-hydroxytamoxifen-induced apoptosis. Int J Mol Epidemiol Genet 1:92–103

41. Yu X, Dhakal IB, Beggs M, Edavana VK, Williams S, Zhang X, Mercer K, Ning B, Lang NP, Kadlubar FF, Kadlubar S (2010) Functional genetic variants in the 3′-untranslated region of sulfotransferase isoform 1A1 (SULT1A1) and their effect on enzymatic activity. Toxicol Sci 118:391–403

42. Moriya Y, Nohata N, Kinoshita T, Mutallip M, Okamoto T, Yoshida S, Suzuki M, Yoshino I, Seki N (2012) Tumor suppressive microRNA-133a regulates novel molecular networks in lung squamous cell carcinoma. J Hum Genet 57:38–45

43. McLellan LI, Wolf CR (1999) Glutathione and glutathione-dependent enzymes in cancer drug resistance. Drug Resist Updat 2:153–164

44. Shea TC, Kelley SL, Henner WD (1988) Identification of an anionic form of glutathione transferase present in many human tumors and human tumor cell lines. Cancer Res 48:527–533

45. Sawers L, Ferguson MJ, Ihrig BR, Young HC, Chakravarty P, Wolf CR, Smith G (2014) Glutathione S-transferase P1 (GSTP1) directly influences platinum drug chemosensitivity in ovarian tumour cell lines. Br J Cancer 111:1150–1158

46. Gottesman MM, Fojo T, Bates SE (2002) Multidrug resistance in cancer: role of ATP-dependent transporters. Nat Rev Cancer 2:48–58

47. Kathawala RJ, Gupta P, Ashby CR Jr, Chen Z (2014) The modulation of ABC transporter-mediated multidrug resistance in cancer: a review of the past decade. Drug Resist Updat 18:1–17

48. Dean M, Hamon Y, Chimini G (2001) The human ATP-binding cassette (ABC) transporter superfamily. J Lipid Res 42:1007–1017

49. Haenisch S, Werk AN, Cascorbi I (2014) MicroRNAs and their relevance to ABC transporters. Br J Clin Pharmacol 77:587–596

50. Gromicho M, Dinis J, Magalhães M, Fernandes A, Tavares P, Laires A, Rueff J, Rodrigues A (2011) Development of Imatinib and Dasatinib resistance: dynamics of the drug transporters expression ABCB1, ABCC1, ABCG2, MVP and SLC22A1. Leuk Lymphoma 52:1980–1990

51. Kovalchuk O, Filkowski J, Meservy J, Ilnytskyy Y, Tryndyak VP, Chekhun VF, Pogribny IP (2008) Involvement of microRNA-451 in resistance of the MCF-7 breast cancer cells to chemotherapeutic drug doxorubicin. Mol Cancer Ther 7:2152–2159

52. Chen J, Tian W, Cai H, He H, Deng Y (2012) Down-regulation of microRNA-200c is associated with drug resistance in human breast cancer. Med Oncol 29:2527–2534

53. Bao L, Hazari S, Mehra S, Kaushal D, Moroz K, Dash S (2012) Increased expression of P-glycoprotein and doxorubicin chemoresistance of metastatic breast cancer is regulated by miR-298. Am J Pathol 180:2490–2503

54. Gromicho M, Magalhaes M, Torres F, Dinis J, Fernandes AR, Rendeiro P, Tavares P, Laires A, Rueff J, Sebastiao Rodrigues A (2013) Instability of mRNA expression signatures of drug transporters in chronic myeloid leukemia patients resistant to imatinib. Oncol Rep 29:741–750

55. van Jaarsveld MT, Helleman J, Berns EM, Wiemer EA (2010) MicroRNAs in ovarian cancer biology and therapy resistance. Int J Biochem Cell Biol 42:1282–1290

56. Zhu H, Wu H, Liu X, Evans BR, Medina DJ, Liu CG, Yang JM (2008) Role of MicroRNA miR-27a and miR-451 in the regulation of MDR1/P-glycoprotein expression in human cancer cells. Biochem Pharmacol 76:582–588

57. Ikemura K, Yamamoto M, Miyazaki S, Mizutani H, Iwamoto T, Okuda M (2013) MicroRNA-145 post-transcriptionally regulates the expression and function of P-glycoprotein in intestinal epithelial cells. Mol Pharmacol 83:399–405

58. Xu Y, Ohms SJ, Li Z, Wang Q, Gong G, Hu Y, Mao Z, Shannon MF, Fan JY (2013) Changes in the expression of miR-381 and miR-495 are inversely associated with the expression of the

MDR1 gene and development of multi-drug resistance. PLoS One 8, e82062

59. Natarajan K, Xie Y, Baer MR, Ross DD (2012) Role of breast cancer resistance protein (BCRP/ABCG2) in cancer drug resistance. Biochem Pharmacol 83:1084–1103

60. Shiozawa K, Oka M, Soda H, Yoshikawa M, Ikegami Y, Tsurutani J, Nakatomi K, Nakamura Y, Doi S, Kitazaki T, Mizuta Y, Murase K, Yoshida H, Ross DD, Kohno S (2004) Reversal of breast cancer resistance protein (BCRP/ABCG2)-mediated drug resistance by novobiocin, a coumermycin antibiotic. Int J Cancer 108:146–151

61. Selever J, Gu G, Lewis MT, Beyer A, Herynk MH, Covington KR, Tsimelzon A, Dontu G, Provost P, Di Pietro A, Boumendjel A, Albain K, Miele L, Weiss H, Barone I, Ando S, Fuqua SA (2011) Dicer-mediated upregulation of BCRP confers tamoxifen resistance in human breast cancer cells. Clin Cancer Res 17:6510–6521

62. Burger H, Foekens JA, Look MP, Meijer-van Gelder ME, Klijn JG, Wiemer EA, Stoter G, Nooter K (2003) RNA expression of breast cancer resistance protein, lung resistance-related protein, multidrug resistance-associated proteins 1 and 2, and multidrug resistance gene 1 in breast cancer: correlation with chemotherapeutic response. Clin Cancer Res 9:827–836

63. Jiao X, Zhao L, Ma M, Bai X, He M, Yan Y, Wang Y, Chen Q, Zhao X, Zhou M, Cui Z, Zheng Z, Wang E, Wei M (2013) MiR-181a enhances drug sensitivity in mitoxantone-resistant breast cancer cells by targeting breast cancer resistance protein (BCRP/ABCG2). Breast Cancer Res Treat 139:717–730

64. Pan YZ, Morris ME, Yu AM (2009) MicroRNA-328 negatively regulates the expression of breast cancer resistance protein (BCRP/ABCG2) in human cancer cells. Mol Pharmacol 75:1374–1379

65. Ma MT, He M, Wang Y, Jiao XY, Zhao L, Bai XF, Yu ZJ, Wu HZ, Sun ML, Song ZG, Wei MJ (2013) MiR-487a resensitizes mitoxantrone (MX)-resistant breast cancer cells (MCF-7/MX) to MX by targeting breast cancer resistance protein (BCRP/ABCG2). Cancer Lett 339:107–115

66. Li X, Pan YZ, Seigel GM, Hu ZH, Huang M, Yu AM (2011) Breast cancer resistance protein BCRP/ABCG2 regulatory microRNAs (hsa-miR-328, -519c and -520h) and their differential expression in stem-like ABCG2+ cancer cells. Biochem Pharmacol 81:783–792

67. Cole SP (2014) Targeting multidrug resistance protein 1 (MRP1, ABCC1): past, present, and future. Annu Rev Pharmacol Toxicol 54:95–117

68. Pogribny IP, Filkowski JN, Tryndyak VP, Golubov A, Shpyleva SI, Kovalchuk O (2010) Alterations of microRNAs and their targets are associated with acquired resistance of MCF-7 breast cancer cells to cisplatin. Int J Cancer 127:1785–1794

69. Liang Z, Wu H, Xia J, Li Y, Zhang Y, Huang K, Wagar N, Yoon Y, Cho HT, Scala S, Shim H (2010) Involvement of miR-326 in chemotherapy resistance of breast cancer through modulating expression of multidrug resistance-associated protein 1. Biochem Pharmacol 79:817–824

70. Pan YZ, Zhou A, Hu Z, Yu AM (2013) Small nucleolar RNA-derived microRNA hsa-miR-1291 modulates cellular drug disposition through direct targeting of ABC transporter ABCC1. Drug Metab Dispos 41:1744–1751

71. Chen ZS, Tiwari AK (2011) Multidrug resistance proteins (MRPs/ABCCs) in cancer chemotherapy and genetic diseases. FEBS J 278:3226–3245

72. Xu K, Liang X, Shen K, Cui D, Zheng Y, Xu J, Fan Z, Qiu Y, Li Q, Ni L, Liu J (2012) miR-297 modulates multidrug resistance in human colorectal carcinoma by down-regulating MRP-2. Biochem J 446:291–300

73. McKenna LB, Schug J, Vourekas A, McKenna JB, Bramswig NC, Friedman JR, Kaestner KH (2010) MicroRNAs control intestinal epithelial differentiation, architecture, and barrier function. Gastroenterology 139:1654–1664, 1664–1651

74. Dalmasso G, Nguyen HT, Yan Y, Laroui H, Charania MA, Obertone TS, Sitaraman SV, Merlin D (2011) MicroRNA-92b regulates expression of the oligopeptide transporter PepT1 in intestinal epithelial cells. Am J Physiol Gastrointest Liver Physiol 300:G52–G59

75. Pullen TJ, da Silva Xavier G, Kelsey G, Rutter GA (2011) miR-29a and miR-29b contribute to pancreatic beta-cell-specific silencing of monocarboxylate transporter 1 (Mct1). Mol Cell Biol 31:3182–3194

76. Jackson SP, Bartek J (2009) The DNA-damage response in human biology and disease. Nature 461:1071–1078

77. Harper JW, Elledge SJ (2007) The DNA damage response: ten years after. Mol Cell 28:739–745

78. Pearl LH, Schierz AC, Ward SE, Al-Lazikani B, Pearl FMG (2015) Therapeutic opportunities within the DNA damage response. Nat Rev Cancer 15:166–180

79. d'Adda di Fagagna F (2008) Living on a break: cellular senescence as a DNA-damage response. Nat Rev Cancer 8:512–522

80. Kelley MR, Fishel ML (2008) DNA repair proteins as molecular targets for cancer therapeutics. Anticancer Agents Med Chem 8:417–425

81. Kelley MR (2011) DNA repair inhibitors: where do we go from here? DNA Repair (Amst) 10:1183–1185

82. Kelley MR (2012) Future directions with DNA repair inhibitors: a roadmap for disruptive approaches to cancer therapy (Chapter 14). In: Mark RK (ed) DNA repair in cancer therapy. Academic, San Diego, CA, pp 301–310. doi:10.1016/b978-0-12-384999-1.10014-9

83. Helleday T, Petermann E, Lundin C, Hodgson B, Sharma RA (2008) DNA repair pathways as targets for cancer therapy. Nat Rev Cancer 8:193–204

84. Dinis J, Silva V, Gromicho M, Martins C, Laires A, Tavares P, Rendeiro P, Torres F, Rueff J, Rodrigues A (2012) DNA damage response in imatinib resistant chronic myeloid leukemia K562 cells. Leuk Lymphoma 53:2004–2014

85. Rodrigues AS, Gomes BC, Martins C, Gromicho M, Oliveira NG, Guerreiro PS, Rueff J (2013) DNA repair and resistance to cancer therapy. In: Chen C (ed) DNA repair and resistance to cancer therapy, new research directions in DNA repair. Intech. doi:10.5772/53952

86. Hoeijmakers J (2001) Genome maintenance mechanisms for preventing cancer. Nature 411:366–374

87. Hoeijmakers JH (2009) DNA damage, aging, and cancer. N Engl J Med 361:1475–1485

88. Sharma V, Misteli T (2013) Non-coding RNAs in DNA damage and repair. FEBS Lett 587:1832–1839

89. Templin T, Paul S, Amundson SA, Young EF, Barker CA, Wolden SL, Smilenov LB (2011) Radiation-induced micro-RNA expression changes in peripheral blood cells of radiotherapy patients. Int J Radiat Oncol Biol Phys 80:549–557

90. Pothof J, Verkaik NS, Van IW, Ta VT, van der Horst GT, Jaspers NG, van Gent DC, Hoeijmakers JH, Persengiev SP (2009) MicroRNA-mediated gene silencing modulates the UV-induced DNA-damage response. EMBO J 28:2090–2099

91. van Jaarsveld MT, Wouters MD, Boersma AW, Smid M, van Ijcken WF, Mathijssen RH, Hoeijmakers JH, Martens JW, van Laere S, Wiemer EA, Pothof J (2014) DNA damage responsive microRNAs misexpressed in human cancer modulate therapy sensitivity. Mol Oncol 8:458–468

92. d'Adda di Fagagna F (2014) A direct role for small non-coding RNAs in DNA damage response. Trends Cell Biol 24:171–178

93. Chowdhury D, Choi YE, Brault ME (2013) Charity begins at home: non-coding RNA functions in DNA repair. Nat Rev Mol Cell Biol 14:181–189

94. Bottai G, Pasculli B, Calin GA, Santarpia L (2014) Targeting the microRNA-regulating DNA damage/repair pathways in cancer. Expert Opin Biol Ther 14:1667–1683

95. Wouters MD, van Gent DC, Hoeijmakers JHJ, Pothof J (2011) MicroRNAs, the DNA damage response and cancer. Mutat Res 717:54–66

96. Weidhaas JB, Babar I, Nallur SM, Trang P, Roush S, Boehm M, Gillespie E, Slack FJ (2007) MicroRNAs as potential agents to alter resistance to cytotoxic anticancer therapy. Cancer Res 67:11111–11116

97. Blower PE, Chung JH, Verducci JS, Lin S, Park JK, Dai Z, Liu CG, Schmittgen TD, Reinhold WC, Croce CM, Weinstein JN, Sadee W (2008) MicroRNAs modulate the chemosensitivity of tumor cells. Mol Cancer Ther 7:1–9

98. Wang Q, Zhong M, Liu W, Li J, Huang J, Zheng L (2011) Alterations of microRNAs in cisplatin-resistant human non-small cell lung cancer cells (A549/DDP). Exp Lung Res 37:427–434

99. Valeri N, Gasparini P, Braconi C, Paone A, Lovat F, Fabbri M, Sumani KM, Alder H, Amadori D, Patel T, Nuovo GJ, Fishel R, Croce CM (2010) MicroRNA-21 induces resistance to 5-fluorouracil by down-regulating human DNA MutS homolog 2 (hMSH2). Proc Natl Acad Sci 107:21098–21103

100. Wang Y, Huang JW, Calses P, Kemp CJ, Taniguchi T (2012) MiR-96 downregulates REV1 and RAD51 to promote cellular sensitivity to cisplatin and PARP inhibition. Cancer Res 72:4037–4046

101. Gasparini P, Lovat F, Fassan M, Casadei L, Cascione L, Jacob NK, Carasi S, Palmieri D, Costinean S, Shapiro CL, Huebner K, Croce CM (2014) Protective role of miR-155 in breast cancer through RAD51 targeting impairs homologous recombination after irradiation. Proc Natl Acad Sci U S A 111:4536–4541

102. Savage KI, Harkin DP (2015) BRCA1, a 'complex' protein involved in the maintenance of genomic stability. FEBS J 282:630–646

103. Moskwa P, Buffa FM, Pan Y, Panchakshari R, Gottipati P, Muschel RJ, Beech J, Kulshrestha

R, Abdelmohsen K, Weinstock DM, Gorospe M, Harris AL, Helleday T, Chowdhury D (2011) miR-182-mediated downregulation of BRCA1 impacts DNA repair and sensitivity to PARP inhibitors. Mol Cell 41:210–220

104. Garcia AI, Buisson M, Bertrand P, Rimokh R, Rouleau E, Lopez BS, Lidereau R, Mikaelian I, Mazoyer S (2011) Down-regulation of BRCA1 expression by miR-146a and miR-146b-5p in triple negative sporadic breast cancers. EMBO Mol Med 3:279–290

105. Li L, Li W (2015) Epithelial-mesenchymal transition in human cancer: comprehensive reprogramming of metabolism, epigenetics, and differentiation. Pharmacol Ther 150:33–46

106. Zielinska HA, Bahl A, Holly JM, Perks CM (2015) Epithelial-to-mesenchymal transition in breast cancer: a role for insulin-like growth factor I and insulin-like growth factor-binding protein 3? Breast Cancer (Dove Med Press) 7:9–19

107. Lindsey S, Langhans SA (2014) Crosstalk of oncogenic signaling pathways during epithelial-mesenchymal transition. Front Oncol 4:358

108. Chen Y, Sun Y, Chen L, Xu X, Zhang X, Wang B, Min L, Liu W (2013) miRNA-200c increases the sensitivity of breast cancer cells to doxorubicin through the suppression of E-cadherin-mediated PTEN/Akt signaling. Mol Med Rep 7:1579–1584

109. Manavalan TT, Teng Y, Litchfield LM, Muluhngwi P, Al-Rayyan N, Klinge CM (2013) Reduced expression of miR-200 family members contributes to antiestrogen resistance in LY2 human breast cancer cells. PLoS One 8, e62334

110. Bai WD, Ye XM, Zhang MY, Zhu HY, Xi WJ, Huang X, Zhao J, Gu B, Zheng GX, Yang AG, Jia LT (2014) MiR-200c suppresses TGF-beta signaling and counteracts trastuzumab resistance and metastasis by targeting ZNF217 and ZEB1 in breast cancer. Int J Cancer 135:1356–1368

111. Izumchenko E, Chang X, Michailidi C, Kagohara L, Ravi R, Paz K, Brait M, Hoque M, Ling S, Bedi A, Sidransky D (2014) The TGFbeta-miR200-MIG6 pathway orchestrates the EMT-associated kinase switch that induces resistance to EGFR inhibitors. Cancer Res 74:3995–4005

112. Kitamura K, Seike M, Okano T, Matsuda K, Miyanaga A, Mizutani H, Noro R, Minegishi Y, Kubota K, Gemma A (2014) MiR-134/487b/655 cluster regulates TGF-beta-induced epithelial-mesenchymal transition and drug resistance to gefitinib by targeting MAGI2 in lung adenocarcinoma cells. Mol Cancer Ther 13:444–453

113. Lee CG, McCarthy S, Gruidl M, Timme C, Yeatman TJ (2014) MicroRNA-147 induces a mesenchymal-to-epithelial transition (MET) and reverses EGFR inhibitor resistance. PLoS One 9, e84597

114. Jiang L, He D, Yang D, Chen Z, Pan Q, Mao A, Cai Y, Li X, Xing H, Shi M, Chen Y, Bruce IC, Wang T, Jin L, Qi X, Hua D, Jin J, Ma X (2014) MiR-489 regulates chemoresistance in breast cancer via epithelial mesenchymal transition pathway. FEBS Lett 588:2009–2015

115. Ma J, Fang B, Zeng F, Ma C, Pang H, Cheng L, Shi Y, Wang H, Yin B, Xia J, Wang Z (2015) Down-regulation of miR-223 reverses epithelial-mesenchymal transition in gemcitabine-resistant pancreatic cancer cells. Oncotarget 6:1740–1749

116. Nguyen LV, Vanner R, Dirks P, Eaves CJ (2012) Cancer stem cells: an evolving concept. Nat Rev Cancer 12:133–143

117. Wiseman DH, Greystoke BF, Somervaille TC (2014) The variety of leukemic stem cells in myeloid malignancy. Oncogene 33: 3091–3098

118. Pattabiraman DR, Weinberg RA (2014) Tackling the cancer stem cells – what challenges do they pose? Nat Rev Drug Discov 13: 497–512

119. Li C, Heidt DG, Dalerba P, Burant CF, Zhang L, Adsay V, Wicha M, Clarke MF, Simeone DM (2007) Identification of pancreatic cancer stem cells. Cancer Res 67:1030–1037

120. O'Brien CA, Pollett A, Gallinger S, Dick JE (2007) A human colon cancer cell capable of initiating tumour growth in immunodeficient mice. Nature 445:106–110

121. Al-Hajj M, Wicha MS, Benito-Hernandez A, Morrison SJ, Clarke MF (2003) Prospective identification of tumorigenic breast cancer cells. Proc Natl Acad Sci U S A 100: 3983–3988

122. Singh SK, Clarke ID, Terasaki M, Bonn VE, Hawkins C, Squire J, Dirks PB (2003) Identification of a cancer stem cell in human brain tumors. Cancer Res 63:5821–5828

123. Antoniou A, Hebrant A, Dom G, Dumont JE, Maenhaut C (2013) Cancer stem cells, a fuzzy evolving concept: a cell population or a cell property? Cell Cycle 12:3743–3748

124. Liu C, Tang DG (2011) MicroRNA regulation of cancer stem cells. Cancer Res 71:5950–5954

125. Shimono Y, Zabala M, Cho R, Lobo N, Dalerba P, Qian D, Diehn M, Liu H, Panula S,

Chiao E (2009) Downregulation of miRNA-200c links breast cancer stem cells with normal stem cells. Cell 138:592–603

126. Ji Q, Hao X, Zhang M, Tang W, Yang M, Li L, Xiang D, Desano JT, Bommer GT, Fan D, Fearon ER, Lawrence TS, Xu L (2009) MicroRNA miR-34 inhibits human pancreatic cancer tumor-initiating cells. PLoS One 4, e6816

127. Raza U, Zhang JD, Sahin O (2014) MicroRNAs: master regulators of drug resistance, stemness, and metastasis. J Mol Med (Berl) 92:321–336

128. Yu CC, Chen YW, Chiou GY, Tsai LL, Huang PI, Chang CY, Tseng LM, Chiou SH, Yen SH, Chou MY, Chu PY, Lo WL (2011) MicroRNA let-7a represses chemoresistance and tumourigenicity in head and neck cancer via stem-like properties ablation. Oral Oncol 47:202–210

129. Yang YP, Chien Y, Chiou GY, Cherng JY, Wang ML, Lo WL, Chang YL, Huang PI, Chen YW, Shih YH, Chen MT, Chiou SH (2012) Inhibition of cancer stem cell-like properties and reduced chemoradioresistance of glioblastoma using microRNA145 with cationic polyurethane-short branch PEI. Biomaterials 33:1462–1476

130. Ma S, Tang KH, Chan YP, Lee TK, Kwan PS, Castilho A, Ng I, Man K, Wong N, To KF, Zheng BJ, Lai PB, Lo CM, Chan KW, Guan XY (2010) miR-130b Promotes CD133(+) liver tumor-initiating cell growth and self-renewal via tumor protein 53-induced nuclear protein 1. Cell Stem Cell 7:694–707

131. Xu CX, Xu M, Tan L, Yang H, Permuth-Wey J, Kruk PA, Wenham RM, Nicosia SV, Lancaster JM, Sellers TA, Cheng JQ (2012) MicroRNA miR-214 regulates ovarian cancer cell stemness by targeting p53/Nanog. J Biol Chem 287:34970–34978

132. Bitarte N, Bandres E, Boni V, Zarate R, Rodriguez J, Gonzalez-Huarriz M, Lopez I, Javier Sola J, Alonso MM, Fortes P, Garcia-Foncillas J (2011) MicroRNA-451 is involved in the self-renewal, tumorigenicity, and chemoresistance of colorectal cancer stem cells. Stem Cells 29:1661–1671

133. Bourguignon LY, Wong G, Earle C, Chen L (2012) Hyaluronan-CD44v3 interaction with Oct4-Sox2-Nanog promotes miR-302 expression leading to self-renewal, clonal formation, and cisplatin resistance in cancer stem cells from head and neck squamous cell carcinoma. J Biol Chem 287:32800–32824

Chapter 10

The Role of MicroRNAs in Resistance to Current Pancreatic Cancer Treatment: Translational Studies and Basic Protocols for Extraction and PCR Analysis

Ingrid Garajová, Tessa Y.S. Le Large, Elisa Giovannetti, Geert Kazemier, Guido Biasco, and Godefridus J. Peters

Abstract

Pancreatic ductal adenocarcinoma (PDAC) is a common cause of cancer death and has the worst prognosis of any major malignancy, with less than 5 % of patients alive 5-years after diagnosis. The therapeutic options for metastatic PDAC have changed in the past few years from single agent gemcitabine treatment to combination regimens. Nowadays, FOLFIRINOX or gemcitabine with nab-paclitaxel are new standard combinations in frontline metastatic setting in PDAC patients with good performance status. MicroRNAs (miRNA) are small, noncoding RNA molecules affecting important cellular processes such as inhibition of apoptosis, cell proliferation, epithelial-to-mesenchymal transition (EMT), metastases, and resistance to common cytotoxic and anti-signaling therapy in PDAC. A functional association between miRNAs and chemoresistance has been described for several common therapies. Therefore, in this review, we summarize the current knowledge on the role of miRNAs in the resistance to current anticancer treatment used for patients affected by metastatic PDAC.

Key words MicroRNAs, Pancreatic cancer, Drug resistance, Prognosis

1 Introduction

A surprising revelation from the human genome project was that 75 % of the genome is transcribed into RNA, but less than 2 % is composed of protein-coding genes [1–3]. The non-coding (nc) portion of the genome is of crucial functional importance in relation to both normal physiology and diseases [3]. They can be divided into three major types, according to the size. Short ncRNAs are <50 nucleotides (nt), those between 50 and 200 nt are referred to as mid-size ncRNAs, while long ncRNAs (lncRNAs) are >200 nt [4] (Table 1). MiRNAs are a class of short ncRNAs containing approximately 19–24 nt. They play an important role in cellular processes, such as proliferation, differentiation, apoptosis, stress

José Rueff and António Sebastião Rodrigues (eds.), *Cancer Drug Resistance: Overviews and Methods*, Methods in Molecular Biology, vol. 1395, DOI 10.1007/978-1-4939-3347-1_10, © Springer Science+Business Media New York 2016

Table 1
Noncoding RNAs and their functions

Type of noncoding RNA	Abbreviation	Related biological function	Reference
Short (<50 nt) and mid-size RNAs (50–200 nt)			
MicroRNAs	miRNAs	Posttranscriptional regulation of gene expression	[4]
PIWI-interacting RNAs	piRNAs	Transposon repression and DNA methylation	[5]
Small nucleolar RNAs	snoRNAs	Role in the modification of ribosomal RNAs	[6]
Endogenous small interfering RNAs	endo-siRNAs	Repress transposon transcripts or endogenous mRNAs	[4, 7]
Sno-derived RNAs	sdRNAs	Guide RNA modifications	[8]
Transcription initiation RNAs	tiRNAs	Regulation of chromatin modifications, protein recruitment involved in transcription initiation	[9]
miRNA-offset RNAs	moRNAs	Guide RISC to complementary target mRNAs	[10]
Long-size RNAs (>200 nt)			
Long intergenic noncoding RNAs	lincRNAs	Embryonic development	[11]
Long noncoding RNAs	lncRNA	Regulate neighboring protein-coding genes	[12]
NATs	NATs	Regulation of gene expression	[13]
Circular RNAs	circRNAs	Regulate microRNA function, role in transcriptional control	[14]
Long enhancer noncoding RNAs	–	Regulate neighboring protein-coding genes	[12]
Transcribed ultraconserved regions	T-UCRs	Interact with miRNAs and overlap with genomic ultraconserved regions	[12]
Pseudogenes	–	Regulate RNA expression	[4]

response, and drug resistance [4]. Remarkably, miRNAs exhibit tissue-specific and disease-specific expression that could provide the basis for their development as novel diagnostic, prognostic, and/or predictive biomarkers, as well as therapeutic targets [4].

1.1 The Role of MicroRNA Role in Cancer

The first miRNA molecule, lin-4, was identified in 1993 by Lee and collaborators [15]. The involvement of miRNAs in cancer was first discovered in a quest to identify tumor suppressors in the frequently deleted 13q14 region in chronic lymphocytic leukemia (CLL). The miRNA cluster miR-15a–miR-16-1 was deleted or downregulated via epigenetic silencing in 69 % of the patients [16]. MiRNAs regulate more than one-third of all human genes, which suggest their impressive influence on human biology [17]. It is known that more than 50 % of miRNA genes are localized within genomic regions that are either frequently amplified or deleted in different tumor types, resulting in miRNAs deregulation and aberrant expression [4, 17]. One of the most striking themes in the study of miRNAs and cancer is indeed the large alteration of miRNA expression in malignant cells compared to their normal counterparts. Most cancers have a specific miRNA signature, or "miRNome," that characterizes the malignant state and defines some of the clinico-pathological features of the tumors (e.g., grade, stage, aggressiveness, vascular invasion, and/or proliferation index) [4]. MiRNAs have a variety of roles in cancer development and progression, acting not only as tumor suppressors or oncogenes, but also as key activators or suppressors of tumor metastasis [4]. Variations in miRNA genes and their precursors, as well as the target sites and genes encoding components of the miRNA processing machinery can affect the cell phenotype and disease susceptibility [18]. Finally, a subclass of miRNAs, known as epi-miRNAs can directly control the epigenetic variations while miRNA expression can also be downregulated via promoter hypermethylation [19], adding another piece to the puzzle of regulatory gene expression networks.

MiRNA expression levels can be detected in a variety of human specimens including both fresh and formalin-fixed paraffin-embedded tissues, fine-needle aspirates, and in almost all human body fluids, including serum, plasma, saliva, urine, and amniotic fluid [4]. The impressive stability of miRNAs in tissues and biofluids is a key advantage over proteins and mRNAs. The present review summarizes the current knowledge on the role of miRNAs in resistance to current anticancer treatment used in patients affected by metastatic PDAC.

1.2 Current Pancreatic Cancer Treatment Regimens and the Role of miRNAs

PDAC is a highly aggressive malignancy and fourth leading cause of cancer-related death in developed countries [20]. The median survival after diagnosis is 2–8 months, and less than 5 % of all patients with PDAC survive 5 years after diagnosis [21]. This is due to advanced stage disease at initial diagnosis, frequent recurrence and the absence of treatment strategies that specifically and effectively target these tumors. In fact, a very limited response to most treatments can be achieved [22], including both conventional and signal transduction targeted therapies. The main mechanisms responsible for drug resistance in PDAC include molecular aberrations in key regulatory genes or signaling pathways, the desmoplastic reaction which characterizes PDAC microenvironment and the presence of resistant cancer stem cells (CSCs) as described elsewhere [23]. miRNAs appear to be involved in all these mechanisms and therefore can be considered as critical regulators of drug resistance in PDAC patients [5].

1.3 Conventional Chemotherapy

Nowadays, there are only a few therapeutic options for PDAC patients in the metastatic setting. They include gemcitabine as monotherapy or in combinations (mostly with nab-paclitaxel) or the combination of 5-FU, leucovorin, irinotecan, and oxaliplatin (FOLFIRINOX). It was uncommon that 2 years survival was observed in patients with metastatic PDAC but now it is observed in approximately 10 % of patients who received either FOLFIRINOX

Table 2
Current standard therapies for metastatic PDAC

	Gemcitabine	Gemcitabine + nab-paclitaxel	FOLFIRINOX
Response rate	6–11 %	23 %	31.6 %
PFS (months)	3.7	5.5	6.4
OS (months)	6.2	8.5	11.1
Toxicity G3-G4			
Hematologic			
Neutropenia	21.0 %	38 %	45.7 %
Febrile neutropenia	1.2 %	3 %	5.4 %
Thrombocytopenia	3.6 %	13 %	9.1 %
Nonhematologic			
Fatigue	17.8 %	17 %	23.6 %
Diarrhea	1.8 %	6 %	12.7 %
Sensory neuropathy	0 %	17 %	9.0 %
Reference	[25]	[26]	[25]

or gemcitabine and nab-paclitaxel. Unfortunately, direct comparison of these two regimes is lacking [24]. A summary of current standard therapies and ongoing experimental therapies for metastatic PDAC can be found in Tables 2 and 3.

1.3.1 Gemcitabine Monotherapy

Since 1997, gemcitabine is being used in the treatment of metastatic PDAC patients. Gemcitabine is a nucleoside analog with structural similarity to cytarabine. Initial studies suggested a low objective response rate (6–11%) in chemotherapy-naive patients with overall survival (OS) of 6.2 months [27]. MiRNAs can alter cellular response to several anticancer drugs (including gemcitabine) via interference with DNA repair. In particular, inhibition of ribonucleotide reductase (RR) by gemcitabine results in deoxyadenosine triphosphate (dATP) depletion, causing DNA replication errors. Moreover, gemcitabine is incorporated into DNA and arrests DNA replication. Both the mispaired bases and the gemcitabine-modified DNA bases can be the substrates for postreplicative DNA mismatch repair (MMR) machinery [28], which influences cancer cell sensitivity. In several papers we described ribonucleotide (NTP) and deoxyribonucleotide (dNTP) levels and related disturbances with differences in effects between sensitive and resistant cell lines. We demonstrated that gemcitabine induced concentration and combination dependent changes in NTP and dNTP pools [29, 30].

Table 3
Summary of novel therapeutic agents being tested usually in combinations in ongoing clinical studies for metastatic PDAC (according to www.clinicaltrial.gov)

Drug/treatment	Class	ClinicalTrials. go identifier
FOLFOX and Abraxane	Chemotherapy	NCT02080221
NC-6004 (Nanoplatin)	Chemotherapy (micellar cisplatin formulation)	NCT02043288
Irinotecan, oxaliplatin and cetuximab	Chemotherapy and anti-EGFR monoclonal antibody	NCT00871169
Mirtazapine	Antidepressant	NCT01598584
Celecoxib	Cox2 inhibitor	NCT01111591
MK0752	Notch signaling pathway inhibitor	NCT01098344
OMP-59R5	Notch signaling pathway inhibitor	NCT01647828
ABT-888 (Veliparib)	PARP inhibitor	NCT01489865
MLN8237 (Alisertib)	PARP inhibitor	NCT01924260
OMP-54F28 (Ipafricept)	Wnt signaling antagonist	NCT02050178
OMP-18R5 (Vantictumab)	Wnt signaling antagonist	NCT02005315
BKM120	PI3K inhibitor	NCT01571024
Dinaciclib + MK2206	Inhibitor of cyclin-dependent kinases + Akt inhibitor	NCT01783171
XL184 (Cabozantinib)	MET and VEGFR2 Inhibitor	NCT01663272
NPC-1C (Ensituximab)	Chimeric monoclonal antibody	NCT01834235
LCL161	Inhibitor of apoptosis (IAP) antagonist	NCT01934634
TH-302	2-Nitroimidazole triggered hypoxia-activated prodrug (HAP) of bromo-isophosphoramide mustard	NCT01833546
Dasatinib	Multi-BCR/Abl and Src family tyrosine kinase inhibitor	NCT01660971
CPI-613	Tumor-selective α-ketoglutarate dehydrogenase inhibitor	NCT01835041
Enzalutamide	Androgen receptor antagonist	NCT02138383
Ipilimumab	Monoclonal antibody anti-CTLA-4	NCT01473940
Algenpantucel-L	Immunotherapy	NCT01836432

In a retrospective study on laser-microdissected PDAC specimens, high miR-21 expression was associated with shorter patients' OS both in the metastatic and in the adjuvant setting. Multivariate analysis confirmed the prognostic significance of miR-21 [31]. The reduced expression of miR-21 was associated with benefit from gemcitabine treatment in two independent cohorts of PDAC patients [32, 33], as well as in a cohort of intraductal papillary mucinous neoplasms (IPMNs) of the pancreas [34]. These results might be explained by the effects of miR-21 expression on certain phenotypic characteristics in PDAC cell lines [32, 35]. Overexpression of miR-21 promotes cell proliferation, increases the metastatic ability through expression of matrix metalloproteinase-2 and metalloproteinase-9 as well as vascular endothelial growth factor (VEGF), and decreases gemcitabine sensitivity, whereas miR-21 repression delivers the opposite results [36]. Furthermore, Giovannetti et al. [31] and Dong and et al. [37] provided experimental evidence for a role of miR-21 in chemoresistance through modulation of apoptosis by directly regulating Bcl-2 and PTEN expression. Frampton et al. identified three miRNAs (miR-21, miR-23a, and miR-27a) that acted as cooperative repressors of a network of tumor suppressor genes that included PDCD4, BTG2, and NEDD4L [38]. In 91 PDAC samples from PDAC radically resected patients, high levels of a combination of these miRNAs were associated with shorter OS. Thus, high expression levels of this triple miRNA combination may be identified as having a much worse prognosis and might benefit from anti-miRNA therapy, although the best way to deliver such a treatment and potential off-target effects are unknown.

MiR-10b is overexpressed in PDAC patients and reduced expression of miR-10b was associated with improved response to multimodality neoadjuvant therapy, likelihood of surgical resection, better progression-free survival (PFS), and increased OS [39], demonstrating its role as a novel diagnostic and predictive biomarker. Finally, several studies reported that miR-155 is commonly overexpressed in PDACs and their precursor lesions. Its elevated expression correlated with shorter OS [40], while gemcitabine treatment induced the expression of miR-155 in PDAC cells suggesting a role in acquired chemoresistance [41]. miR-200a, miR-200b, and miR-200c are all downregulated in PDAC cells resistant to gemcitabine [42]. Moreover, miRNAs regulate the epithelial–mesenchymal transition (EMT) through the regulation of cadherin1 and other molecules [43]. Many members of the let-7 family are downregulated in EMT-type cells that are resistant to gemcitabine [42]. More miRNAs that have been linked to gemcitabine chemoresistance in PDAC are reported in Table 4.

1.3.2 Gemcitabine and Nab-Paclitaxel

Gemcitabine has been combined with many other active cytotoxic agents including nanoparticle albumin-bound paclitaxel, 5-FU, cisplatin, docetaxel, oxaliplatin, and irinotecan. Despite the large number of randomized trials that have been conducted, very few have demonstrated a benefit for a gemcitabine combination compared to gemcitabine alone. A clear superiority for combined therapy was documented for nanoparticle bound paclitaxel (nab-paclitaxel) plus gemcitabine. This combined therapy was associated with higher response rate (23 %), OS (8.5 months)

Table 4
Selected miRNA candidates which are related to gemcitabine resistance in PDAC

miRNA	Expression	Targets	Reference
miR-21	Upregulated	EGFR, HER2/neu, PDCD4, BCL2, PTEN, TIMP2, TIMP3	[32, 35]
miR-222 and miR-221	Upregulated	p27, PUMA, PTEN, Bim	[40, 44]
miR-10a and miR-10b	Upregulated	HOXB8, HOXA1	[45, 46]
miR-214	Upregulated	PTEN, ING4	[47, 48]
mir-320c	Upregulated	SMARCC1	[49]
miR-155	Upregulated	PI3K SMG-1	[41]
miR-1246	Upregulated	CCNG2	[50]
miR-301b	Upregulated	TP63	[51]
miR-365	Upregulated	SHC1 and BAX	[52]
miR-181b	Upregulated	CYLD	[53]
miR-RNA 330	Upregulated	dCK	[54]
miR-125b	Upregulated	BAP1, BBC3, NEU1, BCL2, STARD13	[55]
miR-34a	Downregulated	BCL-2	[56]
let-7	Downregulated	E2F2, c-Myc, KRAS, MAPK	[42]
miR-142-5p	Downregulated	BTG3	[57]
miR-204	Downregulated	MIC-1	[57]
miR-200a, miR-200b, miR-200c	Downregulated	EP300	[42, 58]
miR-29a	Downregulated	Dkk1, Kremen2, sFRP2	[59]

and PFS (5.5 months), in comparison to gemcitabine alone [26]. Combination treatment with gemcitabine and nab-paclitaxel increased intratumoral gemcitabine levels due to a marked decrease in the primary gemcitabine catabolizing enzyme, cytidine deaminase [60]. Correspondingly, paclitaxel reduced the levels of cytidine deaminase protein in cultured cells through reactive oxygen species-mediated degradation, resulting in the increased stabilization of gemcitabine [60]. Nab-paclitaxel alone or in combination with gemcitabine has been demonstrated to reduce the desmoplastic stroma [61]. Moreover, it is hypothesized that the albumin-bound nab-paclitaxel may selectively accumulate in the pancreatic stroma via its binding to "secreted protein acidic and rich in cysteine" (SPARC) matricellular glycoprotein which binds albumin and is overexpressed in tumor stroma [20]. High SPARC expression has been correlated to poor OS and has been suggested as a

Table 5
Selected miRNA related to anticancer treatment resistance

Drug	miRNA	Expression	Tumor type	Reference
Paclitaxel	miR-200c	Downregulated	Ovarian and endometrial cancer	[63]
Paclitaxel	miR-145	Downregulated	Ovarian cancer	[64]
Paclitaxel	miR-17-5p	Downregulated	Lung cancer	[65]
5-fluorouracil	miR-21	Overexpression	PDAC	[66]
Oxaliplatin	miR-203	Overexpression	Colorectal cancer	[67]
5-fluorouracil and oxaliplatin	miR-106a, miR-484, miR-130b	Overexpression	Colorectal cancer	[68]
Irinotecan and cetuximab	miR-345	Overexpression	Colorectal cancer	[69]
Erlotinib	miR-424	Downregulated	Lung cancer	[70]
Gefinitib	miR-214	Overexpression	Lung cancer	[71]
Erlotinib	miR-21	Downregulated	Lung cancer	[47]
Erlotinib	miR-145	Overexpression	Lung cancer	[72]
Erlotinib	miR-518f, miR-636, miR-301a, miR-34c, miR-224, miR-197, miR-205, miR135b, miR-200b, miR-200c, and miR-141	Overexpression	Lung cancer	[73]
Erlotinib	miR-140-3p, miR-628-5p	Downregulated	Lung cancer	[73]

Note: These miRNAs were not tested in PDAC yet, but have been evaluated in other tumor types

possible predictive biomarker for nab-paclitaxel in the phase-II trial [61]. However, no data on SPARC are available from the phase III trial. Neesse et al. showed that the effects of nab-paclitaxel were largely dose-dependent and that SPARC expression in the tumor stroma did not influence drug accumulation in a PDAC mouse model [62]. Further studies are therefore warranted to evaluate tissue and plasma SPARC expression as a potential predictive biomarker for nab-paclitaxel [62].

No data are available on miRNA affecting nab-paclitaxel, but several miRNAs have been associated to resistance to paclitaxel which is currently used for treatment of various cancers (*see* Table 5). The drug target TUBB3 is associated with miR-200c in ovarian and endometrial cancer cells [63]. The ectopic expression of this miRNA downregulated TUBB3 and enhanced sensitivity to microtubule-targeting agents, including paclitaxel [63]. In another study [64], paclitaxel-resistant ovarian cancer patients and cell lines show decreased miR-145 levels and expressed high levels of transcription factor specificity protein 1 (Sp1) and cyclin-dependent kinase 6 protein (Cdk6). Expression of miR-145 in SKOV3/PTX and A2780/PTX cells led to a reduction in Cdk6 and Sp1 along with downregulation of P-glycoprotein (P-gp) and phosphorylated retinoblastoma protein (pRb). These changes resulted in increased accumulation of antineoplastic drugs and G1 cell cycle arrest, which rendered the cells more sensitive to paclitaxel *in vitro* and *in vivo*. These effects could be reversed by reintroducing Sp1 or Cdk6 into cells expressing high levels of miR-145, resulting in restoration of P-gp and pRb levels [64].

MiRNA arrays were used to screen differentially expressed miRNAs between paclitaxel sensitive lung cancer cells A549 and its paclitaxel-resistant cell variant (A549-T24) [74]. MiR-17-5p was one of most downregulated miRNAs in paclitaxel-resistant lung cancer cells compared to paclitaxel-sensitive parental cells [74]. Overexpression of miR-17-5p sensitized paclitaxel resistant lung cancer cells to paclitaxel induced apoptotic cell death. Moreover, miR-17-5p directly binds to the 3'-UTR of the beclin 1 gene, one of the most important autophagy modulators [74]. Moreover, miR-17-5p, which is a member of the miR-17-92 cluster, is upregulated in PDAC and some of the present findings suggest that miR-17-5p plays an important role in pancreatic carcinogenesis and cancer progression, and is associated with a poor prognosis in PDAC [75].

1.3.3 FOLFIRINOX (5-FU, Leucovorin, Irinotecan, and Oxaliplatin)

The phase III ACCORD 11 trial using the FOLFIRINOX regimen in PDAC patients has shown a response rate of 31.6 %, a median PFS of 6.4 months and a median OS of survival of 11.1 months [25]. Therefore, the FOLFIRINOX protocol confers a significant improvement in the OS in metastatic PDAC patients and can be considered as a novel therapeutic option for patients with a good performance status [25]. In patients with a poor performance status or severe toxicity, a dose reduction of either one or more drugs is often needed. No predictive biomarkers are actually used in clinical practice to predict either response or toxicity, but a few studies suggested the role of candidate miRNAs to predict the sensitivity/resistance to 5-FU, and the other drugs in this regimen.

Donahue et al. demonstrated that miR-21 expression in cancer associated fibroblasts (CAFs) was associated with decreased OS in PDAC patients who received 5-FU [66]. Therefore stromal miR-21 might be used as a marker to guide chemotherapy choice in PDAC patients [66]. In a pharmacogenetic study, 18 polymorphisms were evaluated both in miRNA-containing genomic regions (primary and precursor miRNA) and in genes related to miRNA biogenesis and related with outcome in metastatic CRC patients treated with 5-FU and irinotecan [76]. The genotypes CC and CT in the SNP rs7372209 in pri-miR26a-1 were favorable when compared with the TT variant genotype and associated with a higher response and longer OS. Similarly, the SNP rs1834306, located in the pri-miR-100 gene, correlated with a longer PFS [76].

Several associations between miRNAs and oxaliplatin/irinotecan resistance have been described (*see* Table 5). Oxaliplatin is a common component of combination therapeutic regimen in CRC patients, both in adjuvant and metastatic setting. In oxaliplatin-resistant CRC cell lines, miR-203 was upregulated while exogenous expression of miR-203 in chemo-naïve CRC cells induced oxaliplatin resistance [67]. Moreover, knockdown of miR-203 sensitized chemoresistant CRC cells to oxaliplatin. A main target of miR-203 is ataxia telangiectasia mutated (ATM), a primary mediator of the DNA damage response [67]. ATM mRNA and protein levels were downregulated in CRC cells with acquired resistance to oxaliplatin. Moreover, a reverse correlation between miR-203 and ATM expression was found in CRC tissues [67]. Kjersem et al. investigated the expression of 742 miRNAs in plasma samples from 24 mCRC patients (12 responders and 12 non-responders) before onset and after four cycles of 5-FU/oxaliplatin [68]. The top differentially expressed miRNAs between responders and non-responders were selected for further analysis in a validation cohort of 150 patients. The authors found three miRNAs (miR-106a, miR-484, and miR-130b) to be upregulated in non-responders. Therefore, plasma miRNAs analyzed before treatment may serve as noninvasive markers predicting outcome in mCRC patients treated with 5-FU and oxaliplatin-based chemotherapy [68]. In a prospective phase II study in CRC patients treated in third line therapy with cetuximab and irinotecan, out of 738 pretreatment miRNAs, high miR-345 expression was associated with lack of response [69]. These miRNAs were isolated and profiled from whole blood using the TaqMan MicroRNA Array v2.0 [69].

Taken together, no miRNAs as predictive biomarkers are used or are validated to be used to predict response/resistance to current anticancer conventional chemotherapy (gemcitabine, gemcitabine and nab-paclitaxel, FOLFIRINOX) in clinical practice. Several candidate miRNAs have been correlated to gemcitabine resistance, but no specific miRNAs have been investigated as predictive biomarker for combination therapy (gemcitabine and nab-paclitaxel or FOLFIRINOX) in PDAC patients.

1.4 Signal Transduction Targeted Therapy

From its introduction, cancer therapy has been encumbered by its poor selectivity because most antineoplastic drugs are also toxic to fast-replicating cells of the blood compartment, skin cells, and gastrointestinal tract lining cells. This unsatisfactory situation and the development of technology leading to the sequencing of the genome, have led to the development of a group of potentially more specific and less toxic anticancer drugs supposed to be blocking specific molecules involved in tumor growth and progression. Therefore these drugs are generally called "targeted therapies."

Some of these therapeutic regimens are designed to intercept deregulated dominant oncogenes and proved to be effective treatment in "oncogene addicted" tumors [77]. One of these targets is the epidermal growth factor receptor (EGFR) that can be inhibited either by monoclonal antibodies (mAbs) or small molecules inhibiting the tyrosine kinase domain (TKIs). The mAb cetuximab blocks the extracellular domain of EGFR, thereby competing with the ligands, and resulting in the inhibition of the receptor. Cetuximab is approved for the treatment of metastatic CRC, while the EGFR-TKIs gefitinib and erlotinib have been approved as upfront therapy replacing chemotherapy in metastatic NSCLC patients harboring activating-EGFR mutations. Moreover, erlotinib is registered for treatment of patients with metastatic PDAC.

1.4.1 Anti-EGFR Therapy in PDAC

The SWOG group conducted a randomized Phase III clinical trial randomizing PDAC patients with stages III–IV to receive either gemcitabine alone or in combination with cetuximab. The addition of cetuximab did not improve patients' OS [78]. Negative results for this combination were also observed in the adjuvant setting [79]. Similarly, other EGFR and HER2 targeted therapies, including trastuzumab and lapatinib, have not shown OS benefit in PDAC patients [80]. In contrast, a combination of gemcitabine and erlotinib has been approved for use by the US Food and Drug Administration (FDA), and European Medicines Agency (EMEA) as a treatment for PDAC patients on the basis of a randomized trial, showing an overall gain in median survival of 2 weeks [81]. K-RAS mutational status and EGFR gene copy number could not be used as molecular predictors of response [82]. However, accumulating evidence suggests that dysregulation of specific miRNAs may be involved in the acquisition of cancer cell resistance to EGFR targeted agents (*see* Table 5). In particular, miR-7 emerged as a critical modulator of a regulatory network for EGFR signaling in lung cancer cells, with the ability of coordinately downregulating the expression of several members of the EGFR signaling cascade [83]. The binding of c-Myc to the miR-7 promoter increased expression of mir-7, while ectopic miR-7 promoted cell growth and orthotopic tumor formation in nude mice. In these models, quantitative proteomic analysis revealed that miR-7 decreased levels of the Ets2 transcriptional repression factor ERF, which is a direct target of miR-7. Accordingly, the inhibition of miR-7 expression suppressed EGFR mRNA and protein expression in different lung cancer cell lines as well as the growth of the A549 lung adenocarcinoma cells [84]. Of note, miR-7 is preferentially expressed in endocrine cells of the developing and adult human pancreas [85]. However, its role in the regulation of the insulin growth factor-1 receptor expression might affect the development of diabetes-associated PDAC [70].

Other studies in lung cancer cell lines showed that decreased miR-424 levels were indicative of increased resistance to erlotinib, while the gefitinib resistant cell line-HCC827/GR had a significant upregulation of miR-214 [71]. The inhibition of miR-214 has been also correlated with decreased apoptosis. MiR-214 and PTEN were indeed inversely expressed, while knockdown of miR-214 increased the expression of PTEN and p-AKT, re-sensitizing HCC827/GR to gefitinib. MiR-214 is also aberrantly expressed in PDAC and *in vitro* experiments showed that overexpression of miR-214 decreased the sensitivity of the BxPC-3 cells to gemcitabine [47]. Yan et al. demonstrated that patients with NSCLC with reduced miR-21 expression had longer OS and a poor response rate to erlotinib. In contrary, increased miRNA-145 levels can predict longer OS, PFS and excellent response rate to erlotinib [72]. The sensitivity to erlotinib of lung cancer cell lines was also predicted by a 13-gene miRNA signature. Eleven of the signature miRNAs are upregulated (miR-518f, miR-636, miR-301a, miR-34c, miR-224, miR-197, miR-205, miR135b, miR-200b, miR-200c, and miR-141) and two miRNAs were found to be downregulated (miR-140-3p, miR-628-5p), in erlotinib-sensitive cells. Ontological annotation of these miRNA and their potential targets revealed enrichment of this panel of EMT, including the Wnt pathway. This may explain the ability of this signature to separate primary from metastatic tumor samples as well as why the treatment with TGFβ1 modulated both the expression of these miRNA and cell migration [73]. Interestingly, EMT has been inversely correlated with the response of cancers to EGFR-targeted therapy and the TGFβ-mitogen-inducible gene 6-miR200 network orchestrates the EMT-associated kinase switch that induces resistance to EGFR inhibitors in primary tumor xenografts of patient-derived lung and pancreatic cancers carrying wild type EGFR [86]. These data support the low ratio of Mig6 to miR200 as a promising predictive biomarker of the response of PDAC to EGFR-TKIs.

In conclusion, various anti-EGFR agents were tested alone or in combination with gemcitabine for treatment of metastatic PDAC, though just erlotinib achieved a very modest survival benefit. Several miRNAs have been investigated as possible predictive biomarkers to resistance/response to erlotinib, though, similarly as in the case of conventional chemotherapy, none of them has been validated for use in clinical practice.

1.5 Conclusions and Future Perspectives

PDAC is a well-known chemorefractory disease. In the past few years, a modest improvement in OS has been observed due to introduction of newer anticancer agents such as nab-paclitaxel and regimens such as FOLFIRINOX. Both showed a superior activity in comparison to gemcitabine monotherapy. Numerous potential biomarkers have been studied to predict the response/resistance to these anticancer treatment, but the clinical benefit has been limited so far. As important regulators of gene expression, miRNAs possess high potential as predictive markers for therapeutic response to chemotherapeutic drugs. In particular, circulating miRNAs have high translational potential as noninvasive biomarkers. However, as with previous studies on gene profiling, most emerging miRNA signatures of chemoresistance are not overlapping and no conclusive evidence has been obtained on their clinical utility. Additional studies in larger homogeneous populations are therefore warranted.

1.6 Practical Protocol: The Use of miRNAs Analysis in Chemoresistance

The expression of miRNAs is known to change upon stress, inflammation or disease [87]. This feature makes miRNAs very attractive for analysis of a potential predictive value of chemoresistance. Furthermore, they represent a part of molecular mechanism of chemoresistance. One of the important qualities of miRNAs is that they are extremely stable [88]. This was for instance shown by Verhoeven et al. [89], where they proved that miRNAs in graft preservation solution remained stable at room temperature up to 24 h. Furthermore, miRNAs are present in the cell but are also excreted into body fluids such as serum, bile or urine, either bound to proteins or in exosomes [90–94]. These possibilities to measure miRNAs in fluids emphasize the importance of miRNAs to use for screening or for prediction of response to therapy. Different types of material can be used to extract miRNAs. A comparison can be done on tissue before and after treatment, or for example on serum before and after treatment. Significant differences can be validated in a separate cohort to assess their predictive value, while differentially expressed miRNAs can be analyzed for their role *in vitro*.

There are several techniques to distinguish differential expression of miRNAs. For profiling, high-throughput sequencing or microarray profiling can be used. With real-time PCR specific miRNA expression can be validated [95, 96]. In a recent Nature Methods paper [97], the different platforms for quantitative miRNA expression were evaluated. This article shows that there are great differences between platforms and detection of differentially expressed miRNAs. They conclude that there are weaknesses and strengths for each platform, and the decision for the platform should be made for each specific experiment and research question.

Another problem for the analysis of miRNAs is the normalization of data. For assays with many (>50) miRNAs the "global mean normalizer" is recommended. For analysis of less miRNAs, there is no consensus on what method to use. Data can be normalized with spiked-in miRNA, a set of stably expressed miRNAs or general reference genes of small RNA like U6 [98, 99].

Here, we describe a method of extraction of miRNAs from serum and formalin-fixed paraffin-embedded (FFPE) tissue, quantified with qRT-PCR. Extraction of miRNAs from cells and fresh tissues can also be done, usually following standard protocols for mRNA isolation.

2 Materials

2.1 MicroRNA Extraction from Serum

1. miRCURY RNA Isolation Kit—Biofluids (Exiqon, Vedbaek, Denmark).
2. Bench top microcentrifuge.
3. Pipette with RNase-free tips.
4. RNase-free microcentrifuge tubes 1.5 ml.
5. Vortexer.
6. Isopropanol.
7. 100 % ethanol.
8. Serum stored in aliquots of 1 ml in −80 °C.
9. Optional: RNaseZAP Solution.

2.2 MicroRNA Extraction from FFPE Tissue

1. RecoverAll Total Nucleic Acid Isolation Kit for FFPE (Ambion).
2. 100 % xylene.
3. 100 % ethanol.
4. Microtome for tissue sectioning.
5. Bench top centrifuge.
6. Pipette with RNase-free tips.
7. RNase-free microcentrifuge tubes 1.5 ml.
8. Incubators.
9. Spectrophotometer, e.g., the NanoDrop* 1000 Spectrophotometer.
10. Optional: Laser for dissection.
11. Optional: Centrifugal vacuum concentrator.
12. Optional: RNaseZAP Solution.
13. FFPE tissue from patients.

2.3 Reverse Transcription and Quantitative Real-Time PCR

1. Real-time PCR 96-wells plate.
2. TaqMan MicroRNA Assay (Applied Biosystems, Foster City, California, USA).
3. TaqMan MicroRNA Reverse Transcription Kit (Applied Biosystems, Foster City, California, USA).
4. ABI-Prism-7500HT instrument.
5. C-1000 Thermal Cycler (Bio-Rad, Berkeley, California, USA).
6. Taqman Universal PCR Master Mix.

2.4 Methods

All materials should be handled with gloves to minimize contamination. Clean bench and pipettes with RNase decontamination solution like RNaseZap.

2.4.1 MicroRNA Extraction from Serum

The method described here uses the miRNA extraction protocol from the miRCURY RNA Isolation Kit for biofluids (Exiqon). This kit uses a purification column to collect all RNA molecules smaller than 1000 nt. If the samples are used for PCR, you can optionally add a spike-in.

1. Thaw the samples on ice. Centrifuge the sample 5 min at $3000 \times g$ to remove cell debris and collect supernatant in a new tube.

2. Use 200 µl of serum (*see* **Note 1**).

3. Add 60 µl Lysis Solution (provided by Kit), vortex 5 s to mix well and incubate for 1 min at room temperature.

4. Add 20 µl Protein Precipitation Solution BF (provided by Kit) to the sample, vortex for 5 s, and incubate for 1 min at room temperature.

5. Centrifuge sample for 3 min at $11,000 \times g$ and collect the clear supernatant to a new collection tube of 2 ml (provided by Kit).

6. Add 270 µl isopropanol to each tube and vortex 5 s.

7. Place microRNA Mini Spin column BF (provided by Kit) in a collection tube and load sample onto the column (*see* **Note 2**).

8. Incubate for 2 min at room temperature.

9. Centrifuge for 30 s at $11,000 \times g$ and discard flow-through.

10. Add 100 µl Wash Solution 1 (provided by Kit) to the spin column and repeat **item 9**.

11. Add 700 µl Wash Solution 2 (before the first use, the Wash Solution 2 provided by the Kit should be diluted with 80 ml of ethanol 100 %) to the spin column and repeat **item 9**.

12. Add 250 µl Wash solution 2 (provided by Kit) to the spin column and centrifuge for 2 min at $11,000 \times g$ to dry column completely.

13. Place the column in a new collection tube.

14. Add 50 µl RNase-free H_2O (provided by Kit) onto the membrane to elute the small RNAs (*see* **Note 3**).

15. Incubate for 1 min at RT and centrifuge for 1 min at $11,000 \times g$.

16. Store purified RNA sample in –20 °C when it will be used shortly. To store for long term, store at –80 °C.

2.4.2 MicroRNA Extraction from FFPE Tissue

This protocol is used for extraction of miRNAs from FFPE tissue blocks according to manufacturer's protocol (RecoverAll Total Nucleic Acid Isolation Kit for FFPE, Ambion). It is recommended to use tissue slices of 10 μm thick to ensure that you do not use a mono-layer of cells. This will maximize recovery of miRNAs (*see* **Note 4**).

1. Cut slices of FFPE tissue using a microtome and place the equivalent of about 80 μm in a 1.5 microcentrifuge tube.

2. Add 1 ml 100 % xylene to the sample and vortex (*see* **Note 5**).

3. Centrifuge briefly to submerge the tissue in xylene.

4. Heat the sample for 3 min at 50 °C to melt the paraffin.

5. Centrifuge for 2 min at room temperature at maximum speed to pellet the tissue (*see* **Note 6**).

6. Remove xylene without disturbing the pellet (*see* **Note 7**).

7. Add 1 ml of 100 % ethanol and vortex. The sample should become opaque.

8. Centrifuge for 2 min at maximum speed at room temperature and discard supernatant.

9. Repeat **items 7** and **8**.

10. Air dry the pellet for 15–45 min at room temperature, optionally by use a centrifugal vacuum concentrator.

11. Add Digestion Buffer (provided by Kit) (100 μl for sample <40 μm, 200 μl for sample 40–80 μm) to each sample.

12. Add 4 μl Protease (provided by Kit) to each sample and make sure the tissue is fully submerged.

13. Incubate for 15 min at 50 °C and subsequently 15 min at 80 °C (*see* **Note 8**).

14. Combine the Isolation Additive (provided by Kit) and ethanol, the amount needed is dependent on the amount of digestion buffer used (*see* **Note 9**).

15. Add to sample and pipet to mix the sample.

16. Place a Filter Cartridge in the supplied Collection Tube and add up to 700 μl of mixture to the Filter (*see* **Note 10**).

17. Centrifuge at $10,000 \times g$ for 30 s and discard flow-through.

18. Repeat **items 16** and **17** if more mixture was there to start with.

19. Add 700 μl of Wash 1 (provided by Kit) and repeat **item 17**.

20. Add 500 μl of Wash 2/3 (provided by Kit) and repeat **item 17** twice to remove all residual fluid.

21. Create a DNase mix (6 μl 10× DNase Buffer + 4 μl DNase + 50 μl Nuclease-free water) (provided by Kit) and add 60 μl to the sample (*see* **Note 8**).

22. Incubate for 30 min at room temperature.

23. Add 700 μl of Wash 1 and incubate for 30–60 s at room temperature.

24. Repeat **item 17**.

25. Add 500 μl Wash 2/3 and repeat **item 17**.

26. Repeat **item 25**.

27. Centrifuge 1 min at $10,000 \times g$ to remove residual fluids from the filter.

28. Transfer the filter to a new collection tube and apply 60 μl of Elution Solution (provided by Kit) or nuclease-free water.

29. Incubate for 1 min at room temperature and subsequently centrifuge at maximum speed for 1 min.

30. Store the eluate at −20 °C or colder.

2.4.3 MicroRNA Extraction from Cells

We extracted RNA according to the Trizol-chloroform protocol of the VUMC Micro-array Facility (http://www.vumc.com/afdelingen/microarrays/). For that purpose, cells in exponential phase of culturing are harvested by standard procedures (attached cells by trypsinization or scraping) [31, 32]. The suspended cells are centrifuged and pellets are immediately frozen in liquid nitrogen for later processing or immediately used for RNA isolation. RNA yields and integrity were checked by measuring optical density at 260 nm/280 nm with a NanoDrop® spectrophotometer.

2.4.4 MicroRNA Extraction from Fresh Tissue

For frozen tissues, RNA extraction has to be performed after disruption of the tissue for which various methods can be used. Laser microdissection is more complicated for frozen tissues. Therefore before proceeding, the percentage of tumor cells in the tissue has to be assessed by a pathologist. Usually, we require that the tumor consists of >70 % tumor cells.

The tumor can be pulverized using a micro-dismembrator, a procedure during which the tumor remains frozen during the whole procedure preventing RNA degradation [100]. In this procedure, the deep-frozen tissue (−80 °C or liquid nitrogen-stored) was put in a precooled (liquid nitrogen) teflon shaking vial containing a tungsten carbide ball. The micro-dismembrator was operated for 1 min at the maximal amplitude. Under this procedure the tissue remained frozen and the remaining pulverized tissue was removed with a prechilled spatula into a precooled weighted tube, weighted with tissue, while the vial is rinsed with ice-cold homogenization buffer. From both suspensions, a 600-g supernatant was prepared (5 min, 4 °C), this supernatant was centrifuged at $10,000 \times g$ for 10 min. The pellets containing RNA and DNA were subsequently extracted with a standard procedure, such as Trizol-chloroform protocol.

The disadvantage of this procedure is the relatively large amount of tissue that is required. For smaller pieces of tissue an alternative procedure can be used. Using a microtome, thin slices can be prepared, which after suspension in extraction buffer can be further processed. To get a clean suspension a short sonication step may be necessary. Subsequently, RNA and DNA can be isolated as described above.

2.4.5 Reverse Transcription and Quantitative Real-Time PCR

To validate differentially expressed miRNAs, a commonly used method is the reverse transcription and quantitative real-time PCR (qRT-PCR). The described method is for miRNAs from serum but can be used for other samples as well.

1. Create cDNA from the RNA samples according to manufacturer's protocol (TaqMan MicroRNA Assay) (*see* **Note 11**).

2. For the reverse transcriptase reactions, combine 5 μl of RNA sample, 3 μl of RT primer, 1.5 μl RT buffer, 0.15 μl of dNTPs, 1.0 μl of MultiScribe Reverse Transcriptase, and 0.19 μl of RNase inhibitor (all provided by Kit) to a microcentrifugation tube.

3. Incubate mixture for 30 min at 16 °C.

4. Subsequently, incubate for 30 min at 42 °C.

5. Incubate for 5 min at 85 °C.

6. Cool mixture on ice at 4 °C.

7. For the qRT-PCR 14 μl of total mix is needed. For this you add 3 μl of reverse-transcriptase product (result of **item 6**) with 10 μl of TaqMan Universal PCR Master Mix and 1 μl of primer and probe Mix (from the kit) (*see* **Note 11**).

8. All the samples should be loaded on a 96-well PCR plate.

9. Perform real-time PCR amplification followed by melt curve analysis.

3   Data Analysis

All real-time PCR reactions are run in duplicates or triplicates and average threshold cycles and SD values are calculated. The average expression levels of all analyzed miRNAs have to be normalized. Several methods for normalization of serum and plasma miRNA expression data exist (normalization by plate mean, normalization using commonly expressed miRNA targets and normalization using panel of invariant miRNA). To date, there is no consensus on the type of normalization that should be used. We use the normalization by internal reference genes. Subsequently the $2-\Delta CT$ method is applied and finally, normalized expression data are statistically analyzed.

4   Notes

1. The protocol for serum can be scaled up to 300 μl of starting sample without additional steps. Only the amounts of solution need to be adjusted. If more than 300 μl of starting material is used, adjust the solutions accordingly and load the sample in

separate steps. Remember that when using biofluids not to be use heparin in collecting the fluids, since this will hamper the PCR reaction. Furthermore, ensure the correct collection of samples. Hemolysis of samples will affect the miRNA profile.

2. If more than 300 μl of sample is used, load the sample in different steps onto the column to create optimal extraction.

3. The amount of eluate can be adjusted to the concentration needed. This should be evaluated for different biofluids. Furthermore, to isolate exosomes from serum/plasma it is recommended to use 100 μl of eluate.

4. When using tissue, you need to know the amount of cancer cells that you can harvest from the tissues. Some tumors create a desmoplastic reaction. This results in a lower amount of cancer cells in tissue. This should be taken into account since it can change the miRNA profile. Laser microdissection for FFPE samples can increase the harvested tumor cells but it is time consuming. It is advisable to analyze the surrounding tissue as well, since this can affect the tumor.

5. Be careful since xylene is a harmful substance.

6. If no neat pellet is formed, you can repeat this step.

7. Remove as much xylene as possible, but try not to remove the tissue. It is better to leave a small amount of xylene on the sample that you will wash off with the subsequent washing steps, than to lose tissue.

8. Be careful not to leave the sample heated for too long. If the sample is kept more than 17 min on 80 °C, this can result in RNA degradation.

9. When you combine these solutions, if you added 100 μl Digestion buffer add: 120 μl Isolation Additive with 275 μl 100 % ethanol. If you added 200 μl Digestion buffer add: 240 μl Isolation Additive with 550 μl 100 % ethanol. If you work with multiple samples, it is useful to create a master mix for all the samples with 5 % extra.

10. If you have more than 700 μl of sample, add the sample in multiple steps.

11. For each type of biofluid, a correct amount of sample volume needs to be tested to evaluate how much is needed and to make sure there is no inhibition from lack of enough cDNA.

Acknowledgments

We would like to thank the Bennink Foundation, the CCA foundation and AIRC.

References

1. Pennisi E (2012) Genomics ENCODE project writes eulogy for junk DNA. Science 337(1159):1161

2. Consortium E. P (2012) An integrated encyclopedia of DNA elements in the human genome. Nature 489:57–74

3. Khan S, Ansarullah, Kumar D, Jaggi M, Chauhan SC (2013) Targeting microRNAs in pancreatic cancer: microplayers in the big game. Cancer Res 73:6541–6547

4. Ling H, Fabbri M, Calin GA (2013) MicroRNAs and other non-coding RNAs as targets for anticancer drug development. Nat Rev Drug Discov 12: 847–865

5. Tang YT, Xu XH, Yang XD, Hao J, Cao H, Zhu W, Zhang SY, Cao JP (2014) Role of non-coding RNAs in pancreatic cancer: the bane of the microworld. World J Gastroenterol 20:9405–9417

6. Martin R, Hackert P, Ruprecht M, Simm S, Bruning L, Mirus O, Sloan KE, Kudla G, Schleiff E, Bohnsack MT (2014) A pre-ribosomal RNA interaction network involving snoRNAs and the Rok1 helicase. RNA 20:1173–1182

7. Claycomb JM (2014) Ancient endo-siRNA pathways reveal new tricks. Curr Biol 24:R703–R715

8. Taft RJ, Glazov EA, Lassmann T, Hayashizaki Y, Carninci P, Mattick JS (2009) Small RNAs derived from snoRNAs. RNA 15: 1233–1240

9. Taft RJ, Glazov EA, Cloonan N, Simons C, Stephen S, Faulkner GJ, Lassmann T, Forrest AR, Grimmond SM, Schroder K, Irvine K, Arakawa T, Nakamura M, Kubosaki A, Hayashida K, Kawazu C, Murata M, Nishiyori H, Fukuda S, Kawai J, Daub CO, Hume DA, Suzuki H, Orlando V, Carninci P, Hayashizaki Y, Mattick JS (2009) Tiny RNAs associated with transcription start sites in animals. Nat Genet 41:572–578

10. Langenberger D, Bermudez-Santana C, Hertel J, Hoffmann S, Khaitovich P, Stadler PF (2009) Evidence for human microRNA-offset RNAs in small RNA sequencing data. Bioinformatics 25:2298–2301

11. Guttman M, Amit I, Garber M, French C, Lin MF, Feldser D, Huarte M, Zuk O, Carey BW, Cassady JP, Cabili MN, Jaenisch R, Mikkelsen TS, Jacks T, Hacohen N, Bernstein BE, Kellis M, Regev A, Rinn JL, Lander ES (2009) Chromatin signature reveals over a thousand highly conserved large non-coding RNAs in mammals. Nature 458:223–227

12. Lai F, Orom UA, Cesaroni M, Beringer M, Taatjes DJ, Blobel GA, Shiekhattar R (2013) Activating RNAs associate with Mediator to enhance chromatin architecture and transcription. Nature 494:497–501

13. Khorkova O, Myers AJ, Hsiao J, Wahlestedt C (2014) Natural antisense transcripts. Hum Mol Genet 23:R54–R63

14. Guo JU, Agarwal V, Guo H, Bartel DP (2014) Expanded identification and characterization of mammalian circular RNAs. Genome Biol 15:409

15. Lee RC, Feinbaum RL, Ambros V (1993) The C. elegans heterochronic gene lin-4 encodes small RNAs with antisense complementarity to lin-14. Cell 75:843–854

16. Calin G, Dumitru C, Shimizu M, Bichi R, Zupo S, Noch E, Aldler H, Rattan S, Keating M, Rai K (2002) Frequent deletions and down-regulation of micro- RNA genes miR15 and miR16 at 13q14 in chronic lymphocytic leukemia. Proc Natl Acad Sci U S A 99:15524–15529

17. Bhardwaj A, Singh S, Singh AP (2010) MicroRNA-based cancer therapeutics: big hope from small RNAs. Mol Cell Pharmacol 2:213–219

18. Wojcik SE, Rossi S, Shimizu M, Nicoloso MS, Cimmino A, Alder H, Herlea V, Rassenti LZ, Rai KR, Kipps TJ, Keating MJ, Croce CM, Calin GA (2010) Non-codingRNA sequence variations in human chronic lymphocytic leukemia and colorectal cancer. Carcinogenesis 31:208–215

19. Fabbri M (2008) MicroRNAs and cancer epigenetics. Curr Opin Investig Drugs 9:583–590

20. McCarroll JA, Naim S, Sharbeen G, Russia N, Lee J, Kavallaris M, Goldstein D, Phillips PA (2014) Role of pancreatic stellate cells in chemoresistance in pancreatic cancer. Front Physiol 5:141 doi 10.3389/fphys 2014. 00141

21. Arora S, Bhardwaj A, Singh S, Srivastava SK, McClellan S, Nirodi CS, Piazza GA, Grizzle WE, Owen LB, Singh AP (2013) An undesired effect of chemotherapy: gemcitabine promotes pancreatic cancer cell invasiveness through reactive oxygen species-dependent, nuclear factor kappaB- and hypoxia-inducible factor 1alpha-mediated up-regulation of CXCR4. J Biol Chem 288:21197–21207

22. Kaur S, Kumar S, Momi N, Sasson AR, Batra SK (2013) Mucins in pancreatic cancer and its microenvironment. Nat Rev Gastroenterol Hepatol 10:607–620

23. Garajova I, Le Large TY, Frampton AE, Rolfo C, Voortman J, Giovannetti E (2014) Molecular mechanisms underlying the role of MicroRNAs in the chemoresistance of pancreatic cancer. BioMed Res Int 2014:678401

24. Ryan DP, Hong TS, Bardeesy N (2014) Pancreatic adenocarcinoma. N Engl J Med 371:1039–1049

25. Conroy T, Desseigne F, Ychou M, Bouche O, Guimbaud R, Becouarn Y, Adenis A, Raoul JL, Gourgou-Bourgade S, de la Fouchardiere C, Bennouna J, Bachet JB, Khemissa-Akouz F, Pere-Verge D, Delbaldo C, Assenat E, Chauffert B, Michel P, Montoto-Grillot C, Ducreux M (2011) FOLFIRINOX versus gemcitabine for metastatic pancreatic cancer. N Engl J Med 364:1817–1825

26. Von Hoff DD, Ervin TJ, Arena FP, Chiorean EG, Infante JR, Moore MJ, Seay TE, Tjulandin S, Ma WW, Saleh MN, Harris M, Reni M, Dowden S, Laheru D, Bahary N, Ramanathan RK, Tabernero J, Hidalgo M, Goldstein D, Van Cutsem E, Wei X, Iglesias JL, Renschler MF (2013) Increased survival in pancreatic cancer with weekly nab-paclitaxel plus gemcitabine. N Eng J Med 369: 1691-703

27. Sultana A, Smith CT, Cunningham D, Starling N, Neoptolemos JP, Ghaneh P (2007) Meta-analyses of chemotherapy for locally advanced and metastatic pancreatic cancer. J Clin Oncol 25:2607–2615

28. Matthaios D, Zarogoulidis P, Balgouranidou I, Chatzaki E, Kakolyris S (2011) Molecular pathogenesis of pancreatic cancer and clinical perspectives. Oncology 81:259–272

29. Van Moorsel CJ, Smid K, Voorn DA, Bergman AM, Pinedo HM, Peters GJ (2003) Effect of gemcitabine and cis-platinum combinations on ribonucleotide and deoxyribonucleotide pools in ovarian cancer cell lines. Int J Oncol 22:201–207

30. Peters GJ, Van Moorsel CJ, Lakerveld B, Smid K, Noordhuis P, Comijn EC, Weaver D, Willey JC, Voorn D, Van der Vijgh WJ, Pinedo HM (2006) Effects of gemcitabine on cis-platinum-DNA adduct formation and repair in a panel of gemcitabine and cisplatin-sensitive or -resistant human ovarian cancer cell lines. Int J Oncol 28:237–244

31. Giovannetti E, Funel N, Peters GJ, Del Chiaro M, Erozenci LA, Vasile E, Leon LG, Pollina LE, Groen A, Falcone A, Danesi R, Campani D, Verheul HM, Boggi U (2010) MicroRNA-21 in pancreatic cancer: correlation with clinical outcome and pharmacologic aspects underlying its role in the modulation of gemcitabine activity. Cancer Res 70: 4528–4538

32. Hwang JH, Voortman J, Giovannetti E, Steinberg SM, Leon LG, Kim YT, Funel N, Park JK, Kim MA, Kang GH, Kim SW, Del Chiaro M, Peters GJ, Giaccone G (2010) Identification of microRNA-21 as a biomarker for chemoresistance and clinical outcome following adjuvant therapy in resectable pancreatic cancer. PLoS One 5, e10630

33. Jamieson NB, Morran DC, Morton JP, Ali A, Dickson EJ, Carter CR, Sansom OJ, Evans TR, McKay CJ, Oien KA (2012) MicroRNA molecular profiles associated with diagnosis, clinicopathologic criteria, and overall survival in patients with resectable pancreatic ductal adenocarcinoma. Clin Cancer Res 18: 534–545

34. Caponi S, Funel N, Frampton AE, Mosca F, Santarpia L, Van der Velde AG, Jiao LR, De Lio N, Falcone A, Kazemier G, Meijer GA, Verheul HM, Vasile E, Peters GJ, Boggi U, Giovannetti E (2013) The good, the bad and the ugly: a tale of miR-101, miR-21 and miR-155 in pancreatic intraductal papillary mucinous neoplasms. Ann Oncol 24:734–741

35. Moriyama T, Ohuchida K, Mizumoto K, Yu J, Sato N, Nabae T, Takahata S, Toma H, Nagai E, Tanaka M (2009) MicroRNA-21 modulates biological functions of pancreatic cancer cells including their proliferation, invasion, and chemoresistance. Mol Cancer Ther 8:1067–1074

36. Park JK, Lee EJ, Esau C, Schmittgen TD (2009) Antisense inhibition of microRNA-21 or -221 arrests cell cycle, induces apoptosis, and sensitizes the effects of gemcitabine in pancreatic adenocarcinoma. Pancreas 38:e190–e199

37. Dong J, Zhao YP, Zhou L, Zhang TP, Chen G (2011) Bcl-2 upregulation induced by miR-21 via a direct interaction is associated with apoptosis and chemoresistance in MIA PaCa-2 pancreatic cancer cells. Arch Med Res 42:8–14

38. Frampton AE, Castellano L, Colombo T, Giovannetti E, Krell J, Jacob J, Pellegrino L, Roca-Alonso L, Funel N, Gall TM, De Giorgio A, Pinho FG, Fulci V, Britton DJ, Ahmad R, Habib NA, Coombes RC, Harding V, Knosel T, Stebbing J, Jiao LR (2014) MicroRNAs cooperatively inhibit a network of tumor suppressor genes to promote pancreatic tumor growth and progression. Gastroenterology 146(268–277), e218

39. Preis M, Gardner TB, Gordon SR, Pipas JM, Mackenzie TA, Klein EE, Longnecker DS, Gutmann EJ, Sempere LF, Korc M (2011) MicroRNA-10b expression correlates with response to neoadjuvant therapy and survival

in pancreatic ductal adenocarcinoma. Clin Cancer Res 17:5812–5821

40. Greither T, Grochola LF, Udelnow A, Lautenschlager C, Wurl P, Taubert H (2010) Elevated expression of microRNAs 155, 203, 210 and 222 in pancreatic tumors is associated with poorer survival. Int J Cancer 126:73–80

41. Xia QS, Ishigaki Y, Zhao X, Shimasaki T, Nakajima H, Nakagawa H, Takegami T, Chen ZH, Motoo Y (2011) Human SMG-1 is involved in gemcitabine-induced primary microRNA-155/BIC up-regulation in human pancreatic cancer PANC-1 cells. Pancreas 40:55–60

42. Li Y, VandenBoom TG, Kong D, Wang Z, Ali S, Philip PA, Sarkar FH (2009) Up-regulation of miR-200 and let-7 by Natural Agents Leads to the Reversal of Epithelial-to-Mesenchymal Transition in Gemcitabine-Resistant Pancreatic Cancer Cells. Cancer Res 69:6704–6712

43. Wellner U, Schubert J, Burk U, Schmalhofer O, Zhu F, Sonntag A, Waldvogel B, Vannier C, Darling D, Zur H (2009) The EMT-activator ZEB1 promotes tumorigenicity by repressing stemness-inhibiting microRNAs. Nat Cell Biol 11:1487–1495

44. Papaconstantinou IG, Manta A, Gazouli M, Lyberopoulou A, Lykoudis PM, Polymeneas G, Voros D (2013) Expression of microRNAs in patients with pancreatic cancer and its prognostic significance. Pancreas 42:67–71

45. Ohuchida K, Mizumoto K, Lin C, Yamaguchi H, Ohtsuka T, Sato N, Toma H, Nakamura M, Nagai E, Hashizume M, Tanaka M (2012) MicroRNA-10a is overexpressed in human pancreatic cancer and involved in its invasiveness partially via suppression of the HOXA1 gene. Ann Surg Oncol 19:2394–2402

46. Setoyama T, Zhang X, Natsugoe S, Calin GA (2011) microRNA-10b: a new marker or the marker of pancreatic ductal adenocarcinoma? Clin Cancer Res 17:5527–5529

47. Zhang XJ, Ye H, Zeng CW, He B, Zhang H, Chen YQ (2010) Dysregulation of miR-15a and miR-214 in human pancreatic cancer. J Hematol Oncol 3:46

48. Volinia S, Calin GA, Liu CG, Ambs S, Cimmino A, Petrocca F, Visone R, Iorio M, Roldo C, Ferracin M, Prueitt RL, Yanaihara N, Lanza G, Scarpa A, Vecchione A, Negrini M, Harris CC, Croce CM (2006) A microRNA expression signature of human solid tumors defines cancer gene targets. Proc Natl Acad Sci U S A 103:2257–2261

49. Iwagami Y, Eguchi H, Nagano H, Akita H, Hama N, Wada H, Kawamoto K, Kobayashi

S, Tomokuni A, Tomimaru Y, Mori M, Doki Y (2013) miR-320c regulates gemcitabine-resistance in pancreatic cancer via SMARCC1. Br J Cancer 109:502–511

50. Hasegawa S, Eguchi H, Nagano H, Konno M, Tomimaru Y, Wada H, Hama N, Kawamoto K, Kobayashi S, Nishida N, Koseki J, Nishimura T, Gotoh N, Ohno S, Yabuta N, Nojima H, Mori M, Doki Y, Ishii H (2014) MicroRNA-1246 expression associated with CCNG2-mediated chemoresistance and stemness in pancreatic cancer. Br J Cancer 111:1572–1580

51. Funamizu N, Lacy CR, Parpart ST, Takai A, Hiyoshi Y, Yanaga K (2014) MicroRNA-301b promotes cell invasiveness through targeting TP63 in pancreatic carcinoma cells. Int J Oncol 44:725–734

52. Hamada S, Masamune A, Miura S, Satoh K, Shimosegawa T (2014) MiR-365 induces gemcitabine resistance in pancreatic cancer cells by targeting the adaptor protein SHC1 and pro-apoptotic regulator BAX. Cell Signal 26:179–185

53. Takiuchi D, Eguchi H, Nagano H, Iwagami Y, Tomimaru Y, Wada H, Kawamoto K, Kobayashi S, Marubashi S, Tanemura M, Mori M, Doki Y (2013) Involvement of microRNA-181b in the gemcitabine resistance of pancreatic cancer cells. Pancreatology 13:517–523

54. Hodzic J, Giovannetti E, Diosdado B, Adema AD, Peters GJ (2011) Regulation of deoxycytidine kinase expression and sensitivity to gemcitabine by micro-RNA 330 and promoter methylation in cancer cells. Nucleosides Nucleotides Nucleic Acids 30:1214–1222

55. Bera A, VenkataSubbaRao K, Manoharan MS, Hill P, Freeman JW (2014) A miRNA signature of chemoresistant mesenchymal phenotype identifies novel molecular targets associated with advanced pancreatic cancer. PLoS One 9, e106343

56. Ji Q, Hao X, Zhang M, Tang W, Yang M, Li L, Xiang D, Desano JT, Bommer GT, Fan D, Fearon ER, Lawrence TS, Xu L (2009) MicroRNA miR-34 inhibits human pancreatic cancer tumor-initiating cells. PLoS One 4, e6816

57. Ohuchida K, Mizumoto K, Kayashima T, Fujita H, Moriyama T, Ohtsuka T, Ueda J, Nagai E, Hashizume M, Tanaka M (2011) MicroRNA expression as a predictive marker for gemcitabine response after surgical resection of pancreatic cancer. Ann Surg Oncol 18:2381–2387

58. Yu J, Ohuchida K, Mizumoto K, Sato N, Kayashima T, Fujita H, Nakata K, Tanaka M

(2010) MicroRNA, hsa-miR-200c, is an independent prognostic factor in pancreatic cancer and its upregulation inhibits pancreatic cancer invasion but increases cell proliferation. Mol Cancer 9:169

59. Nagano H, Tomimaru Y, Eguchi H, Hama N, Wada H, Kawamoto K, Kobayashi S, Mori M, Doki Y (2013) MicroRNA-29a induces resistance to gemcitabine through the Wnt/beta-catenin signaling pathway in pancreatic cancer cells. Int J Oncol 43:1066–1072

60. Frese KK, Neesse A, Cook N, Bapiro TE, Lolkema MP, Jodrell DI, Tuveson DA (2012) nab-Paclitaxel potentiates gemcitabine activity by reducing cytidine deaminase levels in a mouse model of pancreatic cancer. Cancer Discov 2:260–269

61. Alvarez R, Musteanu M, Garcia-Garcia E, Lopez-Casas PP, Megias D, Guerra C, Munoz M, Quijano Y, Cubillo A, Rodriguez-Pascual J, Plaza C, de Vicente E, Prados S, Tabernero S, Barbacid M, Lopez-Rios F, Hidalgo M (2013) Stromal disrupting effects of nab-paclitaxel in pancreatic cancer. Br J Cancer 109:926–933

62. Neesse A, Michl P, Frese KK, Feig C, Cook N, Jacobetz MA, Lolkema MP, Buchholz M, Olive KP, Gress TM, Tuveson DA (2011) Stromal biology and therapy in pancreatic cancer. Gut 60:861–868

63. Cochrane DR, Spoelstra NS, Howe EN, Nordeen SK, Richer JK (2009) MicroRNA-200c mitigates invasiveness and restores sensitivity to microtubule-targeting chemotherapeutic agents. Mol Cancer Ther 8:1055–1066

64. Zhu X, Li Y, Xie C, Yin X, Liu Y, Cao Y, Fang Y, Lin X, Xu Y, Xu W, Shen H, Wen J (2014) miR-145 sensitizes ovarian cancer cells to paclitaxel by targeting Sp1 and Cdk6. Int J Cancer 135:1286–1296

65. Aggarwal BB, Kunnumakkara AB, Harikumar KB, Tharakan ST, Sung B, Anand P (2008) Potential of spice-derived phytochemicals for cancer prevention. Planta Med 74:1560–1569

66. Donahue TR, Nguyen AH, Moughan J, Li L, Tatishchev S, Toste P, Farrell JJ (2014) Stromal MicroRNA-21 levels predict response to 5-fluorouracil in patients with pancreatic cancer. J Surg Oncol 110(8):952–959

67. Zhou Y, Wan G, Spizzo R, Ivan C, Mathur R, Hu X, Ye X, Lu J, Fan F, Xia L, Calin GA, Ellis LM, Lu X (2014) miR-203 induces oxaliplatin resistance in colorectal cancer cells by negatively regulating ATM kinase. Mol Oncol 8:83–92

68. Kjersem JB, Ikdahl T, Lingjaerde OC, Guren T, Tveit KM, Kure EH (2014) Plasma microRNAs predicting clinical outcome in metastatic colorectal cancer patients receiving first-line oxaliplatin-based treatment. Mol Oncol 8:59–67

69. Schou JV, Rossi S, Jensen BV, Nielsen DL, Pfeiffer P, Hogdall E, Yilmaz M, Tejpar S, Delorenzi M, Kruhoffer M, Johansen JS (2014) MiR-345 in metastatic colorectal cancer: a non-invasive biomarker for clinical outcome in non-KRAS mutant patients treated with 3rd line cetuximab and irinotecan. PLoS One 9, e99886

70. Chakraborty C, George Priya Doss C, Bandyopadhyay S (2013) miRNAs in insulin resistance and diabetes-associated pancreatic cancer: the 'minute and miracle' molecule moving as a monitor in the 'genomic galaxy'. Curr Drug Targets 14:1110–1117

71. Wang YS, Wang YH, Xia HP, Zhou SW, Schmid-Bindert G, Zhou CC (2012) MicroRNA-214 regulates the acquired resistance to gefitinib via the PTEN/AKT pathway in EGFR-mutant cell lines. Asian Pac J Cancer Prev 13:255–260

72. Yan G, Yao R, Tang D, Qiu T, Shen Y, Jiao W, Ge N, Xuan Y, Wang Y (2014) Prognostic significance of microRNA expression in completely resected lung adenocarcinoma and the associated response to erlotinib. Med Oncol 31:203

73. Bryant JL, Britson J, Balko JM, Willian M, Timmons R, Frolov A, Black EP (2012) A microRNA gene expression signature predicts response to erlotinib in epithelial cancer cell lines and targets EMT. Br J Cancer 106:148–156

74. Chatterjee A, Chattopadhyay D, Chakrabarti G (2014) MiR-17-5p downregulation contributes to paclitaxel resistance of lung cancer cells through altering beclin1 expression. PLoS One 9, e95716

75. Yu J, Ohuchida K, Mizumoto K, Fujita H, Nakata K, Tanaka M (2010) MicroRNA miR-17-5p is overexpressed in pancreatic cancer, associated with a poor prognosis, and involved in cancer cell proliferation and invasion. Cancer Biol Ther 10:748–757

76. Boni V, Zarate R, Villa JC, Bandres E, Gomez MA, Maiello E, Garcia-Foncillas J, Aranda E (2011) Role of primary miRNA polymorphic variants in metastatic colon cancer patients treated with 5-fluorouracil and irinotecan. Pharmacogenomics J 11:429–436

77. Gutierrez ME, Kummar S, Giaccone G (2009) Next generation oncology drug development: opportunities and challenges. Nat Rev Clin Oncol 6:259–265

78. Philip PA, Benedetti J, Corless CL, Wong R, O'Reilly EM, Flynn PJ, Rowland KM, Atkins JN, Mirtsching BC, Rivkin SE, Khorana AA,

Goldman B, Fenoglio-Preiser CM, Abbruzzese JL, Blanke CD (2010) Phase III study comparing gemcitabine plus cetuximab versus gemcitabine in patients with advanced pancreatic adenocarcinoma: Southwest Oncology Group-directed intergroup trial S0205. J Clin Oncol 28:3605–3610

79. Fensterer H, Schade-Brittinger C, Muller HH, Tebbe S, Fass J, Lindig U, Settmacher U, Schmidt WE, Marten A, Ebert MP, Kornmann M, Hofheinz R, Endlicher E, Brendel C, Barth PJ, Bartsch DK, Michl P, Gress TM (2013) Multicenter phase II trial to investigate safety and efficacy of gemcitabine combined with cetuximab as adjuvant therapy in pancreatic cancer (ATIP). Ann Oncol 24:2576–2581

80. Arslan C, Yalcin S (2014) Current and future systemic treatment options in metastatic pancreatic cancer. J Gastrointest Oncol 5:280–295

81. Moore MJ, Goldstein D, Hamm J, Figer A, Hecht JR, Gallinger S, Au HJ, Murawa P, Walde D, Wolff RA, Campos D, Lim R, Ding K, Clark G, Voskoglou-Nomikos T, Ptasynski M, Parulekar W (2007) Erlotinib plus gemcitabine compared with gemcitabine alone in patients with advanced pancreatic cancer: a phase III trial of the National Cancer Institute of Canada Clinical Trials Group. J Clin Oncol 25:1960–1966

82. Jimeno A, Tan AC, Coffa J, Rajeshkumar NV, Kulesza P, Rubio-Viqueira B, Wheelhouse J, Diosdado B, Messersmith WA, Iacobuzio-Donahue C, Maitra A, Varella-Garcia M, Hirsch FR, Meijer GA, Hidalgo M (2008) Coordinated epidermal growth factor receptor pathway gene overexpression predicts epidermal growth factor receptor inhibitor sensitivity in pancreatic cancer. Cancer Res 68:2841–2849

83. Chou YT, Lin HH, Lien YC, Wang YH, Hong CF, Kao YR, Lin SC, Chang YC, Lin SY, Chen SJ, Chen HC, Yeh SD, Wu CW (2010) EGFR promotes lung tumorigenesis by activating miR-7 through a Ras/ERK/Myc pathway that targets the Ets2 transcriptional repressor ERF. Cancer Res 70:8822–8831

84. Webster RJ, Giles KM, Price KJ, Zhang PM, Mattick JS, Leedman PJ (2009) Regulation of epidermal growth factor receptor signaling in human cancer cells by microRNA-7. J Biol Chem 284:5731–5741

85. Bravo-Egana V, Rosero S, Molano RD, Pileggi A, Ricordi C, Dominguez-Bendala J, Pastori RL (2008) Quantitative differential expression analysis reveals miR-7 as major

islet microRNA. Biochem Biophys Res Commun 366:922–926

86. Izumchenko E, Chang X, Michailidi C, Kagohara L, Ravi R, Paz K, Brait M, Hoque M, Ling S, Bedi A, Sidransky D (2014) The TGFbeta-miR200-MIG6 pathway orchestrates the EMT-associated kinase switch that induces resistance to EGFR inhibitors. Cancer Res 74:3995–4005

87. Mendell JT, Olson EN (2012) MicroRNAs in stress signaling and human disease. Cell 148:1172–1187

88. Jung M, Schaefer A, Steiner I, Kempkensteffen C, Stephan C, Erbersdobler A, Jung K (2010) Robust microRNA stability in degraded RNA preparations from human tissue and cell samples. Clin Chem 56:998–1006

89. Verhoeven CJ, Farid WR, de Ruiter PE, Hansen BE, Roest HP, de Jonge J, Kwekkeboom J, Metselaar HJ, Tilanus HW, Kazemier G, van der Laan LJ (2013) MicroRNA profiles in graft preservation solution are predictive of ischemic-type biliary lesions after liver transplantation. J Hepatol 59:1231–1238

90. Yu DC, Li QG, Ding XW, Ding YT (2011) Circulating microRNAs: potential biomarkers for cancer. Int J Mol Sci 12:2055–2063

91. Verweij FJ, van Eijndhoven MA, Middeldorp J, Pegtel DM (2013) Analysis of viral microRNA exchange via exosomes in vitro and in vivo. Methods Mol Biol 1024:53–68

92. Li L, Masica D, Ishida M, Tomuleasa C, Umegaki S, Kalloo AN, Georgiades C, Singh VK, Khashab M, Amateau S, Li Z, Okolo P, Lennon AM, Saxena P, Geschwind JF, Schlachter T, Hong K, Pawlik TM, Canto M, Law J, Sharaiha R, Weiss CR, Thuluvath P, Goggins M, Shin EJ, Peng H, Kumbhari V, Hutfless S, Zhou L, Mezey E, Meltzer SJ, Karchin R, Selaru FM (2014) Human bile contains microRNA-laden extracellular vesicles that can be used for cholangiocarcinoma diagnosis. Hepatology 60:896–907

93. Turchinovich A, Weiz L, Langheinz A, Burwinkel B (2011) Characterization of extracellular circulating microRNA. Nucleic Acids Res 39:7223–7233

94. Mlcochova H, Hezova R, Stanik M, Slaby O (2014) Urine microRNAs as potential noninvasive biomarkers in urologic cancers. Urol Oncol 32(41):e41–e49

95. Cirera S, Busk PK (2014) Quantification of miRNAs by a simple and specific qPCR method. Methods Mol Biol 1182:73–81

96. Ach RA, Wang H, Curry B (2008) Measuring microRNAs: comparisons of microarray and

quantitative PCR measurements, and of different total RNA prep methods. BMC Biotechnol 8:69

97. Mestdagh P, Hartmann N, Baeriswyl L, Andreasen D, Bernard N, Chen C, Cheo D, D'Andrade P, DeMayo M, Dennis L, Derveaux S, Feng Y, Fulmer-Smentek S, Gerstmayer B, Gouffon J, Grimley C, Lader E, Lee KY, Luo S, Mouritzen P, Narayanan A, Patel S, Peiffer S, Ruberg S, Schroth G, Schuster D, Shaffer JM, Shelton EJ, Silveria S, Ulmanella U, Veeramachaneni V, Staedtler F, Peters T, Guettouche T, Wong L, Vandesompele J (2014) Evaluation of quantitative miRNA expression platforms in the microRNA quality control (miRQC) study. Nat Methods 11:809–815

98. Meyer SU, Kaiser S, Wagner C, Thirion C, Pfaffl MW (2012) Profound effect of profiling platform and normalization strategy on detection of differentially expressed microRNAs--a comparative study. PLoS One 7, e38946

99. Mestdagh P, Van Vlierberghe P, De Weer A, Muth D, Westermann F, Speleman F, Vandesompele J (2009) A novel and universal method for microRNA RT-qPCR data normalization. Genome Biol 10:R64

100. Noordhuis P, Holwerda U, Van der Wilt CL, Van Groeningen CJ, Smid K, Meijer S, Pinedo HM, Peters GJ (2004) 5-Fluorouracil incorporation into RNA and DNA in relation to thymidylate synthase inhibition of human colorectal cancers. Ann Oncol 15: 1025–1032

Chapter 11

Methods for Studying MicroRNA Expression and Their Targets in Formalin-Fixed, Paraffin-Embedded (FFPE) Breast Cancer Tissues

Bruno Costa Gomes, Bruno Santos, José Rueff, and António Sebastião Rodrigues

Abstract

Drug resistance remains a burden in cancer treatment. In the past few years molecular genetics brought a new hope with personalized therapy. This individual approach allows the identification of genetic profiles that will respond better to a given treatment and consequently get a better outcome. Recently, physicians received an extra aid with the approval of molecular tools based on gene expression signatures. With these tools, physicians have the capacity to identify the probability of disease recurrence in the first 5 years following diagnosis, a fact that is essential for a more effective adjuvant therapy administration. However, some patients still relapse and acquire drug resistance and aggressive tumors. For that reason, a comprehensive understanding of the molecular players in drug resistance is of extreme importance. MicroRNAs have been described as regulators of various cellular pathways and as predictive and prognostic factors. As broad regulators, microRNAs also interfere with drug metabolism and drug targets. Thus it is of paramount importance to understand which microRNAs are deregulated in breast cancer and try to relate this misexpression with resistance to therapeutics, poor outcomes, and survival. Here, we describe a possible approach to study microRNA expression and respective targets from formalin-fixed, paraffin-embedded (FFPE) breast cancer tissues. FFPE tissues are regularly archived for long periods in pathology departments, and microRNAs are well conserved in these tissues.

Key words Drug resistance, MicroRNAs, Breast cancer, FFPE, Immunohistochemistry, RT-qPCR

1 Introduction

Fresh-frozen tissues are still the best and preferred type of biological samples to study molecular biology. However, collection and storage of these samples are not always easy and sometimes the required amounts of nucleic acids and proteins necessary for certain molecular approaches are not easily obtained. In clinical practice tissues coming from surgery are preserved and fixed in formalin and embedded in paraffin. This not only assures tissue preservation but also allows tissues to be cut in thin sections to perform

José Rueff and António Sebastião Rodrigues (eds.), *Cancer Drug Resistance: Overviews and Methods*, Methods in Molecular Biology, vol. 1395, DOI 10.1007/978-1-4939-3347-1_11, © Springer Science+Business Media New York 2016

ordinary pathology protocols. These tissues are usually referenced as Formaldehyde Fixed-Paraffin Embedded (FFPE). Millions of tissue samples are archived every year in hospitals worldwide. This is not only the biggest available collection of human samples, but also the largest collection of clinical data which allows large retrospective studies, correlating molecular characteristics with therapeutic response and clinical outcome. Unfortunately this preservation also has some detrimental effects on nucleic acids and proteins [1]. Indeed, this procedure tends to form cross-links at primary amino groups and side chains of amino acids, promoting DNA and RNA degradation and shortening and antigen site blocking in proteins. Until recently, this raised some doubts on the usefulness of FFPE tissues. Currently, with new nucleic acid isolation and better antigen retrieval procedures the utility of these samples has added new perspectives to molecular biology [1–6].

Noncoding RNAs, such as microRNAs (miRs), have become important tools to characterize and classify human cancers. Likewise, they have been reported as main regulators of several cellular pathways and diseases [7–11]. MicroRNAs are short RNAs capable of regulating gene expression by posttranscriptional regulation of mRNAs. Indeed, approximately 60 % of protein-coding genes have a conserved binding site for miRs and numerous non-conserved sites. Thus, almost all protein-coding genes may be regulated by miRs. The biogenesis and tight regulation of miRNAs have been comprehensively review by Ha and Kim [12]. Due to their size, miRs are well conserved in FFPE tissues and for that reason they may be better molecular tools than mRNA in biomarker discovery when using FFPE tissues [13, 14]. Indeed, several studies demonstrated the robustness of profiling miRs using FFPE tissues [15–19].

Currently, breast cancer (BC) therapy is selected according to the hormone receptor status, estrogen and progesterone receptor expression, ER and PR, respectively, and amplification of human epidermal growth factor receptor 2 (HER2) (ERBB2) proto-oncogenic receptor. HER2 has an N-terminal extracellular ligand-binding site domain and a C-terminal intracellular domain having a tyrosine kinase (TK). Usually, ER-positive BC is treated with antiestrogen drugs, such as aromatase inhibitors, tamoxifen, or fulvestrant. These comprise about 70–75 % of all breast tumors [20]. HER-positive tumors represent about 20 % of all BC and frequently half of these are ER positive. In these cases the therapy applied is trastuzumab, pertuzumab, and lapatinib. Five to ten percent of all tumors are called "triple negative" and lack the expression of ER, PR, and HER2. These tumors are managed only with genotoxic chemotherapy [21]. Recent molecular tools based on gene signatures have been helping the drug selection by clinicians and guided them into the identification of the likelihood of disease recurrence in the first 5 years following diagnosis [22–25].

However, patients still relapse and acquire drug resistance. For that reason, other markers have been studied in order to understand the biology of drug resistance mechanisms and possibly find new targets for drug development.

Several authors [26–29] have been reporting comprehensive data about correlation of miR expression in the NCI-60 cancer cell panel and drug sensitivity profiles. Indeed, a correlation between the expression pattern of miRs and the growth-inhibitory pattern of several drugs has been shown, which allows the identification of new targets and consequently improved therapies. Although these reports showed association between miRs and drug sensitivity, they do not necessarily identify experimentally miR targets. In this respect, numerous reports have been published giving account of miRs and their respective targets in several mechanisms of drug resistance. Particularly for breast cancer, two review articles [30, 31] summarize these studies describing the influence of miRs in all mechanisms of drug resistance.

The most used methodologies to distinguish variations in miR expression are quantitative PCR, hybridization, and small RNA sequencing. All techniques have their strengths and weaknesses and before starting to study miR expression, investigators must have some considerations in mind and be sure what approach is better to reach their goals. A recent report assessed the differences between these techniques and compared different platforms from several vendors [32]. One of the difficulties in studying miRs is the sample amount. Often the samples are too small and the yields are low, so total RNA (including miRs) from FFPE tissues is valuable and cannot be wasted. There are techniques that require a miR-specific cDNA synthesis prior to quantitative PCR [33]. These methodologies are more sample consuming since a cDNA synthesis must be performed for each miR that one wants to study. Conversely, a cDNA synthesis that allows the investigator to study several miRs with only one reaction is less sample consuming [34]. Another problem regarding miR expression quantification studies is the endogenous control selection to normalize the data. In microarray data analysis a global mean normalization is widely used, where all miRs are normalized according to the global mean threshold cycle of all miRs. However this methodology is not recommended in small studies and individual miR expression quantification. In these cases a previous study of several miRs in the samples of interest is suggested in order to identify miRs that are more stable across all samples. In some cases more than one miR can be used at the same time as an endogenous normalizer. The normalization of the data can also be done with general reference genes like U6 snRNA, probably the most used in miR quantification studies. Nevertheless, this gene can vary according to tissue and type of sample (not recommended to use in serum/plasma samples), so precautions must be taken before its use [35, 36].

Along with miR expression one must study their targets. Protein-coding gene transcripts are not suitable for this type of study since the levels of mRNAs may not diminish with regulation by miRs. Hence, protein levels are appropriate in this approach. There are several ways to study protein expression in human samples. Frequently, protein extracts are used as source material for ELISA, Western blot, or mass spectrometry, to name a few techniques. On the other hand, in these techniques the original in vivo morphology is not preserved; in order to do so immunohistochemistry must be performed. This technique allows the visualization of the proteins in situ so that one can understand which cells do in fact express the protein being studied. The expression may be harder to quantify when compared to a Western blot, but the amount of sample required is rather small. For further readings on immunohistochemistry technique see Leong et al. [37].

Here, we describe a methodology to analyze miRs in FFPE of breast cancer tissues and immunohistochemistry to analyze putative targets of the same miRs in the same samples. These methodologies can be applied in every type of tissue with minor modifications. Since in retrospective studies vast clinical information exists of each patient, the profile of the miRs and targets studied can then be compared with the clinicopathological features including drug resistance.

2 Material

2.1 Total RNA Isolation

1. RecoverAll Total Nucleic Acid Isolation Kit for FFPE (Ambion).
2. FFPE tissue from patients.
3. 100 % Xylene.
4. 100 % Ethanol.
5. Microtome.
6. Benchtop centrifuge.
7. Pipettes with filter tips.
8. RNase-free microcentrifuge tubes 1.5 mL.
9. Pestle.
10. Tweezers.
11. Incubators.

2.2 Reverse Transcription and Quantitative Real-Time PCR

1. Pipettes and filter tips.
2. 96-Well plates.
3. Spectrophotometer.
4. Thermal cycler.

5. Real-time PCR instrument.

6. RNase-free microcentrifuge tubes 0.1 mL.

7. Universal cDNA Synthesis Kit II.

8. ExiLENT SYBR® Green master mix.

9. miRCURY LNA™ Universal RT microRNA PCR, microRNA primer set, microRNA reference gene primer set.

10. ROX (passive reference dye).

2.3 Immunohisto-chemistry

1. FFPE tissue from patients.

2. Bi-distilled water.

3. 100, 96, and 70 % Ethanol.

4. 100 % Xylene.

5. Hydrogen peroxide.

6. Tris-EDTA buffer (10 mM Tris base, 1 mM EDTA solution with 0.05 % Tween 20, pH 9.0).

7. Phosphate-buffered saline (PBS; 137 mM NaCl, 2.7 mM KCl; 10 mM Na_2HPO_4, 1.8 mM KH_2PO_4, with 0.05 % Tween 20, pH 7.4).

8. Primary antibody.

9. Secondary antibody against the primary antibody species.

10. 3,3′-Diaminobenzidine (DAB).

11. Mayer's hematoxylin.

12. Mounting medium.

13. Cover slips.

14. Laboratory oven set to 37 °C.

15. Microwave oven.

16. Superfrost® Plus slides.

17. Coplin staining dish.

18. Staining rack.

19. Glass staining dishes.

20. Staining tray (humidity chamber).

21. Microwave-proof container.

22. Plastic wrap.

23. Pipettes and tips.

24. Tweezers.

25. Hydrophobic barrier pen.

26. Optical microscope coupled with digital camera (Olympus, Nikon, Leica; *see* **Note 1**).

3 Methods

3.1 Total RNA Isolation

RNA is extremely sensitive; thus, there should be some precautions in material handling in order to minimize contamination and degradation. The operator must wear gloves during all the procedures and change them frequently, wear a clean coat and if possible dedicated to RNA purification, RNA-dedicated pipettes and filter tips must be used, and the bench must always be clean with a decontamination solution. Total RNA isolation must include RNAs below 200 nucleotides. Most commercial reagent kits do not allow total RNA isolation which justifies an additional care. The methodology here described is based on the RecoverAll Total Nucleic Acid Isolation Kit for FFPE (Ambion).

3.1.1 Deparaffinization

1. Place approximately 80 μm of tissue slices in a microcentrifuge tube. (Slices must have 5–20 μm thickness (*see* **Note 2**)).

2. Add 1 mL of 100 % xylene (*see* **Note 3**) to the sample and vortex briefly to mix. If tissue sticks to the side of the tube centrifuge briefly. Heat the sample for 3 min at 50 °C.

3. Centrifuge the sample for 3 min at maximum speed.

4. Discard the xylene without disturbing the pellet.

5. Add 1 mL of 100 % ethanol (*see* **Note 4**) and vortex to mix (*see* **Note 5**).

6. Centrifuge the sample for 3 min at maximum speed.

7. Discard the ethanol without disturbing the pellet.

8. Repeat steps 6 to 7.

9. Remove any remaining drops of ethanol and let to air-dry for 15–45 min at room temperature.

3.1.2 Protein Digestion

1. Add 200 μL (*see* **Note 6**) of digestion buffer from the kit and 4 μL of protease from the kit in each sample. Mix it by swirling the tube and be sure that all the tissue is immerse; if not, briefly centrifuge the sample.

2. With a pestle homogenize and macerate the tissue (*see* **Note 7**).

3. Incubate the samples during 15 min at 50 °C, and then 15 min at 80 °C (*see* **Note 8**).

3.1.3 Nucleic Acid Isolation

1. Add an additive (from the kit)/ethanol mixture (240 μL/550 μL) (*see* **Note 9**) per sample and mix by pipetting or inversion.

2. For each sample mount a filter cartridge and a collection tube (provided with the kit).

3. Pipette up to 700 μL of the mixture of step 1 (*see* **Note 10**) and centrifuge at $10,000 \times g$ for 30 s. Discard the flow-through, and reinsert the filter cartridge in the same tube.

4. Repeat **step** 3 until all the mixture has passed through the filter.

5. Add 700 μL of Wash 1 (from the kit) to the filter cartridge and centrifuge at $10,000 \times g$ for 30 s. Discard the flow-through.

6. Add 500 μL of Wash 2/3 (from the kit) and centrifuge at $10,000 \times g$ for 30 s. Discard the flow-through.

7. Centrifuge for an additional 30 s to remove residual fluid.

8. Add 60 μL of the DNase mix (*see* **Note 11**) to the center of each filter cartridge.

9. Incubate for 30 min at room temperature.

10. Add 700 μL of Wash 1.

11. Incubate for 30–60 s at room temperature.

12. Centrifuge at $10,000 \times g$ for 30 s. Discard the flow-through.

13. Wash twice with 500 μL of Wash 2/3 (from the kit). Centrifuge at $10,000 \times g$ for 30 s. Discard the flow-through.

14. Centrifuge for an additional 60 s at $10,000 \times g$ to remove residual fluid from the filter.

15. Transfer the filter cartridge to a new collection tube.

16. Apply 60 μL of elution solution or nuclease-free water.

17. Incubate the sample during 60 s at room temperature and centrifuge at full speed for 60 s.

18. Store the nucleic acid at −20 °C or colder.

3.2 Reverse Transcription and Quantitative Real-Time PCR

Prior to RT-qPCR, total RNA samples must be quantified and purity assessed with a spectrophotometer. RNA must have a ratio 260/280 around 2. Once more, all precautions should be undertaken to avoid RNA contamination and degradation. Samples must thaw on ice. There are several commercial kits available. The procedure here described is based on miRCURY LNA Universal RT microRNA PCR.

3.2.1 cDNA Synthesis

1. Dilute RNA samples to a concentration of 5 ng/μL using nuclease-free water.

2. Thaw on ice the 5× reaction buffer, nuclease-free water, and RNA spike-in.

3. Spin down all reagents.

4. For a 10 μL total reaction volume prepare the following mixture:

    (a) 2 μL 5× Reaction buffer.

    (b) 4.5 μL Nuclease-free water.

    (c) 0.5 μL RNA spike-in (*see* **Note 12**).

5. 1 μL Enzyme mix (*see* **Note 13**).

6. Mix the reaction by pipetting to ensure that all components are well mixed and spin down.

7. Dispense 8 μL of working solution into nuclease-free 0.1 mL tubes.

8. Dispense 2 μL template total RNA (5 ng/μL)

9. In a thermal cycler, incubate for 60 min at 42 °C, inactivate the reverse transcriptase for 5 min at 95 °C, and immediately cool to 4 °C.

10. Store at 4 °C or freeze until next use (*see* **Note 14**).

3.2.2   Real-Time qPCR
(See Note 15)

1. Thaw the cDNA, nuclease-free water, PCR master mix (*see* **Notes 16** and **17**), and ROX (*see* **Notes 18** and **19**).

2. Dilute cDNA template needed for the planned real-time PCR reactions 80× in nuclease-free water (*see* **Note 20**).

3. Prepare the required amount of primer/master mix working solution. For a 10.2 μL total reaction volume prepare the following mixture:

   (a) 5 μL PCR master mix.

   (b) 1 μL PCR primer mix (*see* **Note 21**).

   (c) 0.2 μL ROX.

4. In a 96-well plate add 6 μL of primer/master mix solution and then 4 μL diluted cDNA.

5. Mix the reaction with gentle pipetting up and down and following sealing spin down in a plate centrifuge at $1500 \times g$ for 60 s (*see* **Note 22**).

6. Perform real-time PCR amplification in a real-time instrument with the following conditions:

7. Polymerase activation/denaturation at 95 °C for 10 min.

8. 40 Amplification cycles (*see* **Note 23**) at 95 °C for 10 s and 60 °C for 1 min (*see* **Note 24**).

9. Set manually the threshold cycles (Ct).

3.3 Immunohisto-chemistry

1. Cut 2–3 μm paraffin sections (*see* **Note 25**) onto slides (*see* **Note 26**) and let them dry overnight at 37 °C. For adhesion, put the slides in a microwave oven for 2 min at full potency (*see* **Note 27**).

2. Deparaffinize the sections in xylene (two times 10 min each; *see* **Note 2**) and rehydrate them through decreasing concentrations of ethanol (absolute, 96 and 70 %).

3. Wash in bi-distilled water.

4. For antigen retrieval, put the slides in a microwave-proof container with 400/600 mL of Tris-EDTA buffer (pH 9.0) and take those to a microwave oven for 20 min at 80 % potency

Fig. 1 Antigen retrieval in a microwave oven

Fig. 2 Drop effect after the hydrophobic pen appliance around the tissue section (*see* **Note 32**)

(*see* **Notes 28–30**). Cover the container with perforated plastic wrap to avoid splashes inside the microwave (Fig. 1).

5. Let the solution cool at room temperature for 5 min and then wash the slides in running water for 15 min.

6. Block endogenous peroxidase activity with hydrogen peroxide at 3 % for 10 min, followed by wash in running water (*see* **Note 31**).

7. Dry the slide around the tissue section, apply hydrophobic pen (*see* **Note 32**), and wash the section with PBS with Tween 20 at 0.05 % (Fig. 2).

8. Incubate the tissue with blocking solution from Vector Laboratories detection kit (optional; *see* **Note 33**) for 20 min at room temperature.

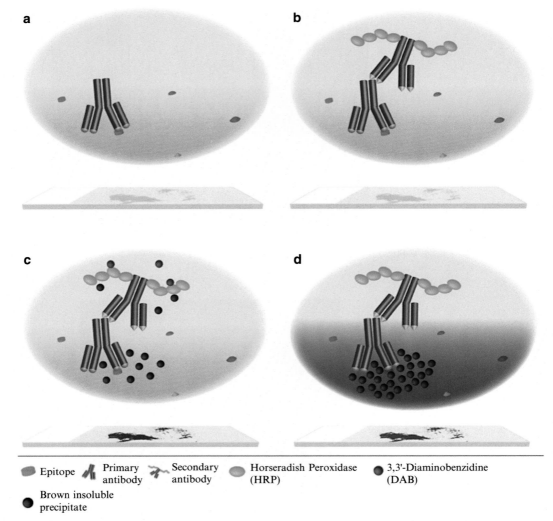

Fig. 3 Schematic representation of what happens during immunohistochemistry of the tumor section, macroscopically and at the molecular level. (**a**) Mouse primary antibody (*red*) specifically recognizes its epitope; (**b**) ImmPRESS™ anti-mouse Ig (*blue*) recognizes the primary antibody FC region; (**c**) DAB (*purple*) is converted by HRP enzymes (*orange*) of the detection system in a insoluble precipitate (*brown*). (**d**) Hematoxylin stains the nuclei giving contrast to a tissue previously translucent

9. Incubate with primary antibody diluted in PBS with Tween 20 at 0.05 %, for example to 1/500, for 30 min at room temperature (Fig. 3a) (*see* **Note 34**).

10. Wash in PBS with Tween 20 at 0.05 %: two times 5 min each.

11. Incubate with Vector Laboratories Ready-to-use ImmPRESS™ anti-mouse Ig secondary antibody at room temperature (Fig. 3b).

12. Wash in PBS with Tween 20 at 0.05 %: two times 5 min each.

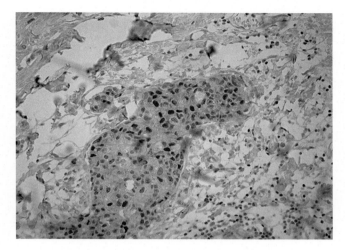

Fig. 4 Example of an SOX2-positive invasive breast carcinoma (20×)

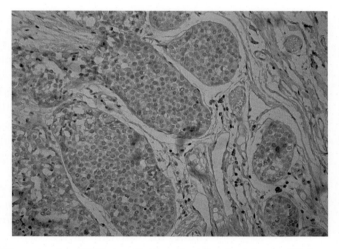

Fig. 5 Example of an SOX2-negative invasive breast carcinoma (20×)

13. Incubate with DAB for 5 min and then wash the slides in running water for 2 min (Fig. 3c) (*see* **Note 35**).

14. Counterstain in Mayer's hematoxylin for 2 min, followed by wash in running water for 5 min (Fig. 3d) (*see* **Note 36**).

15. Dehydrate through increasing concentrations of ethanol and xylene and mount with mounting medium (*see* **Note 37**).

16. Observe the stained tissue section under a microscope (Figs. 4, 5, and 6).

3.4 Data Analysis   All RT-qPCR reactions are run in duplicates and average C_t and standard deviation values are calculated. We normalize microRNA expression to endogenous expression of snRNA U6. Relative

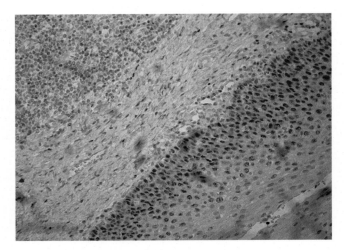

Fig. 6 Example of SOX2 positivity in the positive control used for this antibody: normal tonsil (20×)

expression is calculated according to $2^{-\Delta Ct}$ method. $\Delta C_t = C_t$ of microRNA—C_t of endogenous control. Fold change expression between two related samples can be calculated through $2^{-\Delta Ct}$ (sample 1)/$2^{-\Delta Ct}$ (sample2). Normalized relative expression or fold change expression is then used in statistical approaches to infer the proposed objectives. In this case, normalized relative expression of microRNAs in each tumor sample can be associated with treatment outcome, in order to assess if a specific microRNA can somehow influence treatment and survival.

Slide evaluation can be made in many ways depending on the kind of staining and ultimately on the desired purpose of the evaluation. There are several parameters which may be taken into account; the intensity of the staining and the amount of stained cells are the most preponderant. The intensity is usually assessed qualitatively as absent, weak, moderate, or strong, according to the color intensity of the precipitate obtained (e.g., in case revelation with DAB staining can go from light brown to dark brown). The amount of cells is usually the percentage of stained cells among all the cells that could or should be staining. This assessment can be made by counting cells (e.g., in five high-magnification fields) or by determining an approximate expression via categories (e.g., 1–25 %, 25–50 %). The staining intensity and the number of stained cells can also be combined into a single immunostaining score in order to ease data processing and analysis. In addition to these two parameters it is important to ensure that the tissue section morphology is preserved so that it does not interfere with the staining evaluation, and that the observed staining is in fact specific. In this sense it is imperative to know exactly which cell structures are supposed to stain and which ones should not. The use of positive controls (tissues with known expression of the studied protein) allows inferring about the proper functioning of the antibody and thus validating the technique even in cases in which the tissue under investigation is negative and therefore does not exhibit any kind of staining.

4   Notes

1. Several commercial imaging packages are available for image capture and analysis.

2. Depending on the tissue slices ≥ 10 μm should be used in order to prevent microRNA loss, since cells are more susceptible to split open.

3. Xylene is a toxic substance. Must be handled with precaution and discarding must be done accordingly.

4. Ethanol will remove traces of xylene and accelerates drying.

5. Tissue will turn opaque.

6. If using ≤ 40 μm add 100 μL of digestion buffer.

7. Work quickly in order to avoid RNA degradation. This step should be used in hard tissues and fatty tissues. In our experience, this additional step improves yield.

8. Do not exceed the incubation period at 80 °C; it may result in RNA degradation.

9. If using 100 μL of digestion buffer prepare the mix by half.

10. Avoid removing undigested tissue to prevent the clog of the filter cartridge.

11. The DNase mix has 6 μL of 10× DNase buffer, 4 μL of DNase, and 50 μL of nuclease-free water (supplied with the kit).

12. RNA spike-in is an internal control that can be used as inter-plate calibrator. If not used, replace the volume with nuclease-free water.

13. Remove the enzyme mix from the freezer immediately before use.

14. Using this kit undiluted cDNA may be kept at −20 °C for up to 5 weeks (optional store at 4 °C for up to 4 days).

15. Always work on ice.

16. Do not vortex. Mix by pipetting up and down.

17. PCR master mix must be protected from light during all the process.

18. Some real-time PCR instruments do not need the addition of ROX. See instrument instruction guide. Use it carefully and following the manufacturer's instructions. ROX is a PCR inhibitor when used in excess.

19. ROX must be protected from light during all the process.

20. Do not store 80× dilutions.

21. Prepare different mixes to microRNA of interest, endogenous control, and RNA spike-in if used.

22. At this time procedure can be paused and the reactions stored at 4 °C for up to 24 h.

23. If using the LC480 instrument from Roche use 45 amplification cycles.

24. Ramp rate 1.6 °C/s and optical read at this step.

25. The thickness of the sections is a key element for a successful outcome. Sections cut thicker than 4 μm tend to detach from the slide while sections thinner than 2 μm may give false-negative results due to insufficient amount of detectable protein.

26. The type of slides used is very important to assure a right adhesion. As far as we know, superfrost slides are the best performing. Additionally, there are many ways to promote tissue adhesion onto the slide. In our experience it works best to first let the section dry completely and then melt the paraffin so that the tissue binds to the slide. Doing this step in a microwave oven allows doing it faster.

27. In order to validate the results of samples that eventually did not express the studied protein, a slide containing a section of a tissue with known expression should be included, per run, to serve as control.

28. Antigen retrieval is meant to subvert the effect of formalin fixation in the tissues which promote the formation of methylene bridges that can mask antigens. Antigen retrieval can be achieved by high-temperature processes with the use of steam baths, pressure cookers, or microwaves, and it can be done through enzymatic digestion with trypsin or pepsin. These methods can be combined if needed. Some antigens may not need antigen retrieval which could in fact lead to loss of antigenicity; antigen retrieval should be selected in a case-by-case manner.

29. For heat-induced antigen retrieval there are several buffers that can be used. Choice of buffer used depends on its aggressiveness; solutions with higher pH allows a more intense recovery. The most common buffers used are citrate (pH 6) and Tris-EDTA (pH 9).

30. When performing heat-induced antigen retrieval in a microwave oven, 400–600 mL of volume of buffer should be used to ensure a correct distribution of heat through all the slides. For the same reason, the same amount of slides should always be used, even if

blank slides. As the tissue sections should never dry after the antigen retrieval step, we must ensure that there is enough buffer to cover the slides and because of that, optimize the potency of the microwave, and the time and number of cycles to be carried out, to ensure a desired antigen exposure.

31. Depending on the detection system used it could be necessary to block endogenous tissue constituents. In our case, we used an HRP-based detection system, so we need do block all endogenous peroxidase activity; otherwise every peroxidase would react with DAB and form an unspecific precipitate (unspecific because it does not associate with the antibody-binding site). A 3 % hydrogen peroxide dilution is commonly used, but instead, water/methanol could be used for a more efficient blocking if necessary.

32. Application of a hydrophobic pen is not a crucial step; this is only meant to reduce the volume of reagent applied to the tissue section.

33. Protein blocking is required in case of extensive background or unspecific staining, so it is optional. Several detection system kits include a protein block reagent but this could be made by preparing a 4 % serum and 1 % BSA in PBS (the serum should be the same species in which the secondary antibody was produced).

34. Binding of antibodies is influenced by its dilution, incubation time, temperature, and pH. Therefore, every single antibody should be assessed individually regarding which conditions fit the best.

35. DAB is a hazardous substance. Must be handled with precaution and discarding must be done accordingly.

36. Counterstain works as a second stain. It is often applied to provide contrast that helps the immunohistochemical staining to stand out. Mayer's hematoxylin is the most widely used and stains the nuclei blue.

37. Mounting is the last step and allows the tissue section to be preserved permanently so that anytime in the future the tissue can be observed under a microscope over and over again.

Acknowledgments

This work was supported by grant PEst-OE/SAU/UI0009/2014 from Fundação de Ciência e Tecnologia (FCT). B.C.G. was supported by SFRH/BD/64131/2009 from FCT.

References

1. Srinivasan M, Sedmak D, Jewell S (2002) Effect of fixatives and tissue processing on the content and integrity of nucleic acids. Am J Pathol 161:1961–1971

2. Maes E, Broeckx V, Mertens I, Sagaert X, Prenen H, Landuyt B, Schoofs L (2013) Analysis of the formalin-fixed paraffin-embedded tissue proteome: pitfalls, challenges, and future prospectives. Amino Acids 45:205–218

3. Gnanapragasam VJ (2010) Unlocking the molecular archive: the emerging use of formalin-fixed paraffin-embedded tissue for biomarker research in urological cancer. BJU Int 105:274–278

4. Frankel A (2012) Formalin fixation in the "-omics" era: a primer for the surgeon-scientist. ANZ J Surg 82:395–402

5. Shi SR, Shi Y, Taylor CR (2011) Antigen retrieval immunohistochemistry: review and future prospects in research and diagnosis over two decades. J Histochem Cytochem 59:13–32

6. Macabeo-Ong M, Ginzinger DG, Dekker N, McMillan A, Regezi JA, Wong DT, Jordan RC (2002) Effect of duration of fixation on quantitative reverse transcription polymerase chain reaction analyses. Mod Pathol 15:979–987

7. Di Leva G, Garofalo M, Croce CM (2014) MicroRNAs in cancer. Annu Rev Pathol 9:287–314

8. Bronze-da-Rocha E (2014) MicroRNAs expression profiles in cardiovascular diseases. BioMed Res Int 2014:985408

9. Van den Hove DL, Kompotis K, Lardenoije R, Kenis G, Mill J, Steinbusch HW, Lesch KP, Fitzsimons CP, De Strooper B, Rutten BP (2014) Epigenetically regulated microRNAs in Alzheimer's disease. Neurobiol Aging 35:731–745

10. Singh RP, Massachi I, Manickavel S, Singh S, Rao NP, Hasan S, Mc Curdy DK, Sharma S, Wong D, Hahn BH, Rehimi H (2013) The role of miRNA in inflammation and autoimmunity. Autoimmun Rev 12:1160–1165

11. Menghini R, Casagrande V, Federici M (2013) MicroRNAs in endothelial senescence and atherosclerosis. J Cardiovasc Transl Res 6:924–930 ·

12. Ha M, Kim VN (2014) Regulation of microRNA biogenesis. Nat Rev Mol Cell Biol 15:509–524

13. Liu A, Tetzlaff MT, Vanbelle P, Elder D, Feldman M, Tobias JW, Sepulveda AR, Xu X (2009) MicroRNA expression profiling outperforms mRNA expression profiling in formalin-fixed paraffin-embedded tissues. Int J Clin Exp Pathol 2:519–527

14. Hui A, How C, Ito E, Liu FF (2011) MicroRNAs as diagnostic or prognostic markers in human epithelial malignancies. BMC Cancer 11:500

15. Hui AB, Shi W, Boutros PC, Miller N, Pintilie M, Fyles T, McCready D, Wong D, Gerster K, Waldron L, Jurisica I, Penn LZ, Liu FF (2009) Robust global micro-RNA profiling with formalin-fixed paraffin-embedded breast cancer tissues. Lab Invest 89:597–606

16. Chen L, Li Y, Fu Y, Peng J, Mo MH, Stamatakos M, Teal CB, Brem RF, Stojadinovic A, Grinkemeyer M, McCaffrey TA, Man YG, Fu SW (2013) Role of deregulated microRNAs in breast cancer progression using FFPE tissue. PLoS One 8:e54213

17. Lee TS, Jeon HW, Kim YB, Kim YA, Kim MA, Kang SB (2013) Aberrant microRNA expression in endometrial carcinoma using formalin-fixed paraffin-embedded (FFPE) tissues. PLoS One 8:e81421

18. Meng W, McElroy JP, Volinia S, Palatini J, Warner S, Ayers LW, Palanichamy K, Chakravarti A, Lautenschlaeger T (2013) Comparison of microRNA deep sequencing of matched formalin-fixed paraffin-embedded and fresh frozen cancer tissues. PLoS One 8:e64393

19. Kolbert CP, Feddersen RM, Rakhshan F, Grill DE, Simon G, Middha S, Jang JS, Simon V, Schultz DA, Zschunke M, Lingle W, Carr JM, Thompson EA, Oberg AL, Eckloff BW, Wieben ED, Li P, Yang P, Jen J (2013) Multi-platform analysis of microRNA expression measurements in RNA from fresh frozen and FFPE tissues. PLoS One 8:e52517

20. Groenendijk FH, Bernards R (2014) Drug resistance to targeted therapies: Deja vu all over again. Mol Oncol 8:1067–1083

21. Rexer BN, Arteaga CL (2012) Intrinsic and acquired resistance to HER2-targeted therapies in HER2 gene-amplified breast cancer: mechanisms and clinical implications. Crit Rev Oncog 17:1–16

22. Whitworth P, Stork-Sloots L, de Snoo FA, Richards P, Rotkis M, Beatty J, Mislowsky A, Pellicane JV, Nguyen B, Lee L, Nash C, Gittleman M, Akbari S, Beitsch PD (2014) Chemosensitivity predicted by BluePrint 80-gene functional subtype and MammaPrint

in the Prospective Neoadjuvant Breast Registry Symphony Trial (NBRST). Ann Surg Oncol 21:3261–3267

23. Cusumano PG, Generali D, Ciruelos E, Manso L, Ghanem I, Lifrange E, Jerusalem G, Klaase J, de Snoo F, Stork-Sloots L, Dekker-Vroling L, Lutke Holzik M (2014) European inter-institutional impact study of MammaPrint. Breast 23:423–428

24. Viale G, Slaets L, Bogaerts J, Rutgers E, Van't Veer L, Piccart-Gebhart MJ, de Snoo FA, Stork-Sloots L, Russo L, Dell'Orto P, van den Akker J, Glas A, Cardoso F (2014) High concordance of protein (by IHC), gene (by FISH; HER2 only), and microarray readout (by TargetPrint) of ER, PgR, and HER2: results from the EORTC 10041/BIG 03-04 MINDACT trial. Ann Oncol 25:816–823

25. McVeigh TP, Hughes LM, Miller N, Sheehan M, Keane M, Sweeney KJ, Kerin MJ (2014) The impact of Oncotype DX testing on breast cancer management and chemotherapy pre-scribing patterns in a tertiary referral centre. Eur J Cancer 50:2763–2770

26. Blower PE, Verducci JS, Lin S, Zhou J, Chung JH, Dai Z, Liu CG, Reinhold W, Lorenzi PL, Kaldjian EP, Croce CM, Weinstein JN, Sadee W (2007) MicroRNA expression profiles for the NCI-60 cancer cell panel. Mol Cancer Ther 6:1483–1491

27. Sokilde R, Kaczkowski B, Podolska A, Cirera S, Gorodkin J, Moller S, Litman T (2011) Global microRNA analysis of the NCI-60 cancer cell panel. Mol Cancer Ther 10:375–384

28. Gaur A, Jewell DA, Liang Y, Ridzon D, Moore JH, Chen C, Ambros VR, Israel MA (2007) Characterization of microRNA expres-sion levels and their biological correlates in human cancer cell lines. Cancer Res 67: 2456–2468

29. Blower PE, Chung JH, Verducci JS, Lin S, Park JK, Dai Z, Liu CG, Schmittgen TD, Reinhold WC, Croce CM, Weinstein JN, Sadee W (2008) MicroRNAs modulate the chemosensitivity of tumor cells. Mol Cancer Ther 7:1–9

30. Kutanzi KR, Yurchenko OV, Beland FA, Checkhun VF, Pogribny IP (2011) MicroRNA-mediated drug resistance in breast cancer. Clin Epigenetics 2:171–185

31. Shah NR, Chen H (2014) MicroRNAs in pathogenesis of breast cancer: implications in diagnosis and treatment. World J Clin Oncol 5:48–60

32. Mestdagh P, Hartmann N, Baeriswyl L, Andreasen D, Bernard N, Chen C, Cheo D, D'Andrade P, DeMayo M, Dennis L, Derveaux S, Feng Y, Fulmer-Smentek S, Gerstmayer B, Gouffon J, Grimley C, Lader E, Lee KY, Luo S, Mouritzen P, Narayanan A, Patel S, Peiffer S, Ruberg S, Schroth G, Schuster D, Shaffer JM, Shelton EJ, Silveria S, Ulmanella U, Veeramachaneni V, Staedtler F, Peters T, Guettouche T, Wong L, Vandesompele J (2014) Evaluation of quantitative miRNA expression platforms in the microRNA quality control (miRQC) study. Nat Methods 11: 809–815

33. Chen C, Ridzon DA, Broomer AJ, Zhou Z, Lee DH, Nguyen JT, Barbisin M, Xu NL, Mahuvakar VR, Andersen MR, Lao KQ, Livak KJ, Guegler KJ (2005) Real-time quantifica-tion of microRNAs by stem-loop RT-PCR. Nucleic Acids Res 33:e179

34. Andreasen D, Fog JU, Biggs W, Salomon J, Dahslveen IK, Baker A, Mouritzen P (2010) Improved microRNA quantification in total RNA from clinical samples. Methods 50:S6–S9

35. Han HS, Jo YN, Lee JY, Choi SY, Jeong Y, Yun J, Lee OJ (2014) Identification of suitable ref-erence genes for the relative quantification of microRNAs in pleural effusion. Oncol Lett 8:1889–1895

36. McDermott AM, Kerin MJ, Miller N (2013) Identification and validation of miRNAs as endogenous controls for RQ-PCR in blood specimens for breast cancer studies. PLoS One 8:e83718

37. Leong TY, Cooper K, Leong AS (2010) Immunohistology – past, present, and future. Adv Anat Pathol 17:404–418

Chapter 12

The Regulatory Role of Long Noncoding RNAs in Cancer Drug Resistance

Marjan E. Askarian-Amiri, Euphemia Leung, Graeme Finlay, and Bruce C. Baguley

Abstract

Recent genomic and transcriptomic analysis has revealed that the majority of the human genome is transcribed as nonprotein-coding RNA. These transcripts, known as long noncoding RNA, have structures similar to those of mRNA. Many of these transcripts are now thought to have regulatory roles in different biological pathways which provide cells with an additional layer of regulatory complexity in gene expression and proteome function in response to stimuli. A wide variety of cellular functions may thus depend on the fine-tuning of interactions between noncoding RNAs and other key molecules in cell signaling networks. Deregulation of many noncoding RNAs is thought to occur in a variety of human diseases, including neoplasia and cancer drug resistance. Here we discuss recent findings on the molecular functions of long noncoding RNAs in cellular pathways mediating resistance to anticancer drugs.

Key words Long noncoding RNAs, Gene regulation, Drug resistance, Epigenetics, Gene expression

1 Introduction

Over the last decade, analysis of the genome and the transcriptome has revealed that the majority of the human genome is transcribed as noncoding RNAs (ncRNAs). NcRNAs are highly abundant; some of the best known ncRNAs, such as transfer RNA (tRNA), ribosomal RNA (rRNA), and spliceosomal RNA, mainly act as housekeeping molecules while others such as snoRNAs, microRNAs, siRNAs, snRNAs, exRNAs, piRNAs, and the long ncRNAs (lncRNAs) mainly act as regulatory RNA [1]. A number of ncRNAs have been identified as major regulatory molecules in normal development and disease progression (reviewed in [2]), providing another layer of complexity to gene regulation and implying that such transcripts might act as targets for drug therapy. NcRNAs are often divided into two major classes based on their size; small ncRNAs are transcripts shorter than 200 bases while long ncRNAs (lncRNAs) are longer than 200 bases. Although we do not know

José Rueff and António Sebastião Rodrigues (eds.), *Cancer Drug Resistance: Overviews and Methods*, Methods in Molecular Biology, vol. 1395, DOI 10.1007/978-1-4939-3347-1_12, © Springer Science+Business Media New York 2016

the exact function of all of these ncRNAs, many have emerged as crucial regulators of gene expression. For example, alternative splicing provides cells with a capacity for exquisite fine-tuning of their transcriptomes and proteomes, thus exerting a vital function in normal cell behavior. In the case of lncRNAs, the complexity and diversity of their sequences suggest functions that are distinct from those of small ncRNAs or structural RNAs. Coordinated interactions between ncRNAs and cellular signaling networks have been demonstrated in a wide variety of biological functions. Differential expression of many lncRNAs potentially contributes to the progression of many diseases including cancer.

Resistance to cancer therapy normally arises from deregulation of components of signaling pathways. Since ncRNAs are involved in the function of many, perhaps all, of these pathways, it is important to consider their role in the development of resistance. Many biological events are dependent on fine-tuning of ncRNA expression and function and recent developments in transcriptome analysis have made it possible to identify genes that are differentially expressed in drug resistance, either in cell lines or in tumor samples. These findings have the potential to improve the efficacy of therapeutic intervention.

The role of lncRNA in drug resistance forms the major focus of this chapter; we commence with a summary of common mechanisms of resistance, as outlined in Fig. 1. We then review some of the evidence available for the involvement of individual lncRNA species in these mechanisms. We have also compiled a list of the lncRNAs involved in cancer drug resistance (Table 1) and discuss their roles in the regulation of signaling pathways.

1.1 Mechanisms of Resistance to Cancer Drugs

The most common current approach to anticancer drug development involves the discovery of genetic and/or epigenetic defects that lead to aberrant signaling pathways, and the subsequent development of therapeutic agents that target these defects. Targeted therapy has now become one of the principal modes of treatment for cancer and resistance to such therapy is a major obstacle. Cellular signaling pathways are woven into a complex network of cross talk, feedback, and feed-forward communications, and biological dysfunction may arise from miscommunication along any of the routes of information flow. Current cancer therapy cannot effectively target such complex networks.

Resistance to chemotherapeutics can be either intrinsic or acquired. Intrinsic resistance makes therapy ineffective from the commencement of treatment, but in acquired cases drug resistance develops during drug treatment [3]. Variation in drug efficacy and toxicity can be affected by intrinsic gene polymorphisms that lead to differential gene expression patterns, splice variation, as well as functional defects. Acquired drug resistance is developed as a consequence of the genetic instability of tumor cells, as well as the

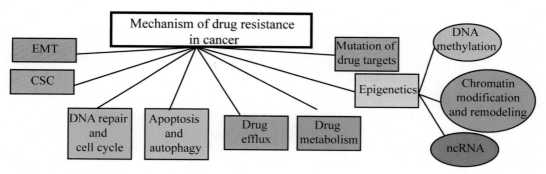

Fig. 1 Mechanisms of drug resistance. *Cancer Stem Cells* (CSC) are undifferentiated cells with the capacity for self-maintenance and possibly for production of differentiated cells characteristic of particular tissues [72]. CSCs have the potential for indefinite, tissue-maintaining proliferation and for causing relapse following response to therapy. Many pathways involved in cancer progression may regulate stem cell development [73]. Drug-resistant tumor cells share properties of CSCs including intrinsic or acquired resistance mechanisms that lead to recurrence of tumors of increased invasiveness. Stem cell-like properties have been observed in tumor cells that have become resistant to erlotinib or lapatinib, giving them the capacity to survive treatment. *Epithelial to mesenchymal transitions (EMT)*: As they self-renew, CSCs are known to generate cellular heterogeneity, a histological hallmark in different cancers. Cancer cells within the same tumor respond differently to therapeutic reagents and this heterogeneity leads to drug resistance. Adaptive changes such as EMT and its reverse process (MET) may be observed during cancer therapy and can lead to resistance. EMT can trigger cells to revert to a CSC-like phenotype and acquire drug resistance [74], suggesting the close inter-weaving of different mechanisms in drug resistance. *Drug efflux*: A major mechanism of drug resistance to cancer therapy is increased efflux of (often multiple) chemotherapeutic agents from cells, so reducing intracellular drug concentrations. Increased expression of ATP-binding cassette (ABC) transporter proteins is a well-known cause of multidrug resistance (MDR) (reviewed in [75]). *Secondary mutation in drug target protein*: One of the most common mechanisms of acquired resistance is genetic mutation of drug target sites. Mutants exist at low frequencies before drug treatment, and undergo positive selection during therapy (reviewed in [76]). Amplification of genes encoding drug target proteins is also observed as an additional somatic mutation that makes tumors resistant to therapy [77]. These observations suggest that the application of molecular targeted cancer therapy can lead to drug resistance because of emerging mutations altering the target site. *Different drug metabolism*: Cancer cells have altered drug metabolism relative to normal cells and this could lead to drug resistance. Two major classes of enzyme used in drug metabolism are cytochrome P450s and glutathione-S-transferases [78]. *DNA repair mechanisms and the cell cycle*: Cancer cells may repair damage more efficiently, making them more tolerant of chemotherapy and enabling them to antagonize the effects of DNA-damaging drugs [79]. DNA damage activates the checkpoint pathways that block CDKs and change cell cycle progression [80]. *Apoptosis and autophagy*: Aberrant regulation of these pathways following therapy can lead to drug resistance. Cancer cells can acquire apoptosis resistance during treatment [81]. Autophagy on the other hand facilitates the degradation of cellular components through lysosomes; it is used by cells as a survival mechanism in stress conditions such as hypoxia, nutrient deprivation, and chemo/radiotherapy [82] *Epigenetics*: Different epigenetic events such as DNA methylation, chromatin modification and ncRNAs regulate the expression of genes associated with drug resistance in cancer

effects of drug treatment. Cancer cells utilize many different mechanisms to resist drug therapy. The known mechanisms include changes in drug efflux and metabolism, the acquisition of mutations in target genes, changes in the cell cycle, in apoptosis, and in the rate of DNA repair, as well as the generation of more cancer stem cells and the induction of epithelial-mesenchymal transition (EMT).

Table 1
LncRNAs and drugs affecting their expression in cancers

LncRNA	Cancer	Drug	Reference
UCA1	Bladder	?	[7]
Linc-ROR	Hepatocellular carcinoma	Sorafenib and doxorubicin	[12]
XIST	Ovarian, breast	Cisplatin, abexinostat	[13–15]
MALAT-1	Pancreas	Gemcitabine	[18, 19]
URHC	Hepatocellular carcinoma	PD98059	[20]
HOTAIR	Lung	Cisplatin	[25]
PCGEM1	Prostate	Doxorubicin	[26, 27]
GAS5	Lung	Gefitinib	[32, 33]
AK126698	Lung	Cisplatin	[35]
ERIC	Bone osteosarcoma	Etoposide	[37]
PANDA	Breast	Doxorubicin	[38]
PDAM	Oligodendroglial	Cisplatin	[39]
HOTTIP	Pancreas	Gemcitabine	[42]
vRNA		Mitoxantrone	[46]
H19	Hepatocellular carcinoma cells	Paclitaxel, doxorubicin, etoposide, and vincristine	[51]
MRUL	Gastric	Doxorubicin and vincristine	[52]
ARA	Breast, hepatocellular carcinoma cells	Doxorubicin	[53]
PVT1	Pancreas, gastric cancer	Gemcitabine, paclitaxel	[61, 64]
BCAR4	Breast	Tamoxifen	[66–68]

Another key factor in drug resistance development is the acquisition of abnormal intracellular signaling pathways due to altered regulatory mechanisms. More recently, epigenetic elements have been implicated in drug resistance (reviewed in [4, 5]) and these are shown diagrammatically in Fig. 1.

Recent advances in genomic and transcriptomic analyses, in combination with bioinformatics and systems biology, have increased our capacity to identify novel genes, signaling pathways, molecular signatures, and genotypes that are involved in tumor responsiveness to certain chemotherapies. Such advances have also helped to identify novel targets for therapy. Different molecular mechanisms increase the rate of drug efflux, the rate and nature of drug metabolism, and also the rate of mutation of genes that encode drug target proteins. In most current therapies, patients respond to the initial cancer treatment but relapse occurs when cancer cells become resistant to the drug. Most studies investigating drug resistance have been built on the hypothesis that there are biochemical differences between the resistant and sensitive cells.

1.2 Role of lncRNA in Drug Resistance

Various mechanisms of transcriptional regulation of gene expression by lncRNAs have been proposed and although the differential expression of many lncRNAs in cancer and drug resistance has been reported, the mechanisms of action of the majority of them are unknown. Among the known functions of lncRNAs, regulation of transcription via chromatin-modifying complexes or epigenetic modifications of DNA are the best studied [2]. Recently the existence of a large proportion of lncRNAs in the cytoplasm, and their association with ribosomes, has been reported [6]. This finding suggests post-transcriptional regulatory roles for lncRNAs, which are yet to be studied. Many of the published studies have used high-throughput analyses to identify the lncRNAs that are differentially expressed in drug-resistant cancer cells, but detailed analyses of underlying mechanisms have not yet been carried out. However, despite a lack of information about the functions of lncRNAs in the development of resistance to cancer therapy, there is abundant evidence to evince their importance. Here we discuss the role of some of the specific lncRNAs involved in drug resistance in different cancers.

1.3 Involvement of Stem Cells

1.3.1 Urothelial Carcinoma-Associated 1 (UCA1; Also Known as CUDR)

The lncRNA *UCA1* is expressed during embryogenesis, repressed in adult tissues, and derepressed in cancer cells, suggesting a role in carcinogenesis. Ectopic expression of *UCA1* increased the invasiveness of cancer cells and also led to resistance to cytotoxic drugs such as cisplatin. Cisplatin is one of the most effective chemotherapeutic drugs against bladder cancer, though resistance to therapy is a major obstacle to treatment. Up-regulation of *UCA1* in cisplatin-treated tumors was observed and cisplatin-resistant tumors also had higher levels of *UCA1*. Inactivation of *UCA1* was associated with increased cisplatin sensitivity. It was proposed that *UCA1* promoted tumor progression via activation of target genes such as *MBD3* and genes encoding Wnt signaling pathway components, transcriptional regulators, and cell division regulators [7].

The Wnt signaling pathway is involved in regulating self-renewal in a variety of epithelial and hematopoietic stem cells, as well as in oncogenesis. Wnt proteins contribute to cancer progression when deregulated (reviewed in [8]). Activation of Wnt signaling also contributes to the development of chemoresistance in different human cancers including neuroblastoma, multiple myeloma, hepatocellular carcinoma, and bladder cancer. *UCA1* is positively up-regulated in the WNT6 pathway and leads to cisplatin resistance through WNT6 expression [9]. *UCA1* has also been reported to promote the Warburg effect [10], which is associated with a high rate of glycolysis and consumption of glucose regardless of oxygen availability. *UCA1* appears to promote glycolysis by activating mTOR to up-regulate hexokinase 2 (HK2) through activation of STAT3 [10].

1.3.2 Long Intergenic ncRNA Regulator of Reprogramming (linc-ROR)

The long intergenic ncRNA *ROR* regulates the expression of the mRNAs of the key core transcription factors OCT4, NANOG, and SOX2 by posttranscriptional mechanisms. The *linc-ROR* and the mRNAs encoding the above transcription factors compete directly for miRNA-145 binding. Therefore the presence of *linc-ROR* prevents the binding of miRNA-145 to the mRNAs and leads to the pluripotency factors being translated, ensuring the stem cell status of these cells [11].

Sorafenib and doxorubicin are used to treat hepatocellular carcinoma (HCC), although they have limited efficacy due to cellular toxicity and stress, and eventually cells become resistant to these agents. TGFβ changes the local microenvironment and induces cellular reprogramming, leading to chemoresistance. In HCC, the release of extracellular vesicles (EV), which contain protein, miRNA, and lncRNA components, allows cells to communicate with each other. Chemotherapeutic agents up-regulated *linc-ROR* in EV derived from HepG2 cells and also increased EV released by these cells. This observation suggested that a TGFβ-mediated pathway led to enrichment of *linc-ROR* within EV and to chemoresistance. *CASP8* (*gene for caspase 8*) and *GADD45B* genes, associated with apoptosis and DNA damage, are induced upon downregulation of *linc-ROR* in HepG2 cells, while cell cycle progression is not affected. In addition, siRNA to *linc-RoR* increased the activity of p53, suggesting that the effects of *linc-ROR* are mediated through p53-dependent signaling [12].

The stem cell-like phenotype is a factor that increases resistance to therapy, and the involvement of *linc-ROR* in regulating core pluripotent factors has been shown in stem cells. Takahashi et al. [12] considered the effect of this RNA on the self-renewal capacity of HCC and showed that TGFβ enhanced the growth of tumor-initiating cells and that *linc-ROR* induced expression of CD133, a surface marker in those cells.

1.3.3 X-Inactive-Specific Transcript (XIST)

Xist lncRNA is expressed from the X inactivation center (XIC) on the inactive X chromosome and acts as a major effector of the X inactivation process. *XIST* is among the best studied lncRNAs. Huang et al. [13] studied the sensitivity of cell lines derived from ovarian cancers to anticancer drugs including carboplatin and paclitaxel. Expression of *XIST* lncRNA showed a significant association with paclitaxel sensitivity, but not to carboplatin or cisplatin sensitivity. Loss of the inactive X chromosome is one mechanism for the loss of *XIST* expression in some cell lines [14] and to confirm the clinical significance of *XIST* expression, ovarian carcinoma tissue taken from the patients who were treated with paclitaxel and platinum was examined. Similar associations between expression of *XIST* and patient responses to chemotherapy were found and a significant correlation between *XIST* expression and paclitaxel-treated breast cancer cell lines was detected [13]. Low expression of *XIST* also showed significant correlation with cisplatin hyper-sensitivity in mammary tumors [14], suggesting the use of *XIST* expression as a biomarker for platinum-based chemotherapy in human breast cancer [14]. More recently it was shown that low expression of *XIST* predicts response to the histone deacetylase inhibitor abexinostat, which induces a significant reduction of the breast CSC population [15]. It was suggested that *XIST* could be used as a novel therapeutic strategy for breast cancer through the induction of CSC differentiation [15].

1.4 Involvement of EMT/CSC

1.4.1 Metastasis-Associated Lung Adenocarcinoma Transcript 1 (MALAT-1)

MALAT-1 is known as an oncogenic lncRNA in many different types of cancer. In pancreatic cancer, *MALAT-1* contributes to EMT and to the regulation of expression of CSC markers [16, 17]. Up-regulation of *MALAT-1* in pancreatic carcinoma has been reported and it was proposed that *MALAT-1* can suppress G2/M cell cycle arrest and apoptosis, and that it induces EMT [18]. The EMT process may give rise to CSCs, or to cells with stem cell-like properties. Downregulation of *MALAT-1* suppressed EMT and decreased the expression of cancer stem cell markers including CD44, CD24, and ALDH [18, 19]. High levels of *MALAT-1* promoted spheroid formation, anchorage-independent growth, angiogenesis, and tumorigenicity in pancreatic cancer cells. *MALAT-1* can play important roles in the expression of self-renewal proteins such as SOX2 [19]. These data suggest that up-regulation of *MALAT-1* induces CSC properties in pancreatic carcinoma.

Gemcitabine is commonly used as anticancer drug in pancreatic carcinoma. High levels of *MALAT-1* reduce chemo-sensitivity to this drug in pancreatic cancer. Although the exact role of *MALAT-1* in drug resistance is not clear, its role in EMT and CSC suggests a critical regulatory function during chemotherapy.

1.5 Cell Cycle and Apoptosis

1.5.1 Up-Regulated in Hepatocellular Carcinoma (URHC)

A novel lncRNA, termed *URHC*, was identified as one of the most frequently up-regulated lncRNAs in malignant hepatocytes. It was found that *URHC* levels were significantly increased in HCC tissues and cell lines, suggesting that increases in the expression levels play significant roles in HCC carcinogenesis, in tumor growth, and in the poor survival of HCC patients. Moreover, the downregulation of *URHC* expression inhibited hepatoma cell proliferation and induced apoptosis in vitro. ZAK, also known as ZAK-α or MLK-MAP triple kinase-α, functions as a tumor-suppressor gene, and the gene product inhibits human lung cancer growth by regulating cell proliferation. It was identified as a functional target of *URHC* and a negative correlation between *ZAK* and *URHC* expression in HCC was detected. It was suggested that *URHC* regulated cell proliferation and apoptosis via downregulating *ZAK*. The inactivation of the ERK/MAPK pathway functioned downstream of the *URHC* ZAK axis [20].

1.5.2 HOX Transcript Antisense RNA (HOTAIR)

HOTAIR is up-regulated in a variety of cancers and is a negative prognostic indicator in breast [21], colorectal [22], hepatocellular [23], gastrointestinal stromal tumors [24], and pancreatic cancer. In breast cancer, *HOTAIR* expression is up-regulated in both primary and metastatic tumors, and its expression in primary tumors is strongly correlated with later metastases, patient prognosis, and death [21]. The up-regulation of *HOTAIR* has been reported to promote cisplatin resistance in part by downregulating p21^{Cip1} in lung cancer cell lines [25].

1.5.3 Prostate-Specific Transcript (Nonprotein Coding) PCGEM1

The expression of *PCGEM1* lncRNA is prostate tissue-specific and prostate cancer-associated. In African–American prostate cancer patients, high expression of *PCGEM1* was significantly associated with high mortality rates [26]. In LNCaP prostate cancer cells, overexpression of *PCGEM1* resulted in increased cell proliferation rates, induction of pRB phosphorylation, and resistance to apoptosis induced by doxorubicin. Induction of p53 and p21^{Cip1} by doxorubicin was delayed in the LNCaP cells stably overexpressing *PCGEM1* but not in androgen-independent variants of LNCaP. Cleavage of caspase 7 was suppressed in cells expressing *PCGEM1* upon doxorubicin treatment, suggesting that *PCGEM1* attenuates the induction of apoptosis in doxorubicin-treated cells. LNCaP cells overexpressing *PCGEM1* were resistant to doxorubicin-induced apoptosis, as evinced by delays in p53 and p21^{Cip1} induction, suppression of caspase 7 cleavage, and suppression of PARP cleavage at 8 and 24 h. The inhibition of PARP cleavage by *PCGEM1* overexpression was also observed in LNCaP-*PCGEM1* cells treated with sodium selenite and etoposide; these agents are reported to induce apoptosis in LNCaP cells [27].

1.5.4 Growth Arrest-Specific 5 (GAS5)

GAS5 is a 5′-terminal oligopyrimidine tract (5′-TOP) gene, containing up to 12 exons which are alternatively spliced to yield two possible mature lncRNAs [28]. It is believed that *GAS5* promotes interplay between the mTOR pathway and the nonsense-mediated decay (NMD) pathway, explaining why transcripts accumulate in growth-arrested cells [28–30]. The expression of *GAS5* is downregulated in breast cancer, and patient prognosis is related to *GAS5* transcript levels (reviewed in [31]). In actively proliferating cells, in which mTOR activity is high, translation of the short reading frame is promoted, resulting in degradation of transcripts by the NMD pathway and consequently low cellular *GAS5* lncRNA levels. Suppression of cell growth and mTOR activity prevents the active translation of *GAS5* transcripts and their consequent degradation by NMD, resulting in the accumulation of *GAS5* lncRNA [28–30]. Direct inhibition of mTOR activity by rapamycin and rapalogues produces the predicted increase in cellular *GAS5* lncRNA [32, 33].

GAS5 is downregulated in lung adenocarcinoma tissue, especially in tumors that are larger and more poorly differentiated. Experimental overexpression of *GAS5* in a gefitinib-resistant lung adenocarcinoma cell line led to increased cell death following drug treatment. Therefore it has been suggested that *GAS5* sensitizes resistant cells to the cytotoxic effects of gefitinib. In this study, *GAS5* overexpression, together with gefitinib treatment, decreased the abundance of phospho-EGFR and of downstream signaling proteins (phospho-ERK and phospho-Akt) in the lung cancer cells [34].

1.5.5 AK126698

Cisplatin is widely used as a chemotherapeutic agent and acts by inducing DNA damage in proliferating cancer cells; resistance to cisplatin is a significant obstacle to therapy. Yang et al. [35] investigated the differential expression of mRNA, lncRNA, and miRNA in lung cancer cells resistant to cisplatin. More than 1300 lncRNAs showed differential expression, eight of which correlated with mRNA expression levels. These were investigated further and the results suggested that lncRNAs and miRNAs epigenetically regulate mRNA in cisplatin resistance. It was suggested that the correlation between lncRNA and mRNA reflects a specific biological function. The lncRNA *AK126698* was identified as being co-expressed with many members of the Wnt pathway, which is involved in cell growth, differentiation, embryogenesis, and oncogenesis. *AK126698* concordantly regulates the expression of NKD2, a negative regulator of Wnt signaling. The downregulation of this lncRNA increases expression of β-catenin and it was proposed that induction is regulated posttranscriptionally [35]. Though the initial finding suggests that *AK126698* lncRNA acts in cisplatin resistance, further investigation is required to identify the mechanism of this action.

1.5.6 E2F1-Regulated Inhibitor of Cell Death (ERIC)

E2Fs are transcription factors best known for their involvement in the timely regulation of gene expression required for cell cycle progression. Some of the E2F family members are downstream effectors of the tumor suppressor, pRB. Deregulation of E2F activity results in uncontrolled cell proliferation, a hallmark of tumor cells [36].

E2F1, a member of E2F family, transcriptionally up-regulates the lncRNA *ERIC* which is composed of two exons and has a transcript size of 1745 bp. *ERIC* is also induced by DNA damage, and inhibition of *ERIC* enhanced apoptotic cell death induced by either endogenous E2F1 or a DNA-damaging agent. It was suggested that endogenous E2F1 regulates *ERIC* expression both in cycling cells and in arrested cells that resume growth after serum starvation. Knockdown of *ERIC* significantly enhanced

E2F1-induced apoptosis. This observation suggests a negative feedback loop in which the E2F1-regulated *ERIC* inhibits E2F1-induced apoptosis [37].

Administration of the chemotherapeutic drug etoposide to bone osteosarcoma U2OS cells induces apoptotic cell death as well as induction in *ERIC* RNA levels. It is proposed that *ERIC* induction is E2F1 independent since a decrease in E2F1 protein levels was detected after drug administration. Silencing of *ERIC* by siRNAs showed no effect on cell viability, although when combined with the chemotherapeutic drug etoposide it led to a significant increase in etoposide-induced apoptosis. It was suggested that *ERIC* inhibits etoposide-induced apoptosis by an as yet unknown mechanism [37].

1.5.7 P21-Associated ncRNA DNA Damage Activated (PANDA)

PANDA is a 1.5 kb lncRNA with varied expression levels in different types of cancer. *PANDA* is found to be induced in a p53-dependent manner, and its activation is totally dependent on p53. Upon DNA damage, p53 directly binds to the *CDKN1A* locus and activates *PANDA*. *PANDA* then interacts with the transcription factor NF-YA to inhibit the expression of pro-apoptotic genes and enables cell cycle arrest by sequestering the transcription factor NF-YA. *PANDA* knockdown in human fetal fibroblasts induced expression of genes involved in apoptosis and sensitized fibroblasts to DNA damage-induced apoptosis. *PANDA* depletion also markedly sensitized human fibroblasts to apoptosis by doxorubicin [38].

1.5.8 p53-Dependent Apoptosis Modulator (PDAM)

Co-deletion of chromosomes 1p and 19q is a common event in oligodendroglial tumors, suggesting the involvement of these regions in tumor progression. The absence of *PDAM* located in 1p36.31–p36.32 was reported in oligodendroglial tumors. Loss of lncRNA *PDAM* expression in oligodendroglial tumors was caused not only by gene deletion but also by hyper-methylation of its promoter [39].

The chromosome 1p deletion predicts chemoresistance in oligodendroglial tumors [40]. *PDAM* does not play a role in mediating sensitivity to the commonly used drugs (vincristine, lomustine, and temozolomide) and paclitaxel in oligodendroglial tumors, but knockdown of *PDAM* led to elevated IC_{50} values in two out of five cell lines when treated with cisplatin [39].

Suppression of *PDAM* led to increased expression of p53 and of its downstream target p21^{Cip1}. Cisplatin suppressed the expression of the antiapoptotic proteins BCL2L1 and BCL2, but this repression was prevented by the knockdown of *PDAM*. This suggests that the loss of *PDAM* expression acts to maintain the levels of anti-apoptotic proteins to induce cisplatin resistance

1.5.9 HOXA Transcript at the Distal Tip (HOTTIP)

HOTTIP is a lincRNA, the gene of which is located about 330 bases upstream of the *HOX13* gene and is transcribed only as an antisense RNA. *HOX13* and *HOTTIP* are concordantly expressed in mouse embryonic cells. The *HOTTIP* gene is occupied by polycomb repressive complex 2 (PRC2) and histone-lysine N-methyltransferase 2A (MLL). *HOTTIP* is necessary to coordinate the expression of *HOXA* genes [41]. It is upregulated in pancreatic cancer, and may deregulate the cell cycle, promote invasion, and induce EMT, at least partly through *HOX13* up-regulation [42]. In a xenograft model of pancreatic ductal adenocarcinoma, it was shown that knockdown of *HOTTIP,* as well as gemcitabine treatment, inhibits tumor growth, and that combined knockdown of *HOTTIP* and gemcitabine treatment further reduced tumor size [42].

1.6 Drug Transport and Metabolism

1.6.1 Vault Noncoding RNAs

The functions of ncRNAs such as ribosomal and spliceosomal RNAs are vital for cell survival. These RNAs function in large ribonucleoprotein complexes and their ribozyme activity is essential for normal cellular function. The largest ribonucleoprotein complexes in eukaryotic cells are known as vaults, and they act as shuttles between nucleus and cytoplasm. Low-resolution X-ray crystal structure and electron microscopy of vault particles revealed their structure as cagelike with two protruding caps and an invaginated waist [43, 44]. Each vault particle is composed of one major vault protein (MVP), two minor proteins (VPARP and TEP1), and three ncRNAs (*hvg-1, hvg-2,* and *hvg-3*). Though these ncRNAs are not lncRNAs, their functions are distinct from those of small ncRNAs such as miRNA. Vault RNAs (vRNAs) are transcribed by RNA polymerase III located at the caps of vault particles. Their location suggests an interaction between the interior and exterior of vault particles. It is believed that these RNAs play functional roles rather than being simple structural components of the vault. The high level of conservation of vault structure and of its protein composition among different species suggests a crucial cellular function in eukaryotes. Despite the variation in size of vRNAs in rat, mouse, and human, the RNA is folded into similar secondary structures. Significant portions of vRNAs are not associated with vault particles suggesting dynamic interactions between the vRNAs and the particles [45].

It has been shown that vRNAs play roles in the development of drug resistance in cancer therapy. vRNA *hvg-1* and *hvg-2* interact with chemotherapeutic drugs through specific binding sites. The vault particle is up-regulated in MDR with elevated levels of vault proteins. It was suggested that vRNAs interact with the chemotherapeutic compound mitoxantrone and that these interactions play important roles in vault functions, possibly by exporting toxic compounds. Vault noncoding RNAs may also bind and sequester drugs, preventing them from reaching their sites of action [46].

1.6.2 H19

H19 is paternally imprinted and transcribed only from the maternal allele. The 2.3 kb RNA transcript does not contain any open reading frame despite its being capped, spliced, polyadenylated, and transported to the cytoplasm [47]. The expression of *H19* is tightly regulated during normal development; it is abundantly expressed in human placenta and embryonic tissue but repressed in almost all adult tissues. This observation suggests a role in fetal development. There is also evidence that *H19* may play a role in cancer progression [48, 49]. Initially the tumor-suppressor activity of *H19* was detected in two embryonal tumor cell lines [48], while its oncogenic potential in breast cancer was reported later [49]. Differential expression of *H19* was detected in different cancers including breast, liver, bladder, ovarian, head and neck, endometrial, lung, esophageal, acute T cell leukemia /lymphoma, and testicular germ cell cancer (reviewed in [50]).

Overexpression of *H19* RNA is observed in multidrug-resistant human breast carcinoma MCF-7/AdrVp cells and lung carcinoma NCI-H1688 cells, accompanied with an up-regulation of a 95 kDa membrane glycoprotein termed p95, which was not P-glycoprotein or another multidrug resistance protein [51]. Increased levels of *H19* were found in hepatocellular carcinoma cells that were resistant to drugs such as paclitaxel, doxorubicin, etoposide, and vincristine. It was suggested that *H19* acted through the *MDR1* gene (formally known as *ABCB1*) to suppress drug sensitivity of hepatocellular carcinoma cells. It has been shown that a reduced level of *H19* in these cells increases the cellular concentration of doxorubicin, suggesting that *H19* may regulate P-glycoprotein or other transporters that modulate drug accumulation in the cells [51].

Hypomethylation of the *ABCB1* gene in drug-resistant cells has also been detected; this could lead to activation of the *ABCB1* gene and resistance to therapy. Reduced levels of *H19* were associated with increased methylation at the *ABCB1* promoter region and it was suggested that *H19* upregulates *ABCB1* and its associated drug resistance by altering the degree of *ABCB1* promoter methylation in HepG2 cells, possibly through the DNA methyltransferases [51].

1.7 Drug Resistance-Related Protein

1.7.1 MDR-Related and Up-Regulated lncRNA (MRUL)

P-glycoprotein is the most well-studied MDR-related protein; its upregulation reduces intracellular chemotherapeutic drug concentrations by transporting drugs out of the cytoplasm, but the mechanisms involved in its induction are not clear. Since lncRNAs function in many biological processes including gene regulation, further investigation into the relationship between lncRNAs and MDR in cancer development is required. Wang et al. [52] have proposed that lncRNAs have enhancer-like functions in the induction of MDR in gastric cancer. A lncRNA, *MRUL*, is located 400 kb downstream of *ABCB1* and is significantly up-regulated in two multidrug-resistant gastric cancer cell sublines, SGC7901/ADR (resistant to doxorubicin) and SGC7901/VCR (resistant to vincristine), and demonstrated an enhancer-like role [52]. In the same study, a total of 32 lncRNAs in various tumor types were found to be differentially expressed, and of these 14 were located upstream or downstream of drug resistance-related protein-coding genes.

1.7.2 Adriamycin Resistance Associated (ARA)

Genome-wide lncRNA microarray analysis in MCF-7 breast cancer cell lines resistant to doxorubicin (Adriamycin) revealed differential expression of lncRNAs. A specific differentially expressed lncRNA *ARA* was validated in breast and liver cancer cell lines. The *ARA* gene is located in the intronic region of the *PAK3* gene and has conserved sequences in primates and secondary structure prediction of the transcript suggested several stable hairpins. *ARA* expression was increased in doxorubicin-resistant breast and liver cancer cell lines and was markedly up-regulated in parental sensitive MCF-7 and HepG2 cell lines treated with doxorubicin. Downregulation of *ARA* using siRNA in vitro suppressed proliferation, induced cell death, caused G2/M arrest, induced migration defects, and reduced resistance to doxorubicin. *ARA* modulates multiple signaling components, including those associated with MAPK, metabolism, cell cycle control, and cell adhesion; it regulates transcription and protein-binding functions. Overall, these data provide novel insights of doxorubicin resistance at the level of lncRNA [53].

1.8 Secondary Mutation or Translocation

1.8.1 Plasmacytoma Variant Translocation 1 (PVT1)

The human and mouse *PVT1* loci direct the synthesis of a large transcript with no protein products. The *PVT1* locus is the site of both tumorigenic translocations and (in mouse) retroviral insertions. Knockout of the *Pvt1* locus in mouse is embryonic lethal; this indicates that *PVT1* plays a critical role in normal development [54] and that aberrant expression might lead to the development of disease. In Burkitt's lymphoma the so-called variant translocations juxtapose immunoglobulin kappa or lambda light-chain genes to the *PVT1* locus. The chimeric transcripts are of 0.9–1.2 kb, containing the first exon of *PVT1* on chromosome 8 and the constant region of kappa or lambda [55]. *PVT1* is rearranged in 16 % of multiple myeloma patients. The location of *PVT1* near *MYC* at human chromosome 8q24 forms a cluster of *MYC*-activating chromosomal translocation breakpoints in Burkitt's lymphoma, plasmacytoma, ovarian, breast, and lung cancer [56–59]. Co-amplification of human *MYC* and *PVT1* correlates with rapid progression of breast cancer and with poor clinical survival in postmenopausal or HER2-positive breast cancer patients [60].

Gemcitabine is a first-line chemotherapeutic agent for the treatment of advanced pancreatic cancer, either alone or in combination with other chemotherapeutic agents; inherent resistance of pancreatic cancer to these agents presents a major challenge [61]. Genome-wide mutagenesis causing *PVT1* gene inactivation resulted in increased sensitivity to gemcitabine in human pancreatic cancer ASPC-1 cells. The functional perturbation of *PVT1* through DNA rearrangements or amplifications is known to contribute to carcinogenesis [58].

Pvt1 is also a common retroviral integration site in mouse and rat. Retroviruses can act as insertional mutagens and tag integration sites; therefore sites commonly integrated would identify protoon-cogenes and tumor-suppressor genes [62]. The *pvt1* locus also encodes miRNAs, and retroviral insertion can lead to altered expression of at least one of these miRNAs [63].

In a study by Ding et al. [64], the expression of lncRNA in gastric cancer tissue was analyzed by microarray and compared with that of normal adjacent tissue. This group also investigated expression of lncRNAs in paclitaxel-resistant SGC7901 cells and compared it with that of normal SGC7901 cells, finding that 252 and 545 lncRNAs were up-regulated in gastric cancer tissues and resistant cells, respectively, when compared to control. Among those 28 that had an obvious fold change in the gastric cancer tissues, *PVT1* expression levels in the gastric cancer tissues and paclitaxel-resistant SGC7901 cells were more than five and three times higher, respectively, than those in control. *PVT1* expression was also correlated with lymph node invasion of gastric cancer.

1.9 Endocrine Resistance

LncRNAs act as key regulatory molecules in the cell, suggesting that aberrant expression of lncRNA may lead to resistance to endocrine therapy. Hormonal regulation of *H19* is important, as estradiol is able to stimulate *H19* transcription in MCF-7 breast cancer cells, while tamoxifen inhibits this transcription [65]. The presence of estrogen response elements in the *HOTAIR* promoter could result in estradiol-induced overexpression of *HOTAIR* in breast cancer. These observations suggest that the aberrant expression of lncRNA can play critical roles in endocrine resistance cells.

1.9.1 Breast Cancer Anti-estrogen Resistance 4 (BCAR4)

The lncRNA *BCAR4* was discovered in a functional screen of estrogen-dependent ZR-75-1 breast cancer cells that aimed to identify genes responsible for factors that are key players in resistance to anti-estrogenic agents [66]. Ectopic expression of *BCAR4* was induced in tamoxifen-resistant human ZR-75-1 and MCF-7 breast cancer cells, and blocked the antiproliferative effects of tamoxifen. *BCAR4* is a clinically relevant biomarker of increased invasiveness and tamoxifen resistance [66–68]. The function of *BCAR4* is independent of the estrogen receptor and relies on the co-expression of HER2 [68]. A HER2 inhibitor may thus be useful for those patients whose tumors are resistant to endocrine therapy due to high levels of *BCAR4*, and are accompanied by HER2 overexpression. In silico analysis showed that *BCAR4* lncRNA is highly expressed in human placenta and oocytes but is absent in other normal tissues [68]. Due to the tissue specificity and time-restricted expression of *BCAR4*, it may be a good target for treating anti-estrogen-resistant breast cancer.

2   Conclusion

The data gathered here strongly implicate lncRNAs as key players in the regulation of protein-coding genes whose products are associated with cellular resistance to different chemotherapeutic agents. It is not surprising, then, that the dysregulation of lncRNAs appears to be a primary feature of drug resistance in cancer. Here we described some of the cases in which lncRNA affects critical pathways involved in drug resistance. In most of the studies reported here, the lncRNA was identified through the treatment of cancer cell lines by chemotherapeutic drugs followed by transcription analysis. In some cases the expression of the lncRNA identified was examined in tumor samples. The roles of identified lncRNAs in different cellular pathways were then examined. The technology available to analyze the transcriptome is growing rapidly and in the near future better facilities will be available to examine gene expression in tumor samples. That will provide invaluable information that would be used for better diagnosis and therapy as well as for identifying the pathways involved in drug resistance. More detailed studies are required to elucidate the mechanisms underlying therapy-induced deregulation of lncRNA.

Some of these lncRNAs may act through different targets in the same pathway (Fig. 2). As discussed here, most of the lncRNAs involved in drug resistance seem to regulate cell cycle progression and apoptosis. DNA-damaging drugs such as doxorubicin and cisplatin affect the deregulation of lncRNAs such as *HOTAIR*, *PCGEM*, and *PDAM*. Those lncRNAs in turn affect drug resistance through p21^{Cip} or p53 pathways. The same drugs could affect different lncRNAs and their downstream pathways in different cancer types (Fig. 2 and Table 1).

Antiapoptotic signaling pathways and EMT are also involved in CSC-mediated drug resistance. Thus, nanodelivery of selective genetic inhibitors, such as miRNA and siRNA, and small inhibitor molecules provide potential combination strategies for suppressing the drug resistance and viability of CSCs. The niche is defined as the microenvironment where CSCs are located, and where they interact with other types of cells. The normal stem cell niche functions in specific vascular locations to regulate stem cell activity during tissue generation, maintenance, and repair. The niche-associated vasculature can provide physical and physiological protection from stem cell depletion. Evidently, the CSC niche is a dynamic supportive system with specific anatomic and functional features that contains a variety of cell types, cytokines, and signaling pathways [69–71].

As discussed earlier, lncRNAs regulate many protein-coding genes at transcriptional and posttranscriptional levels, as well as

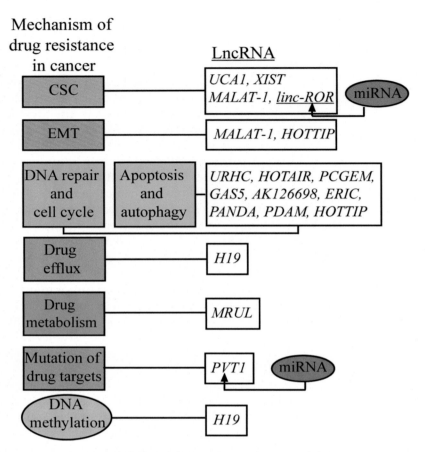

Fig. 2 Mechanisms by which lncRNAs effect drug resistance. Mechanisms are shown in colored boxes, while lncRNAs are in plain boxes. *Linc-ROR, MALAT-1,* and *H19* act through two pathways to regulate downstream genes

acting as vehicles in transporting drugs, as in the case of vRNA. Therefore it seems logical that lncRNAs may be targets for therapy for the reversal of drug sensitivity in the future. Recently, lncRNAs have also become a frontier for drug development. Current RNA therapy is focused on the attempt to suppress the inhibitory effects of lncRNAs on gene transcription, although this approach can be extended to other lncRNAs that affect gene regulation in different ways (Fig. 3). The best known function of lncRNA is its interaction with polycomb repressor complex (PRC) and suppression of transcription of protein-coding genes, though more recently the presence of about 40 % of transcripts in cytoplasmic extracts has been reported. The association of lncRNA with ribosomes suggests a new era of posttranscriptional regulation for protein synthesis that may affect drug resistance in cancer therapy.

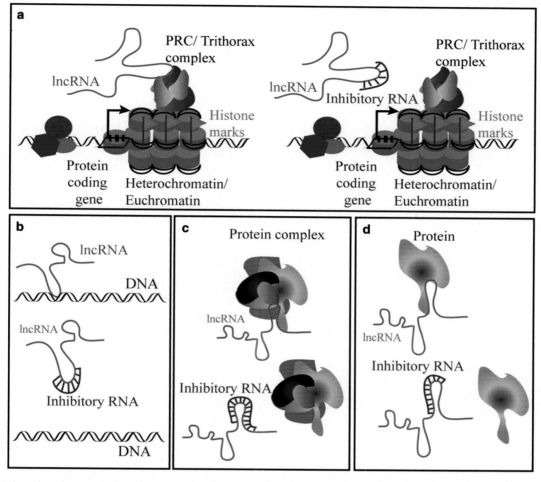

Fig. 3 Inhibitory RNAs bind to target lncRNAs to suppress their functions. (a) LncRNAs regulate gene expression through the transcriptional machinery. Inhibitory RNAs bind to lncRNAs to prevent their interactions with PRC, trithorax, or other regulatory complexes. (b) Interaction between lncRNAs and DNA can be prevented by inhibitory RNAs. (c, d) LncRNAs can bind to protein complexes (c) or proteins (d) in the nucleus or cytoplasm. RNA–protein interactions can be inhibited by inhibitory RNAs, leading to suppression or induction of protein activities

References

1. Morris KV, Mattick JS (2014) The rise of regulatory RNA. Nat Rev Genet 15:423–437
2. Esteller M (2011) Non-coding RNAs in human disease. Nat Rev Genet 12:861–874
3. Holohan C, Van Schaeybroeck S, Longley DB, Johnston PG (2013) Cancer drug resistance: an evolving paradigm. Nat Rev Cancer 13:714–726
4. Xia H, Hui KM (2014) Mechanism of cancer drug resistance and the involvement of noncoding RNAs. Curr Med Chem 21:3029–3041
5. Balch C, Huang TH, Brown R, Nephew KP (2004) The epigenetics of ovarian cancer drug resistance and resensitization. Am J Obstet Gynecol 191:1552–1572
6. van Heesch S, van Iterson M, Jacobi J, Boymans S, Essers PB, de Bruijn E, Hao W, Macinnes AW, Cuppen E, Simonis M (2014) Extensive localization of long noncoding RNAs to the cytosol and mono- and polyribosomal complexes. Genome Biol 15:R6

7. Wang F, Li X, Xie X, Zhao L, Chen W (2008) UCA1, a non-protein-coding RNA up-regulated in bladder carcinoma and embryo, influencing cell growth and promoting invasion. FEBS Lett 582:1919–1927

8. Taipale J, Beachy PA (2001) The Hedgehog and Wnt signalling pathways in cancer. Nature 411:349–354

9. Fan Y, Shen B, Tan M, Mu X, Qin Y, Zhang F, Liu Y (2014) Long non-coding RNA UCA1 increases chemoresistance of bladder cancer cells by regulating Wnt signaling. FEBS J 281:1750–1758

10. Li Z, Li X, Wu S, Xue M, Chen W (2014) Long non-coding RNA UCA1 promotes glycolysis by upregulating hexokinase 2 through the mTOR-STAT3/microRNA143 pathway. Cancer Sci 105:951–955

11. Wang Y, Xu Z, Jiang J, Xu C, Kang J, Xiao L, Wu M, Xiong J, Guo X, Liu H (2013) Endogenous miRNA sponge lincRNA-RoR regulates Oct4, Nanog, and Sox2 in human embryonic stem cell self-renewal. Dev Cell 25:69–80

12. Takahashi K, Yan IK, Kogure T, Haga H, Patel T (2014) Extracellular vesicle-mediated transfer of long non-coding RNA ROR modulates chemosensitivity in human hepatocellular cancer. FEBS Open Bio 4:458–467

13. Huang KC, Rao PH, Lau CC, Heard E, Ng SK, Brown C, Mok SC, Berkowitz RS, Ng SW (2002) Relationship of XIST expression and responses of ovarian cancer to chemotherapy. Mol Cancer Ther 1:769–776

14. Rottenberg S, Vollebergh MA, de Hoon B, de Ronde J, Schouten PC, Kersbergen A, Zander SA, Pajic M, Jaspers JE, Jonkers M, Loden M, Sol W, van der Burg E, Wesseling J, Gillet JP, Gottesman MM, Gribnau J, Wessels L, Linn SC, Jonkers J, Borst P (2012) Impact of intertumoral heterogeneity on predicting chemotherapy response of BRCA1-deficient mammary tumors. Cancer Res 72:2350–2361

15. Salvador MA, Wicinski J, Cabaud O, Toiron Y, Finetti P, Josselin E, Lelievre H, Kraus-Berthier L, Depil S, Bertucci F, Collette Y, Birnbaum D, Charafe-Jauffret E, Ginestier C (2013) The histone deacetylase inhibitor abexinostat induces cancer stem cells differentiation in breast cancer with low Xist expression. Clin Cancer Res 19:6520–6531

16. Fan Y, Shen B, Tan M, Mu X, Qin Y, Zhang F, Liu Y (2014) TGF-beta-induced upregulation of malat1 promotes bladder cancer metastasis by associating with suz12. Clin Cancer Res 20:1531–1541

17. Ying L, Chen Q, Wang Y, Zhou Z, Huang Y, Qiu F (2012) Upregulated MALAT-1 contributes to bladder cancer cell migration by inducing epithelial-to-mesenchymal transition. Mol Biosyst 8:2289–2294

18. Jiao F, Hu H, Yuan C, Wang L, Jiang W, Jin Z, Guo Z (2014) Elevated expression level of long noncoding RNA MALAT-1 facilitates cell growth, migration and invasion in pancreatic cancer. Oncol Rep 32:2485–2492

19. Jiao F, Hu H, Han T, Yuan C, Wang L, Jin Z, Guo Z (2015) Long noncoding RNA MALAT-1 enhances stem cell-like phenotypes in pancreatic cancer cells. Int J Mol Sci 16:6677–6693

20. Xu WH, Zhang JB, Dang Z, Li X, Zhou T, Liu J, Wang DS, Song WJ, Dou KF (2014) Long non-coding RNA URHC regulates cell proliferation and apoptosis via ZAK through the ERK/MAPK signaling pathway in hepatocellular carcinoma. Int J Biol Sci 10:664–676

21. Gupta R, Shah N, Wang KC, Kim J, Horlings HM, Wong DJ, Tsai MC, Hung T, Argani P, Rinn JL, Wang Y, Brzoska P, Kong B, Li R, West RB, van de Vijver MJ, Sukumar S, Chang HY (2010) Long non-coding RNA HOTAIR reprograms chromatin state to promote cancer metastasis. Nature 464:1071–1076

22. Kogo RS, Mimori T, Kawahara K, Imoto K, Sudo S, Tanaka T, Shibata F, Suzuki K, Komune A, Miyano S, Mori M (2011) Long non-coding RNA HOTAIR regulates polycomb-dependent chromatin modification and is associated with poor prognosis in colorectal cancers. Cancer Res 71:6320–6326

23. Geng YJ, Xie SL, Li Q, Ma J, Wang GY (2011) Large intervening non-coding RNA HOTAIR is associated with hepatocellular carcinoma progression. J Int Med Res 39:2119–2128

24. Niinuma T, Suzuki H, Nojima M, Nosho K, Yamamoto H, Takamaru H, Yamamoto E, Maruyama R, Nobuoka T, Miyazaki Y, Nishida T, Bamba T, Kanda T, Ajioka Y, Taguchi T, Okahara S, Takahashi H, Nishida Y, Hosokawa M, Hasegawa T, Tokino T, Hirata K, Imai K, Toyota M, Shinomura Y (2012) Upregulation of miR-196a and HOTAIR drive malignant character in gastrointestinal stromal tumors. Cancer Res 72:1126–1136

25. Liu Z, Sun M, Lu K, Liu J, Zhang M, Wu W, De W, Wang Z, Wang R (2013) The long noncoding RNA HOTAIR contributes to cisplatin resistance of human lung adenocarcinoma cells via downregualtion of p21(WAF1/CIP1) expression. PLoS One 8:e77293

26. Srikantan V, Zou Z, Petrovics G, Xu L, Augustus M, Davis L, Livezey JR, Connell T, Sesterhenn IA, Yoshino K, Buzard GS, Mostofi FK, McLeod DG, Moul JW, Srivastava S (2000) PCGEM1, a prostate-specific gene, is overexpressed in prostate cancer. Proc Natl Acad Sci U S A 97:12216–12221

27. Fu X, Ravindranath L, Tran N, Petrovics G, Srivastava S (2006) Regulation of apoptosis by a prostate-specific and prostate cancer-associated noncoding gene, PCGEM1. DNA Cell Biol 25:135–141

28. Smith C, Steitz JA (1998) Classification of gas5 as a multi-small-nucleolar-RNA (snoRNA) host gene and a member of the 5'-terminal oligopyrimidine gene family reveals common features of snoRNA host genes. Mol Cell Biol 18:6897–6909

29. Schneider C, King RM, Philipson L (1988) Genes specifically expressed at growth arrest of mammalian cells. Cell 54:787–793

30. Williams GT, Farzaneh F (2012) Are snoRNAs and snoRNA host genes new players in cancer? Nat Rev Cancer 12:84–88

31. Hansji H, Leung EY, Baguley BC, Finlay GJ, Askarian-Amiri ME (2014) Keeping abreast with long non-coding RNAs in mammary gland development and breast cancer. Front Genet 5:379

32. Mourtada-Maarabouni M, Hasan AM, Farzaneh F, Williams GT (2010) Inhibition of human T-cell proliferation by mammalian target of rapamycin (mTOR) antagonists requires noncoding RNA growth-arrest-specific transcript 5 (GAS5). Mol Pharmacol 78:19–28

33. Williams GT, Mourtada-Maarabouni M, Farzaneh F (2011) A critical role for non-coding RNA GAS5 in growth arrest and rapamycin inhibition in human T-lymphocytes. Biochem Soc Trans 39:482–486

34. Dong S, Qu X, Li W, Zhong X, Li P, Yang S, Chen X, Shao M, Zhang L (2015) The long non-coding RNA, GAS5, enhances gefitinib-induced cell death in innate EGFR tyrosine kinase inhibitor-resistant lung adenocarcinoma cells with wide-type EGFR via downregulation of the IGF-1R expression. J Hematol Oncol 8:43

35. Yang Y, Li H, Hou S, Hu B, Liu J, Wang J (2013) The noncoding RNA expression profile and the effect of lncRNA AK126698 on cisplatin resistance in non-small-cell lung cancer cell. PLoS One 8:e65309

36. Polager S, Ginsberg D (2008) E2F – at the crossroads of life and death. Trends Cell Biol 18:528–535

37. Feldstein O, Nizri T, Doniger T, Jacob J, Rechavi G, Ginsberg D (2013) The long non-coding RNA ERIC is regulated by E2F and modulates the cellular response to DNA damage. Mol Cancer 12:131

38. Hung T, Wang Y, Lin MF, Koegel AK, Kotake Y, Grant GD, Horlings HM, Shah N, Umbricht C, Wang P, Wang Y, Kong B, Langerod A, Borresen-Dale AL, Kim SK, van de Vijver M, Sukumar S, Whitfield ML, Kellis M, Xiong Y, Wong DJ, Chang HY (2011) Extensive and coordinated transcription of noncoding RNAs within cell-cycle promoters. Nat Genet 43:621–629

39. Pang JC, Li KK, Lau KM, Ng YL, Wong J, Chung NY, Li HM, Chui YL, Lui VW, Chen ZP, Chan DT, Poon WS, Wang Y, Mao Y, Zhou L, Ng HK (2010) KIAA0495/PDAM is frequently downregulated in oligodendroglial tumors and its knockdown by siRNA induces cisplatin resistance in glioma cells. Brain Pathol 20:1021–1032

40. Cairncross JG, Ueki K, Zlatescu MC, Lisle DK, Finkelstein DM, Hammond RR, Silver JS, Stark PC, Macdonald DR, Ino Y, Ramsay DA, Louis DN (1998) Specific genetic predictors of chemotherapeutic response and survival in patients with anaplastic oligodendrogliomas. J Natl Cancer Inst 90:1473–1479

41. Wang KC, Yang YW, Liu B, Sanyal A, Corces-Zimmerman R, Chen Y, Lajoie BR, Protacio A, Flynn RA, Gupta RA, Wysocka J, Lei M, Dekker J, Helms JA, Chang HY (2011) A long noncoding RNA maintains active chromatin to coordinate homeotic gene expression. Nature 472:120–124

42. Li Z, Zhao X, Zhou Y, Liu Y, Zhou Q, Ye H, Wang Y, Zeng J, Song Y, Gao W, Zheng S, Zhuang B, Chen H, Li W, Li H, Fu Z, Chen R (2015) The long non-coding RNA HOTTIP promotes progression and gemcitabine resistance by regulating HOXA13 in pancreatic cancer. J Transl Med 13:84

43. Mikyas Y, Makabi M, Raval-Fernandes S, Harrington L, Kickhoefer VA, Rome LH, Stewart PL (2004) Cryoelectron microscopy imaging of recombinant and tissue derived vaults: localization of the MVP N termini and VPARP. J Mol Biol 344:91–105

44. Anderson DH, Kickhoefer VA, Sievers SA, Rome LH, Eisenberg D (2007) Draft crystal structure of the vault shell at 9-A resolution. PLoS Biol 5:e318

45. Kickhoefer VA, Searles RP, Kedersha NL, Garber ME, Johnson DL, Rome LH (1993) Vault ribonucleoprotein particles from rat and bullfrog contain a related small RNA that is transcribed by RNA polymerase III. J Biol Chem 268:7868–7873

46. Gopinath SC, Wadhwa R, Kumar PK (2010) Expression of noncoding vault RNA in human malignant cells and its importance in mitoxantrone resistance. Mol Cancer Res 8:1536–1546

47. Eddy SR (2001) Non-coding RNA genes and the modern RNA world. Nat Rev Genet 2:919–929

48. Hao Y, Crenshaw T, Moulton T, Newcomb E, Tycko B (1993) Tumour-suppressor activity of H19 RNA. Nature 365:764–767

49. Lottin S, Adriaenssens E, Dupressoir T, Berteaux N, Montpellier C, Coll J, Dugimont T, Curgy JJ (2002) Overexpression of an ectopic H19 gene enhances the tumorigenic properties of breast cancer cells. Carcinogenesis 23:1885–1895

50. Matouk I, Raveh E, Ohana P, Lail RA, Gershtain E, Gilon M, De Groot N, Czerniak A, Hochberg A (2013) The increasing complexity of the oncofetal h19 gene locus: functional dissection and therapeutic intervention. Int J Mol Sci 14:4298–4316

51. Tsang WP, Kwok TT (2007) Riboregulator H19 induction of MDR1-associated drug resistance in human hepatocellular carcinoma cells. Oncogene 26:4877–4881

52. Wang Y, Zhang D, Wu K, Zhao Q, Nie Y, Fan D (2014) Long noncoding RNA MRUL promotes ABCB1 expression in multidrug-resistant gastric cancer cell sublines. Mol Cell Biol 34:3182–3193

53. Jiang M, Huang O, Xie Z, Wu S, Zhang X, Shen A, Liu H, Chen X, Wu J, Lou Y, Mao Y, Sun K, Hu S, Geng M, Shen K (2014) A novel long non-coding RNA-ARA: adriamycin resistance associated. Biochem Pharmacol 87:254–283

54. Marcu KB, Bossone SA, Patel AJ (1992) myc function and regulation. Annu Rev Biochem 61:809–860

55. Shtivelman E, Bishop JM (1990) Effects of translocations on transcription from PVT. Mol Cell Biol 10:1835–1839

56. Pleasance ED, Stephens PJ, O'Meara S, McBride DJ, Meynert A, Jones D, Lin ML, Beare D, Lau KW, Greenman C, Varela I, Nik-Zainal S, Davies HR, Ordonez GR, Mudie LJ, Latimer C, Edkins S, Stebbings L, Chen L, Jia M, Leroy C, Marshall J, Menzies A, Butler A, Teague JW, Mangion J, Sun YA, McLaughlin SF, Peckham HE, Tsung EF, Costa GL, Lee CC, Minna JD, Gazdar A, Birney E, Rhodes MD, McKernan KJ, Stratton MR, Futreal PA, Campbell PJ (2010) A small-cell lung cancer genome with complex signatures of tobacco exposure. Nature 463:184–190

57. Pomerantz MM, Beckwith CA, Regan MM, Wyman SK, Petrovics G, Chen Y, Hawksworth DJ, Schumacher FR, Mucci L, Penney KL, Stampfer MJ, Chan JA, Ardlie KG, Fritz BR, Parkin RK, Lin DW, Dyke M, Herman P, Lee S, Oh WK, Kantoff PW, Tewari M, McLeod DG, Srivastava S, Freedman ML (2009) Evaluation of the 8q24 prostate cancer risk locus and MYC expression. Cancer Res 69:5568–5574

58. Enciso-Mora V, Broderick P, Ma Y, Jarrett RF, Hjalgrim H, Hemminki K, van den Berg A, Olver B, Lloyd A, Dobbins SE, Lightfoot T, van Leeuwen FE, Forsti A, Diepstra A, Broeks A, Vijayakrishnan J, Shield L, Lake A, Montgomery D, Roman E, Engert A, von Strandmann EP, Reiners KS, Nolte IM, Smedby KE, Adami HO, Russell NS, Glimelius B, Hamilton-Dutoit S, de Bruin M, Ryder LP, Molin D, Sorensen KM, Chang ET, Taylor M, Cooke R, Hofstra R, Westers H, van Wezel T, van Eijk R, Ashworth A, Rostgaard K, Melbye M, Swerdlow AJ, Houlston RS (2010) A genome-wide association study of Hodgkin's lymphoma identifies new susceptibility loci at 2p16.1 (REL), 8q24.21 and 10p14 (GATA3). Nat Genet 42:1126–1130

59. Shao RG, Cao CX, Shimizu T, O'Connor PM, Kohn KW, Pommier Y (1997) Abrogation of an S-phase checkpoint and potentiation of camptothecin cytotoxicity by 7-hydroxystaurosporine (UCN-01) in human cancer cell lines, possibly influenced by p53 function. Cancer Res 57:4029–4035

60. Borg A, Baldetorp B, Ferno M, Olsson H, Sigurdsson H (1992) c-myc amplification is an independent prognostic factor in postmenopausal breast cancer. Int J Cancer 51:687–691

61. O'Reilly EM, Abou-Alfa GK (2007) Cytotoxic therapy for advanced pancreatic adenocarcinoma. Semin Oncol 34:347–353

62. Hayward WS, Neel BG, Astrin SM (1981) Activation of a cellular onc gene by promoter insertion in ALV-induced lymphoid leukosis. Nature 290:475–480

63. Beck-Engeser GB, Lum AM, Huppi K, Caplen NJ, Wang BB, Wabl M (2008) Pvt1-encoded microRNAs in oncogenesis. Retrovirology 5:4

64. Ding J, Li D, Gong M, Wang J, Huang X, Wu T, Wang C (2014) Expression and clinical significance of the long non-coding RNA PVT1 in human gastric cancer. Onco Targets Ther 7:1625–1630

65. Adriaenssens E, Lottin S, Dugimont T, Fauquette W, Coll J, Dupouy JP, Boilly B, Curgy JJ (1999) Steroid hormones modulate H19 gene expression in both mammary gland and uterus. Oncogene 18:4460

66. Meijer D, van Agthoven T, Bosma PT, Nooter K, Dorssers LC (2006) Functional screen for genes responsible for tamoxifen resistance in human breast cancer cells. Mol Cancer Res 4:379–386

67. Godinho MF, Sieuwerts AM, Look MP, Meijer D, Foekens JA, Dorssers LC, van Agthoven T (2010) Relevance of BCAR4 in tamoxifen resistance and tumour aggressiveness of human breast cancer. Br J Cancer 103:1284–1291

68. Godinho M, Meijer D, Setyono-Han B, Dorssers LC, van Agthoven T (2011) Characterization of BCAR4, a novel oncogene causing endocrine resistance in human breast cancer cells. J Cell Physiol 226:1741–1749

69. Scadden DT (2006) The stem-cell niche as an entity of action. Nature 441:1075–1079

70. Takeshita H, Kusuzaki K, Ashihara T, Gebhardt MC, Mankin HJ, Hirasawa Y (2000) Intrinsic resistance to chemotherapeutic agents in murine osteosarcoma cells. J Bone Joint Surg Am 82-A: 963–969

71. Vinogradov S, Wei X (2012) Cancer stem cells and drug resistance: the potential of nanomedicine. Nanomedicine (Lond) 7:597–615

72. Reya T, Morrison SJ, Clarke MF, Weissman IL (2001) Stem cells, cancer, and cancer stem cells. Nature 414:105–111

73. Tam WL, Ng HH (2014) Sox2: masterminding the root of cancer. Cancer Cell 26:3–5

74. Polyak K, Weinberg RA (2009) Transitions between epithelial and mesenchymal states: acquisition of malignant and stem cell traits. Nat Rev Cancer 9:265–273

75. Fletcher JI, Haber M, Henderson MJ, Norris MD (2010) ABC transporters in cancer: more than just drug efflux pumps. Nat Rev Cancer 10:147–156

76. Garraway LA, Janne PA (2012) Circumventing cancer drug resistance in the era of personalized medicine. Cancer Discov 2:214–226

77. Gorre ME, Mohammed M, Ellwood K, Hsu N, Paquette R, Rao PN, Sawyers CL (2001) Clinical resistance to STI-571 cancer therapy caused by BCR-ABL gene mutation or amplification. Science 293:876–880

78. Basseville A, Preisser L, de Carne Trecesson S, Boisdron-Celle M, Gamelin E, Coqueret O, Morel A (2011) Irinotecan induces steroid and xenobiotic receptor (SXR) signaling to detoxification pathway in colon cancer cells. Mol Cancer 10:80

79. Salehan MR, Morse HR (2013) DNA damage repair and tolerance: a role in chemotherapeutic drug resistance. Br J Biomed Sci 70: 31–40

80. Shah MA, Schwartz GK (2001) Cell cycle-mediated drug resistance: an emerging concept in cancer therapy. Clin Cancer Res 7: 2168–2181

81. Marquez RT, Tsao BW, Faust NF, Xu L (2013) Drug resistance and molecular cancer therapy: apoptosis versus autophagy. In: Apoptosis. Runder, J (Ed.). ISBN: 978-953-51-1133-7, InTech, doi: 10.5772/55415. www.intechopen.com/books/apoptosis/ drug-resistance-and-molecular-cancer-therapy-apoptosis-versus-autophagy. InTech; 2013

82. Mathew R, Karantza-Wadsworth V, White E (2007) Role of autophagy in cancer. Nat Rev Cancer 7:961–967

Chapter 13

Cancer Exosomes as Mediators of Drug Resistance

Maria do Rosário André, Ana Pedro, and David Lyden

Abstract

In the last decades, several studies demonstrated that the tumor microenvironment is a critical determinant not only of tumor progression and metastasis, but also of resistance to therapy. Exosomes are small membrane vesicles of endocytic origin, which contain mRNAs, DNA fragments, and proteins, and are released by many different cell types, including cancer cells. Mounting evidence has shown that cancer-derived exosomes contribute to the recruitment and reprogramming of constituents associated with the tumor microenvironment. Understanding how exosomes and the tumor microenvironment impact drug resistance will allow novel and better strategies to overcome drug resistance and treat cancer.

Here, we describe a technique for exosome purification from cell culture, and fresh and frozen plasma, and further analysis by electron microscopy, NanoSight microscope, and Western blot.

Key words Tumor microenvironment, Pre-metastatic niche, Tumor-derived exosomes, Drug resistance, Cell culture, Plasma, NanoSight microscope, Electron microscopy

1 Introduction

1.1 The Tumor Microenvironment, the Pre-metastatic Niche, and the Role of Tumor Exosomes in Tumor Progression and Metastasis Development

Solid tumors are complex, organ-like structures, consisting of cancer cells along with a supportive stroma composed of multiple nonmalignant cell types, such as fibroblasts, endothelial cells, mesenchymal stem cells, and immune cells, sustained by an extracellular matrix and a vascular network [1]. The tumor-associated microenvironment differs from normal tissues by an increased number of fibroblasts and an altered extracellular matrix, and the most frequently found immune cells within the tumor microenvironment are tumor-associated macrophages (TAMs) [2]. TAMs have been shown to be involved in multiple steps of tumor development, namely angiogenesis, invasion, and intravasation [3]. In past years, an association between the presence of these cells and poor prognosis in several types of cancers, including breast, gastric, urogenital, and head and neck cancer, was demonstrated [4]. On the other hand, the presence of mature dendritic cells in the primary tumor is associated with a good outcome in lung and head and

José Rueff and António Sebastião Rodrigues (eds.), *Cancer Drug Resistance: Overviews and Methods*, Methods in Molecular Biology, vol. 1395, DOI 10.1007/978-1-4939-3347-1_13, © Springer Science+Business Media New York 2016

neck cancers [5, 6]. Fibroblasts are present at increased numbers in the tumor microenvironment, and influence the ability of tumor cells to invade and metastasize through the synthesis of growth factors, chemokines, and adhesion molecules [7].

As important as the primary tumor microenvironment, the future metastatic organ structure and composition are crucial for metastasis to develop. The pre-metastatic niche model suggests that in order for tumor cells to engraft and to form metastatic lesions at secondary sites, a suitable microenvironment must evolve in these pre-metastatic organs [8]. This theory advocates that metastatic proliferation does not depend solely on the characteristics and genetic alterations of the cancer cell itself, but that the formation of this pre-metastatic niche is also essential for metastasis to occur. These niches form as a consequence of growth factors, e.g., VEGF or placental growth factor (PIGF), secreted by the primary tumor [9]. In response to these soluble factors, tumor-associated cells such as hematopoietic progenitor cells or macrophages are mobilized to the pre-metastatic niches. Besides immature myeloid cells, other cells are also involved actively in the formation of the pre-metastatic niche. Platelets, resident fibroblasts, and endothelial cells are also important in this process. At the pre-metastatic niche, the mobilized bone marrow-derived cells together with resident cells produce chemokines, growth factors, and matrix-degrading proteins (e.g., MMP9). These alter the surrounding microenvironment, making it more suitable for the engraftment of tumor cells and the formation of metastatic lesions [8, 10].

Besides the multiple cell types, tumor-derived chemokines, and growth factors mentioned above, exosomes are also important mediators of metastasis, being involved in a permanent cross talk between the primary tumor and local/distant host cells. Exosomes are small membranous extracellular vesicles, ranging in size between 40 and 100 nm in diameter, that contain microRNAs, messenger RNAs (mRNA), DNA fragments, and proteins [11]. These small vesicles consist of a lipid bilayer membrane surrounding a small cytosol, are devoid of cellular organelles, and are secreted by many kinds of cells, including tumor cells, reticulocytes, and hematopoietic cells [12]. Exosomes are formed by the inward budding of cells known as multivesicular endosomes. Fusion of these endosomes with the plasma membrane leads to the release of internal vesicles known as exosomes [13]. The major role of exosomes seems to be the transport of bioactive molecules between cells, with consequences in targeted cell phenotypes. Exosomes are involved in the normal physiology of the body, including immune regulation, tissue repair, and communication within the nervous system [14]. In cancer patients, the abundance of secreted exosomes suggests an important role of these mediators in cancer development. In fact, a positive correlation between increased exosome secretion and cancer stage and progression has been shown [15]. Exosomes travel to surrounding cells or distant tissues to execute important functions in tumor biology such as angiogenesis, immune suppression, induction of proliferation, and transfer of genetic material [16]. The transport of oncogenic proteins and miRNAs by exosomes released by tumor cells and the uptake of these oncogenic elements by nonmalignant cells in the tumor microenvironment can result in the transfer of oncogenic activity [17]. Work by Peinado et al. demonstrated that tumor-derived exosomes promote metastatic niche formation by educating bone marrow-derived cells towards a more pro-vasculogenic and pro-metastatic phenotype, through the exosome-mediated transfer of the oncoprotein MET [18]. Further studies supported these results, and confirmed the importance of exosomes in tumor growth, angiogenesis, and metastatic development [19].

1.2 The Tumor Microenvironment and Chemotherapy Response

The acquisition of resistance to chemotherapeutic drugs continues to be a major obstacle in cancer treatment. Although it was believed for several years that drug resistance resulted primarily from selection of mutant tumor cells that were resistant to the cytotoxic effects of certain therapies, mounting evidence suggests that there is more to this story than once believed. Functional gene mutations that alter the expression of proteins involved in the uptake, metabolism, and export of drugs are main causes of drug resistance, as are nonmutational (epigenetic) changes that can be associated to transient drug resistance. However, as discussed above, the tumor cell is only part of a complex group of constituents, and this tumor microenvironment is a critical determinant not only of tumor progression and metastasis, but also of resistance to therapy.

In 1998 Brown and Giaccia proposed that the microenvironment could be a major mechanism of drug resistance through the reduction of drug distribution throughout the tumor, therefore protecting high proportions of cells from damage induced by the drug [20]. In fact, the tumor stromal components contribute to an increase in interstitial fluid pressure, and several studies have shown an association between high interstitial fluid pressure and poor drug penetration, with a suggested association to response to chemotherapy [21]. On the other hand, the increase in interstitial fluid pressure in association with an oncotic pressure gradient of almost zero can lead to the extravasation of macromolecules, which can decrease the effectiveness of the treatment if the administered drug is lost at the tumor periphery [22].

Response to chemotherapy is also influenced by the vasculature, not only because the delivery of cytotoxic drugs can be impaired as a consequence of vascular disorganization [23], but also because this disorganized blood flow results in an abnormal and limited delivery of nutrients to the tumor, and the appearance of hypoxia [24, 25]. The first link between glucose deprivation and drug resistance was reported by Shen et al. in 1987. They showed that in Chinese hamster ovary cells stress conditions that induced the endoplasmic reticulum-resident stress proteins, such as hypoxia or glucose deprivation, were associated with significant resistance to doxorubicin. Moreover, it was shown that the removal of these conditions resulted in the disappearance of drug resistance [26]. Hypoxic conditions can lead to the activation of genes associated with angiogenesis and cell survival [27]. The expression of these genes may result in an expansion of biochemically altered cells, with a drug-resistant phenotype. As an example, transient hypoxia has been shown to be associated with an increased expression of genes encoding P-glycoprotein and dihydrofolate reductase, which induces drug resistance, and with selection for cells that are deficient in DNA mismatch repair which increases their resistance to platinum-based chemotherapy [28]. Furthermore, the limited supply of nutrients induces cell cycle arrest, with a consequent reduction of tumor cell proliferation rate [29]. As most chemotherapeutic drugs are more effective against proliferating cells, the slow-growing cells localized most distant to the tumor vasculature have a high likelihood of becoming resistant to therapy [30].

Another known mechanism by which tumor stroma can influence drug resistance is through the interactions between tumor cells and the extracellular matrix. Work performed by Garrido et al., 1997, demonstrated that confluent cells in culture are more resistant to anticancer drugs than non-confluent cells [31]. Moreover, tumor cell adhesion to extracellular matrix mediated by integrins has been shown to protect small-cell lung cancer cells from drug-induced apoptosis [32].

In recent years, mounting evidence has suggested that certain growth factors and immune suppressor cells within the tumor microenvironment can induce tumor growth and mediate resistance to therapy. Straussman et al. demonstrated that in BRAF-mutant melanoma, hepatocyte growth factor (HGF) secretion by stromal cells was associated with poor response to BRAF inhibition. Furthermore, it was demonstrated that HGF plasmatic levels were inversely related to the response to BRAF inhibition in BRAF-mutant melanoma [33]. Recent work by Sun et al. suggests that microenvironment-mediated therapy resistance in the clinical management of prostate cancer may also arise from an adaptive, reciprocal signaling dialogue between the microenvironment and tumor cells. Specifically, it was shown that WNT16B was increased within fibroblasts exposed to cytotoxic drugs both in vitro and in vivo, and that in human tumors, WNT16B expression was associated with higher rates of disease recurrence after chemotherapy. Furthermore, when high-expressing fibroblasts were co-cultured with epithelial cells or xenograft tumors and then exposed to cytotoxic agents, there was a survival advantage as compared to cultures with low or absent WNT16B-expressing fibroblasts. This work demonstrated that WNT16B signals through a paracrine manner to tumor cells, increasing their proliferation and resistance to apoptosis [34].

More recently, the role of exosomes in drug resistance has begun to be explored. In a study published recently, MCF-7 (breast cancer) cells sensitive to docetaxel were exposed to exosomes extracted from the supernatant of a docetaxel-resistant MCF-7 variant. It was demonstrated that exosomes effectively transferred drug resistance characteristics from drug-resistant breast cancer cells to sensitive ones [35]. Another study using breast cancer-derived exosomes reinforced these results, showing that adriamycin and docetaxel-resistant breast cancer cells may spread resistance capacity to sensitive cells by releasing exosomes and that these effects are attributed to the intercellular transfer of specific miR-NAs [36]. Moreover, it was demonstrated that docetaxel resistance in hormone refractory prostate cancer cells can be acquired by noninvasive cell lines also via exosomes [37]. The addition of cisplatin (DDP) to A549 tumor cells (lung cancer cell line) has been shown to increase exosome secretion and the interaction of these secreted exosomes with other cancer cells increased the resistance of these A549 cells to DDP [38]. This study also demonstrated that when A549 cells were exposed to DDP, the expression levels of several miRNAs and mRNAs reportedly associated with DDP sensitivity change significantly in exosomes, and that these changes probably mediate the DDP resistance of these tumor cells.

Exosomes may also contribute to chemotherapy resistance through drug expulsion. Exosomes released from tumor cells have been shown to contain cisplatin, potentially redirecting the drug away from the nucleus where it would normally act, causing DNA damage, cell cycle arrest, and apoptosis [39]. A recent study identified another method by which exosomes may contribute to chemotherapeutic resistance. It was observed that exosomes released from cancer cells might impede antibody and drug therapies by expressing cancer-derived cell surface proteins that sequester the compound away from the target cell [40].

Taken together, the current data suggests that accurate predictions of response to cancer treatment will be incomplete unless an integrative approach is undertaken. It seems proper to consider that more attention should be given to the role of the microenvironment in drug resistance, namely the role of exosomes in therapy resistance. Understanding how exosomes impact drug resistance will allow novel and better strategies to treat cancer and prevent the emergence of drug resistance.

2   Materials

Materials

1. Cell lines.
2. Fresh plasma.
3. Frozen plasma.
4. 1.2 μm nylon filters.
5. 0.22 μm filter.

Reagents

1. Culture media.
2. 40 % Tris/sucrose/D_2O solution (40 % sucrose cushion).
3. 40 g protease-free sucrose.
    (a) 2.4 g Tris base.
    (b) 50 ml D_2O.
    (c) Adjust pH to 7.4 with 10 N HCl drops.
    (d) Adjust volume to 100 ml with D_2O.
    (e) Sterilize by passing through a 0.22 μm filter.
    (f) Store for up to 2 months at 4 °C.
4. FBS, Hyclone.
5. PBS.
6. RIPA buffer.
7. Protease inhibitor tablet (Roche).
8. Antibodies against CD3, CD9, and MHC-I.
9. 2 % PFA.
10. 200 nm phosphate buffer (pH 7.4).
11. FormVar-carbon-coated grid.
12. 1 % glutaraldehyde.
13. Aqueous uranyl oxalate.
14. 0.4 % w/v uranyl acetate.
15. 1.8 % w/v methyl cellulose.

Equipment

1. Sorvall Surespin 630 rotor.
2. Sorvall S100AT5 rotor.
3. Centrifuge.
4. NanoSight microscope.
5. Electron microscope.
6. SDS-PAGE equipment.

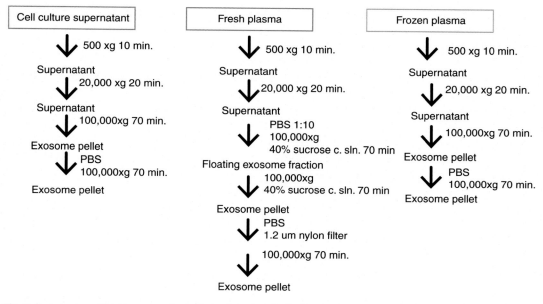

Fig. 1 Exosome purification procedures for cultured cells, fresh plasma, and frozen plasma

3 Methods

3.1 Exosome Purification from Cell Culture

The exosomes are purified by ultracentrifugation: the first steps are designed to eliminate large dead cells and large cell debris by successive centrifugations at increasing speeds. At each of these steps, the pellet is thrown away, and the supernatant is used for the following step (Fig. 1). The final supernatant is then ultracentrifuged at $100,000 \times g$ to pellet the small vesicles that correspond to exosomes. The pellet is washed in a large volume of PBS, to eliminate contaminating proteins, and centrifuged one last time at the same high speed.

1. FBS (Fetal Bovine Serum, Hyclone) is depleted of bovine exosomes by ultracentrifugation at $100,000 \times g$ for 70 min (Sorvall Surespin 630 rotor).

2. Cells are cultured in media supplemented with 10 % exosome-depleted FBS.

3. Supernatant fractions are collected from 48 to 72-h cell cultures and pelleted by centrifugation at $500 \times g$ for 10 min.

4. The supernatant is centrifuged at $20,000 \times g$ for 20 min.

5. Exosomes are then harvested by centrifugation at $100,000 \times g$ for 70 min.

6. The exosome pellet is resuspended in 20 ml of PBS and collected by ultracentrifugation at $100,000 \times g$ for 70 min.

7. The exosome pellet is resuspended in PBS and then stored at 4 °C for short term (1–7 days) or –20 °C for long term.

3.2 Exosome Isolation from Fresh Mouse and Human Plasma

Circulating exosomes are isolated from mouse and human plasma in the same way as from cell culture with an extra purification step with a sucrose cushion and an additional filtration through 1.2 μm nylon filters (GE) before the last step of ultracentrifugation. The extra purification step with a sucrose cushion eliminates more contaminants, such as proteins nonspecifically associated with exosomes, or large protein aggregates, which are sedimented by centrifugation but do not float on a sucrose gradient. The filtration through 1.2 μm nylon filters will eliminate dead cells and large debris while keeping small membranes for further purification by ultracentrifugation.

1. Plasma is pelleted by centrifugation at $500 \times g$ for 10 min.
2. The supernatant is centrifuged at $20,000 \times g$ for 20 min.
3. The supernatant is diluted 1:10 in PBS.
4. Exosomes are then harvested by ultracentrifugation at $100,000 \times g$ for 70 min on a 40 % sucrose cushion solution.
5. The floating exosome fraction is collected again by ultracentrifugation as above.
6. The exosome pellet is resuspended in 20 ml of PBS and filtered through 1.2 μm nylon filters (GE).
7. The exosome pellet is collected by ultracentrifugation at $100,000 \times g$ for 70 min.
8. The exosome pellet is resuspended in PBS and then stored at 4 °C for short term (1–7 days) or –20 °C for long term.

3.3 Exosome Isolation for Retrospective Studies Using Frozen Human Plasma

Plasma for retrospective studies is previously centrifuged at $3000 \times g$ for 20 min before storing at –80 °C.

1. 2 ml of cell-free frozen plasma is centrifuged at $500 \times g$ for 10 min.
2. Then the supernatant is centrifuged at $20,000 \times g$ for 20 min.
3. Exosomes are then harvested by centrifugation at $100,000 \times g$ for 70 min.
4. The exosome pellet is resuspended in PBS and collected by ultracentrifugation at $100,000 \times g$ for 70 min (Sorvall S100AT5 rotor).
5. The exosome pellet is resuspended in PBS and then stored at 4 °C for short term (1–7 days) or –20 °C for long term.

3.4 Electron Microscope Analysis of Exosomes

Exosomes purified as described above are fixed in 2 % PFA (w/v) in 200 mM phosphate buffer (pH 7.4). Fixed exosomes are dropped onto a formvar-carbon-coated grid and left to dry at room temperature for 20 min. After washing in PBS, the exosomes are fixed in 1 % glutaraldehyde for 5 min, washed in water, and stained with saturated aqueous uranyl oxalate for 5 min. Samples are then

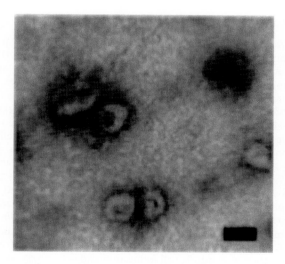

Fig. 2 Representative electron microscopic image of exosomes derived from the plasma of a melanoma subject. Scale bar, 100 nm [2]

embedded in 0.4 % w/v uranyl acetate and 1.8 % w/v methylcellulose and incubated on ice for 10 min. The excess liquid is removed. The grid is dried at room temperature for 10 min and viewed at 20,000 and 50,000 magnification using an electron microscope (model 910, Carl Zeiss) (Fig. 2).

3.5 Identification of Exosome-Specific Markers by Western Blot Analysis

Exosomes are lysed with RIPA buffer containing a complete protease inhibitor tablet (Roche). Lysates are cleared by centrifugation at $14,000 \times g$ for 20 min. Supernatant fractions are used for Western blot. Protein extracts are resolved by SDS-PAGE and probed with the indicated antibodies. For Western Blot analysis the following antibodies are used to identify specific exosome markers: anti-CD3, anti-CD9, and anti-MHC-I.

3.6 Quantification of Exosome Size, Distribution, and Number by LM10 Nanoparticle Characterization System (NanoSight)

The LM10 nanoparticle characterization system (NanoSight) equipped with a blue laser (405 nm) is used for real-time characterization of the vesicles.

Acknowledgements

We thank Prof. Maria de Sousa from the Instituto de Biologia Molecular e Celular, Universidade do Porto, Porto, Portugal, and the Instituto de Ciências Biomédicas de Abel Salazar, Porto, Portugal, for her advice.

References

1. Joyce JA, Pollard JW (2009) Microenvironmental regulation of metastasis. Nat Rev Cancer 9:239–252

2. Grivennikov SI, Greten FR, Karin M (2010) Immunity, inflammation, and cancer. Cell 140:883–899

3. Condeelis J, Pollard JW (2006) Macrophages: obligate partners for tumor cell migration, invasion, and metastasis. Cell 124:263–266

4. Zhang QW, Liu L, Gong CY, Shi HS, Zeng YH, Wang XZ, Zhao YW, Wei YQ (2012) Prognostic significance of tumor-associated macrophages in solid tumor: a meta-analysis of the literature. PLoS One 7:e50946

5. Inoshima N, Nakanishi Y, Minami T, Izumi M, Takayama K, Yoshino I, Hara N (2002) The influence of dendritic cell infiltration and vascular endothelial growth factor expression on the prognosis of non-small cell lung cancer. Clin Cancer Res 8:3480–3486

6. Bron L, Jandus C, Andrejevic-Blant S, Speiser DE, Monnier P, Romero P, Rivals JP (2013) Prognostic value of arginase-II expression and regulatory T-cell infiltration in head and neck squamous cell carcinoma. Int J Cancer 132:E85–E93

7. Kalluri R, Zeisberg M (2006) Fibroblasts in cancer. Nat Rev Cancer 6:392–401

8. Kaplan RN, Riba RD, Zacharoulis S, Bramley AH, Vincent L, Costa C, MacDonald DD, Jin DK, Shido K, Kerns SA, Zhu Z, Hicklin D, Wu Y, Port JL, Altorki N, Port ER, Ruggero D, Shmelkov SV, Jensen KK, Rafii S, Lyden D (2005) VEGFR1-positive haematopoietic bone marrow progenitors initiate the pre-metastatic niche. Nature 438:820–827

9. Hiratsuka S, Watanabe A, Aburatani H, Maru Y (2006) Tumour-mediated upregulation of chemoattractants and recruitment of myeloid cells predetermines lung metastasis. Nat Cell Biol 8:1369–1375

10. Hiratsuka S, Nakamura K, Iwai S, Murakami M, Itoh T, Kijima H, Shipley JM, Senior RM, Shibuya M (2002) MMP9 induction by vascular endothelial growth factor receptor-1 is involved in lung-specific metastasis. Cancer Cell 2:289–300

11. Simpson RJ, Lim JW, Moritz RL, Mathivanan S (2009) Exosomes: proteomic insights and diagnostic potential. Expert Rev Proteomics 6:267–283

12. Keller S, Sanderson MP, Stoeck A, Altevogt P (2006) Exosomes: from biogenesis and secretion to biological function. Immunol Lett 107:102–108

13. Akers JC, Gonda D, Kim R, Carter BS, Chen CC (2013) Biogenesis of extracellular vesicles (EV): exosomes, microvesicles, retrovirus-like vesicles, and apoptotic bodies. J Neurooncol 113:1–11

14. Corrado C, Raimondo S, Chiesi A, Ciccia F, De Leo G, Alessandro R (2013) Exosomes as intercellular signaling organelles involved in health and disease: basic science and clinical applications. Int J Mol Sci 14:5338–5366

15. Taylor DD, Gercel-Taylor C (2005) Tumour-derived exosomes and their role in cancer-associated T-cell signalling defects. Br J Cancer 92:305–311

16. EL Andaloussi S, Mager I, Breakefield XO, Wood MJ (2013) Extracellular vesicles: biology and emerging therapeutic opportunities. Nat Rev Drug Discov 12:347–357

17. Al-Nedawi K, Meehan B, Micallef J, Lhotak V, May L, Guha A, Rak J (2008) Intercellular transfer of the oncogenic receptor EGFRvIII by microvesicles derived from tumour cells. Nat Cell Biol 10:619–624

18. Peinado H, Aleckovic M, Lavotshkin S, Matei I, Costa-Silva B, Moreno-Bueno G, Hergueta-Redondo M, Williams C, Garcia-Santos G, Ghajar C, Nitadori-Hoshino A, Hoffman C, Badal K, Garcia BA, Callahan MK, Yuan J, Martins VR, Skog J, Kaplan RN, Brady MS, Wolchok JD, Chapman PB, Kang Y, Bromberg J, Lyden D (2012) Melanoma exosomes educate bone marrow progenitor cells toward a pro-metastatic phenotype through MET. Nat Med 18:883–891

19. Hood JL, San RS, Wickline SA (2011) Exosomes released by melanoma cells prepare sentinel lymph nodes for tumor metastasis. Cancer Res 71:3792–3801

20. Brown JM, Giaccia AJ (1998) The unique physiology of solid tumors: opportunities (and problems) for cancer therapy. Cancer Res 58:1408–1416

21. Heldin CH, Rubin K, Pietras K, Ostman A (2004) High interstitial fluid pressure - an obstacle in cancer therapy. Nat Rev Cancer 4:806–813

22. Netti PA, Baxter LT, Boucher Y, Skalak R, Jain RK (1995) Time-dependent behavior of interstitial fluid pressure in solid tumors: implications for drug delivery. Cancer Res 55:5451–5458

23. Durand RE (2001) Intermittent blood flow in solid tumours--an under-appreciated source of 'drug resistance'. Cancer Metastasis Rev 20:57–61

24. Galmarini FC, Galmarini CM, Sarchi MI, Abulafia J, Galmarini D (2000) Heterogeneous distribution of tumor blood supply affects the response to chemotherapy in patients with head and neck cancer. Microcirculation 7:405–410

25. Tannock IF, Rotin D (1989) Acid pH in tumors and its potential for therapeutic exploitation. Cancer Res 49:4373–4384

26. Shen J, Hughes C, Chao C, Cai J, Bartels C, Gessner T, Subjeck J (1987) Coinduction of glucose-regulated proteins and doxorubicin resistance in Chinese hamster cells. Proc Natl Acad Sci U S A 84:3278–3282

27. Pouyssegur J, Dayan F, Mazure NM (2006) Hypoxia signalling in cancer and approaches to enforce tumour regression. Nature 441:437–443

28. Rice GC, Hoy C, Schimke RT (1986) Transient hypoxia enhances the frequency of dihydrofolate reductase gene amplification in Chinese hamster ovary cells. Proc Natl Acad Sci U S A 83:5978–5982

29. Hirst DG, Denekamp J (1979) Tumour cell proliferation in relation to the vasculature. Cell Tissue Kinet 12:31–42

30. Valeriote F, van Putten L (1975) Proliferation-dependent cytotoxicity of anticancer agents: a review. Cancer Res 35:2619–2630

31. Garrido C, Ottavi P, Fromentin A, Hammann A, Arrigo AP, Chauffert B, Mehlen P (1997) HSP27 as a mediator of confluence-dependent resistance to cell death induced by anticancer drugs. Cancer Res 57:2661–2667

32. Sethi T, Rintoul RC, Moore SM, MacKinnon AC, Salter D, Choo C, Chilvers ER, Dransfield I, Donnelly SC, Strieter R, Haslett C (1999) Extracellular matrix proteins protect small cell lung cancer cells against apoptosis: a mechanism for small cell lung cancer growth and drug resistance in vivo. Nat Med 5:662–668

33. Straussman R, Morikawa T, Shee K, Barzily-Rokni M, Qian ZR, Du J, Davis A, Mongare MM, Gould J, Frederick DT, Cooper ZA, Chapman PB, Solit DB, Ribas A, Lo RS, Flaherty KT, Ogino S, Wargo JA, Golub TR (2012) Tumour micro-environment elicits innate resistance to RAF inhibitors through HGF secretion. Nature 487:500–504

34. Sun Y, Campisi J, Higano C, Beer TM, Porter P, Coleman I, True L, Nelson PS (2012) Treatment-induced damage to the tumor microenvironment promotes prostate cancer therapy resistance through WNT16B. Nat Med 18:1359–1368

35. Lv MM, Zhu XY, Chen WX, Zhong SL, Hu Q, Ma TF, Zhang J, Chen L, Tang JH, Zhao JH (2014) Exosomes mediate drug resistance transfer in MCF-7 breast cancer cells and a probable mechanism is delivery of P-glycoprotein. Tumour Biol 35:10773–10779

36. Chen WX, Liu XM, Lv MM, Chen L, Zhao JH, Zhong SL, Ji MH, Hu Q, Luo Z, Wu JZ, Tang JH (2014) Exosomes from drug-resistant breast cancer cells transmit chemoresistance by a horizontal transfer of microRNAs. PLoS One 9:e95240

37. Corcoran C, Rani S, O'Brien K, O'Neill A, Prencipe M, Sheikh R, Webb G, McDermott R, Watson W, Crown J, O'Driscoll L (2012) Docetaxel-resistance in prostate cancer: evaluating associated phenotypic changes and potential for resistance transfer via exosomes. PLoS One 7:e50999

38. Xiao X, Yu S, Li S, Wu J, Ma R, Cao H, Zhu Y, Feng J (2014) Exosomes: decreased sensitivity of lung cancer A549 cells to cisplatin. PLoS One 9:e89534

39. Safaei R, Larson BJ, Cheng TC, Gibson MA, Otani S, Naerdemann W, Howell SB (2005) Abnormal lysosomal trafficking and enhanced exosomal export of cisplatin in drug-resistant human ovarian carcinoma cells. Mol Cancer Ther 4:1595–1604

40. Ciravolo V, Huber V, Ghedini GC, Venturelli E, Bianchi F, Campiglio M, Morelli D, Villa A, Della Mina P, Menard S, Filipazzi P, Rivoltini L, Tagliabue E, Pupa SM (2012) Potential role of HER2-overexpressing exosomes in countering trastuzumab-based therapy. J Cell Physiol 227:658–667

Chapter 14

Isolation and Characterization of Cancer Stem Cells from Primary Head and Neck Squamous Cell Carcinoma Tumors

Hong S. Kim, Alexander T. Pearson, and Jacques E. Nör

Abstract

Drug resistance remains a significant problem in the treatment of patients with head and neck squamous cell carcinoma (HNSCC). Recent reports showed that a subpopulation of highly tumorigenic cells, called cancer stem cells (CSCs), is uniquely resistant to chemotherapy, suggesting that these cells play an important role in the relapse of HNSCC. The development of methods for the isolation and culture of cancer stem cells is a key step to enable studies exploring the mechanisms underlying the role of these cells in chemoresistance. Here, we describe a method to isolate cancer stem cells from primary head and neck tumors and for the generation of orospheres.

Key words Head and neck squamous cell carcinoma, Cancer stem cells, Chemoresistance, Self-renewal, Multipotency, Orospheres

1 Introduction

Head and neck squamous cell carcinoma (HNSCC) is the sixth most common cancer worldwide, with approximately 500,000 cases diagnosed annually [1]. In the past, treatment of HNSCC was primarily limited to up-front surgical resection when technically feasible. Owing to the high morbidity often associated with radical surgery, most patients today receive radiation combined with chemotherapy as front-line treatment for locally advanced HNSCC. Chemotherapy alone is implemented for the treatment of distant metastatic disease, but most patients do not exhibit substantial response. A major difficulty in treating HNSCC is the frequency of recurrent disease, with 20–40 % of patients developing loco-regional recurrence and 5–20 % developing distant metastatic disease at 2 years [2]. Chemotherapy for HNSCC typically involves the use of platinum, taxane, or pyrimidine analog agents [3]. While these agents are effective at indiscriminately debulking tumor cells, recent data report that a subpopulation

José Rueff and António Sebastião Rodrigues (eds.), *Cancer Drug Resistance: Overviews and Methods*, Methods in Molecular Biology, vol. 1395, DOI 10.1007/978-1-4939-3347-1_14, © Springer Science+Business Media New York 2016

of cells, called cancer stem cells (CSCs), are resistant and survive these therapies [4–6]. Indeed, mounting evidence suggests that chemotherapy enriches the cancer stem cell population [6, 7]. Notably, the mechanisms exploited by CSCs to resist chemotherapy are largely unknown. Therefore, the development of methods for the isolation and characterization of CSCs from primary tumors is a critical enabling step to improve the mechanistic understanding of the processes mediating chemoresistance, and for the development of therapeutic strategies to overcome this resistance.

The most widely used method to study CSCs is the sphere assay. The sphere assays were originally used for culturing normal neuronal stem cells [8]. Since the initial discovery that culturing cells in ultralow-attachment plates and serum-free (or low serum) conditions maintains their undifferentiated state, the sphere assay evolved as an important tool to study not only normal stem cells but also CSCs. Multiple reports show that CSCs are anoikis resistant and able to grow in suspension as spheres, as for example the breast CSCs [9]. Spheres from head and neck tumors are called orospheres [10, 11]. The orosphere assay is useful to evaluate the stemness, self-renewal, and tumorigenicity of CSCs, but also to study processes involved in the chemoresistance of these cells to drugs [6].

This chapter describes in detail the method for isolation of CSCs from HNSCC and for their culture in suspension as orospheres. There is ongoing discussion about the ideal markers to identify head and neck CSCs. Here, we have used the combination of ALDH activity and CD44 expression to sort these cells from primary tumors [12, 13]. The selection of markers can certainly be adapted to the specific tumor of interest. Nevertheless, the overall principles of the assay described here might be germane to other tumor types.

2 Materials

2.1 Tumor Digestion

1. Supplemented media (*see* **Note 1**).

2. Sterile petri dish.

3. Sterile razor.

4. Tumor digestion reagents:

    (a) Collagenase/hyaluronidase solution (STEMCELL Technologies, Vancouver, BC, Canada). **OR**

    (b) Miltenyi Biotech human tumor dissociation kit (Miltenyi Biotech, San Diego, CA, USA).

5. 40 μm nylon cell strainer for 50 mL conical tube.

6. Serum: Serum neutralizes the digestion process.

7. Ammonium-chloride-potassium (ACK) lysing buffer (Gibco Life Technologies, Grand Island, NY, USA) (*see* **Note 2**).

2.2 Fluorescence-Activated Cell Sorting

1. 1× PBS or 1× PBS with 2 % fetal bovine serum (staining buffer).

2. 5 mL round-bottom flow cytometry tubes.

3. 4-Diethylaminobenzaldehyde (DEAB) (STEMCELL Technologies, Vancouver, BC, Canada).

4. Activated ALDEFLUOR™ (STEMCELL Technologies, Vancouver, BC, Canada).

5. APC CD44 and APC isotype IgG (BD Pharmingen, Franklin Lakes, NJ, USA).

6. Lineage markers: PE-Cy5 CD2, CD3, CD10, CD16, CD18 (BD Pharmingen, Franklin Lakes, NJ, USA).

7. 7-Aminoactinomycin (7-AAD) (BD Pharmingen).

8. Aluminum foil.

2.3 Orosphere Culture

1. Sphere media: DMEM/F12 medium (500 mL) (Gibco Life Technologies, Grand Island, NY, USA), fibroblast growth factor (FGF) (20 ng/mL), epithelial growth factor (EGF) (20 ng/mL), N2 supplement (Gibco Life Technologies, Grand Island, NY, USA) (*see* **Note 3**).

2. 6-well low-attachment plate (Corning, Corning, NY, USA).

3. 0.05 % Trypsin/ethylenediaminetetraacetic acid (EDTA) (Gibco Life Technologies, Grand Island, NY, USA).

4. Trypsin-neutralizing solution (Lonza, Walkersville, MD, USA).

3 Methods

3.1 Preparation for Digestion

1. Place tumor in supplemented media.

2. Pour off the media and add fresh supplemented media. Centrifuge at $130 \times g$-force for 5 min at 4 °C. Wash the tumor 3–4 times (*see* **Note 4**).

3. Place tumor with 5 mL of supplemented media on a petri dish.

4. Cut the tumor into small pieces (approximately 4 mm × 4 mm in size) with sterile razor blade (*see* Fig. 1a) (*see* **Note 5**).

5. Collect minced tumor fragments into a 50 mL conical tube. Add 25–30 mL base media without supplements. Centrifuge at $130 \times g$-force for 5 min at 4 °C.

3.2 Tumor Digestion

1. Collagenase/hyaluronidase method.

   (a) Decant media and place the tumor fragments on a new sterile petri dish. Add 6–10 mL of 1× collagenase/hyaluronidase solution.

   (b) Mix the tumor fragments and digestion solution by pipetting several times with 25 mL pipette. Incubate in 37 °C for 15 min.

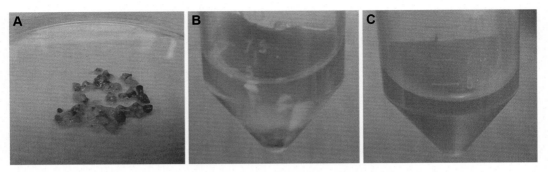

Fig. 1 Tumor digestion steps. (**a**) Cut tumors into small fragments. (**b**) Before tumor digestion. (**c**) After tumor digestion. Notice the digestion media turned opaque after the digestion process.

(c) Take out the petri dish from the incubator. Pipet the mixture 2–3 times to mechanically digest the tumor. Repeat **steps (b)** and **(c)** two more times (*see* **Note 6**).

(d) Prepare a 50 mL tube with 5 mL serum and place 40 μm nylon mesh on top of the tube. Filter the tumor digestion mix and collect cell suspension. Collect and filter remaining cell/fragment mixture with 5 mL of supplemented media (*see* **Note 7**). Centrifuge at $130 \times g$-force for 5 min at 4 °C.

(e) Decant media. Add 1–5 mL of ACK lysing buffer. Incubate in room temperature for 1 min. Centrifuge at $130 \times g$-force for 5 min at 4 °C.

(f) Decant ACK lysing buffer and resuspend the cell pellet in PBS (*see* **Note 8**).

2. GentleMACS method.

(a) Prepare enzyme cocktail mix with RPMI in the appropriate tube for GentleMACS.

(b) Transfer minced tissue fragments into the enzyme cocktail mix. Close the tube tightly (*see* Fig. 1b).

(c) Mechanically dissociate tumors using GentleMACS homogenizer. Select appropriate tumor dissociation program.

(d) Incubate in 37 °C for 30 min on a shaker or rotator.

(e) Repeat **steps (c)** and **(d)**.

(f) Repeat **step (c)** (*see* Fig. 1c).

(g) Transfer the digested tumor to 50 mL tube with 40 μm cell strainer placed on top. Use equal volume of supplemented medium to collect residual tumor cells and undigested tissues. Collect digested cell suspension by filtering with the cell strainer.

(h) Centrifuge at $130 \times g$-force for 5 min at 4 °C.

(i) Decant media. Add 1–5 mL of ACK lysing buffer. Incubate in room temperature for 1 min. Centrifuge at $130 \times g$-force for 5 min at 4 °C.

(j) Decant ACK lysing buffer and resuspend the cell pellet in PBS (*see* **Note 8**).

3.3 Isolation of Cancer Stem Cells

1. Prepare and label tubes.

    Unstained.

    7-AAD.

    Lineage.

    APC IgG.

    APC CD44.

    DEAB.

    ALDH.
    Sample: T1, T2, T3, etc.

2. Count the cells recovered from tumor digestion. Add 1×10^5 cells to each control tube except DEAB tube. Add $\leq 1 \times 10^6$ cells to the sample tube.

3. Resuspend cells in 1 mL PBS or staining buffer.

4. Add 1 µL of DEAB reagent to the control DEAB tube (*see* **Note 9**). Add 1 µL of ALDEFLUOR™ reagent control ALDH tube. Mix by pipetting once and transfer 500 µL of cells immediately to control DEAB tube (*see* **Note 10**). Mix well.

5. Add 5 µL of ALDEFLUOR™ reagent to each sample tube. Mix well. Protect from light by covering the tubes with aluminum foil. Incubate in 37 °C for 30–45 min.

6. Centrifuge at $130 \times g$-force for 5 min at 4 °C. Remove PBS or staining buffer from all the tubes. Resuspend cells in 1 mL PBS or staining buffer.

7. Add 1 µL of APC isotype IgG to control IgG APC tube. Add 1 µL of APC CD44 antibody to control APC tube. Add 5 µL APC CD44 to all the sample tubes.

8. Add 1 µL of CD2, CD3, CD10, CD16, and CD18 to control lineage tube. Add 5 µL of the same antibodies to the sample tubes.

9. Protect from light by covering the tubes with aluminum foil. Incubate in 4 °C for 30 min.

10. Add 1 mL of PBS to all of the tubes. Centrifuge at $130 \times g$-force for 5 min at 4 °C.

11. Aspirate PBS and resuspend cells with 1 mL PBS. Centrifuge at $130 \times g$-force for 5 min at 4 °C.

12. Aspirate PBS. Add 200 μL PBS to all the control tubes except control 7AAD tube.

13. Prepare 7-AAD solution by adding 5 μL 7-AAD for each 1 mL of PBS. Add 200 μL 7-AAD solution to control 7-AAD tube. Resuspend sample tubes with 500 μL 7-AAD solution.

14. Sort cancer stem cells by fluorescence-activated cell sorting (FACS) (*see* Fig. 2) (*see* **Note 11**).

3.4 Orosphere Assay

1. Seed sorted CSCs in 6-well low-attachment plate. 5,000 sorted cells are plated in each well of low-attachment plate (*see* **Note 12**). Add 1.5 mL of sphere media in each well (*see* **Note 13**).

2. Feed spheres every 3–4 days by adding 300 μL of sphere media to each well.

3. Count spheres on days 3 and 7 (*see* **Note 14**) (*see* Fig. 3).

4. On day 7, collect spheres in 15 mL tube. Centrifuge at $130 \times g$-force for 5 min at 4 °C.

5. Wash with PBS. Centrifuge at $130 \times g$-force for 5 min at 4 °C. Remove supernatant.

6. Add 1 mL 0.05 % trypsin/EDTA. Incubate for 5–10 min in room temperature. Mechanically dissociate the spheres by pipetting up and down every 2 min until spheres are invisible.

7. Add 1 mL trypsin-neutralizing solution and mix. Centrifuge at $130 \times g$-force for 5 min at 4 °C.

8. Count the cells.

9. Seed the cells in low-attachment plate. 5,000 cells are plated in each well of low-attachment plate. Add 1.5 mL of sphere media in each well (*see* **Note 13**).

10. Feed spheres every 3–4 days by adding 300 μL of sphere media to each well. Count spheres on days 10 and 14.

4   Notes

1. Depending on the tumor type, add necessary supplements and growth factors to keep tumor fragment. The media should include antibiotics (e.g., amphotericin B, AAA, nystatin) to prevent bacterial/fungal contamination.

2. ACK lysis buffer eliminates red blood cells from the digested cell pellet.

3. CSCs from different tumor type require different supplements needed for sphere formation. Here, we describe the supplements needed for orosphere assay.

4. Washing the tumor prevents potential contamination during cancer stem cell culture following isolation of cancer stem cells.

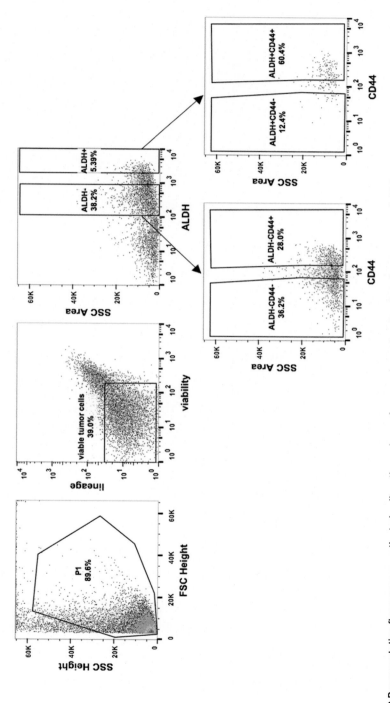

Fig. 2 Representative fluorescence-activated cell sorting scheme from a primary head and neck squamous cell CSC

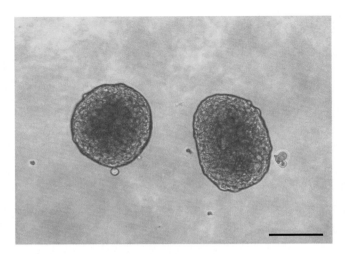

Fig. 3 Examples of orospheres. Scale bar = 100 µm

5. Mincing the tumor into smaller pieces is important for good cell recovery from digestion. Too much cutting results in unhealthy cells, and too little cutting results in poor tumor digestion.

6. As the tumor fragments are digested, the solution will become opaque, and the tumor fragments will be able to pass through 10 mL pipette.

7. Incompletely digested tumor fragments can be stored for future use. Resuspend the chunks in freeze media that contains 5 % DMSO and 5 % serum and move gradually to liquid nitrogen for storage.

8. Serum-containing staining buffer will enhance the cell viability during the staining procedure.

9. After adding DEAB to the tube, close the lid so the solution will not evaporate.

10. Transfer the cell suspension as soon as possible to minimize the background ALDEFLUOR™ signal.

11. CSCs are 7-AAD-negative, lineage-negative, ALDEFLUOR™-positive, CD44-positive cells.

12. Cell density can affect the number and the quality of orosphere. Optimize the number of cells plated in each well.

13. Add chemotherapeutic reagents to the sphere media to study the chemoresistance of CSCs.

14. Orospheres are considered as non-adherent colony with at least 25 cells.

References

1. Kamangar F, Dores GM, Anderson WF (2006) Patterns of cancer incidence, mortality, and prevalence across five continents: defining priorities to reduce cancer disparities in different geographic regions of the world. J Clin Oncol 24:2137–2150

2. Forastiere AA, Goepfert H, Maor M, Pajak TF, Weber R, Morrison W, Glisson B, Trotti A, Ridge JA, Chao C, Peters G, Lee DJ, Leaf A, Ensley J, Cooper J (2003) Concurrent chemotherapy and radiotherapy for organ preservation in advanced laryngeal cancer. N Engl J Med 349:2091–2098

3. Pfister DG, Spencer S, Brizel DM, Burtness B, Busse PM, Caudell JJ, Cmelak AJ, Colevas AD, Dunphy F, Eisele DW, Gilbert J, Gillison ML, Haddad RI, Haughey BH, Hicks WL Jr, Hitchcock YJ, Jimeno A, Kies MS, Lydiatt WM, Maghami E, Martins R, McCaffrey T, Mell LK, Mittal BB, Pinto HA, Ridge JA, Rodriguez CP, Samant S, Schuller DE, Shah JP, Weber RS, Wolf GT, Worden F, Yom SS, McMillian NR, Hughes M (2014) Head and neck cancers, Version 2.2014. Clinical practice guidelines in oncology. J Natl Compr Canc Netw 12:1454–1487

4. Okamoto A, Chikamatsu K, Sakakura K, Hatsushika K, Takahashi G, Masuyama K (2009) Expansion and characterization of cancer stem-like cells in squamous cell carcinoma of the head and neck. Oral Oncol 45:633–639

5. Zhang Q, Shi S, Yen Y, Brown J, Ta JQ, Le AD (2010) A subpopulation of CD133(+) cancer stem-like cells characterized in human oral squamous cell carcinoma confer resistance to chemotherapy. Cancer Lett 289:151–160

6. Nor C, Zhang Z, Warner KA, Bernardi L, Visioli F, Helman JI, Roesler R, Nor JE (2014) Cisplatin induces Bmi-1 and enhances the stem cell fraction in head and neck cancer. Neoplasia 16:137–146

7. Reers S, Pfannerstill AC, Maushagen R, Pries R, Wollenberg B (2014) Stem cell profiling in head and neck cancer reveals an Oct-4 expressing subpopulation with properties of chemoresistance. Oral Oncol 50:155–162

8. Reynolds BA, Weiss S (1992) Generation of neurons and astrocytes from isolated cells of the adult mammalian central nervous system. Science 255:1707–1710

9. Dontu G, Abdallah WM, Foley JM, Jackson KW, Clarke MF, Kawamura MJ, Wicha MS (2003) In vitro propagation and transcriptional profiling of human mammary stem/progenitor cells. Genes Dev 17:1253–1270

10. Krishnamurthy S, Dong Z, Vodopyanov D, Imai A, Helman JI, Prince ME, Wicha MS, Nor JE (2010) Endothelial cell-initiated signaling promotes the survival and self-renewal of cancer stem cells. Cancer Res 70:9969–9978

11. Krishnamurthy S, Nor JE (2013) Orosphere assay: a method for propagation of head and neck cancer stem cells. Head Neck 35:1015–1021

12. Clay MR, Tabor M, Owen JH, Carey TE, Bradford CR, Wolf GT, Wicha MS, Prince ME (2010) Single-marker identification of head and neck squamous cell carcinoma cancer stem cells with aldehyde dehydrogenase. Head Neck 32:1195–1201

13. Prince ME, Sivanandan R, Kaczorowski A, Wolf GT, Kaplan MJ, Dalerba P, Weissman IL, Clarke MF, Ailles LE (2007) Identification of a subpopulation of cells with cancer stem cell properties in head and neck squamous cell carcinoma. Proc Natl Acad Sci U S A 104:973–978

Chapter 15

Clinical and Molecular Methods in Drug Development: Neoadjuvant Systemic Therapy in Breast Cancer as a Model

Sofia Braga

Abstract

Neoadjuvant chemotherapy (NACT), neoadjuvant endocrine therapy (NAET), and neoadjuvant targeted therapy (NATT), more recently, have been adopted worldwide as standard of care in locally advanced and inoperable BC. These modalities, collectively called neoadjuvant systemic therapy (NAST), are also used for organ preservation and for mechanistic biological studies on drug response and resistance, drug development, and clinical trials. Furthermore, the response to NACT is a valuable indicator of long-term survival. In this work, the advantages and pitfalls of using NAST in BC for studying drug response and resistance for drug development and clinical trials are discussed as well as practical points on how to set up a NAST clinical trial in BC.

Key words Neoadjuvant breast cancer, Radiologic response, Metabolic response, ki67 labeling, Pathologic complete response, Overall survival, Polymerase chain reaction, Immunohistochemistry

1 Introduction

Breast carcinoma is the most frequently diagnosed malignancy in women and the second cause of death by cancer in women [1]. Locally advanced breast cancers (LABC) present a clinical challenge because part of the patients will eventually relapse and die despite aggressive multimodality treatments [2]. LABC include stage III disease with T0–T3 breast primaries with clinically detectable, palpable lymph nodes. Some T2 breast primaries with involved, but not palpable, lymph nodes and T3 tumors without involved axillary nodes are called large operable breast carcinomas (LOBC) and can also be managed by multimodality therapy including systemic therapy, surgery, and radiotherapy. Inflammatory breast carcinoma (IBC) is a rare clinical entity accounting for less than 3 % of BC. It is an aggressive form of BC with poor prognosis. Clinically it is characterized by diffuse erythema and edema of

José Rueff and António Sebastião Rodrigues (eds.), *Cancer Drug Resistance: Overviews and Methods*, Methods in Molecular Biology, vol. 1395, DOI 10.1007/978-1-4939-3347-1_15, © Springer Science+Business Media New York 2016

the breast without an underlying palpable mass. IBC must be distinguished from indolent neglected LABC; therefore a short clinical history of less than 6 months is a diagnostic criterion. IBC should be managed with systemic treatment, surgery, and radiotherapy [3]. In these three clinical situations, a consensus has been reached that systemic treatment should be the initial modality as it increases resectability and breast conservation and provides valuable information regarding systemic treatment sensitivity without compromising long-term survival [4]. Patients treated with neoadjuvant systemic therapy (NAST) when compared stage by stage with patients treated solely with adjuvant systemic therapy (AST) have worse outcomes. Matched ypTNM and pTNM BC have worse disease-free survival (DFS) and overall survival (OS); in daily practice, this means that a patient with, for example, six involved lymph nodes that underwent primary surgery followed by AST has the same prognosis as a patient with one involved lymph node after NAST [5]. NAST, chemotherapy, endocrine therapy, or targeted therapy has thus been established as a standard of care in the last three decades.

In drug development, the testing of novel agents in the preoperative (NAST) setting is an efficient strategy, using pathologic complete response (pCR) as the primary endpoint. Molecularly, NAST allows the assessment of drug effects on the target, the pharmacodynamic response, by performing serial biopsies. Molecular profiling of the residual tumor in the surgical specimen may also provide insights into actionable mechanisms of resistance. NAST is the only clinical trial research setting where it is relatively easy to obtain a snapshot of the tumor before and after treatment and therefore study drug activity with biological material. In addition tumors are drug naive, which is a unique opportunity to evaluate drug sensitivity and maximal drug effect as well as resistance development which is the main reason metastatic breast cancer is still an incurable disease. NAST provides a unique opportunity to assess in pretreatment and posttreatment tumor tissue changes in molecular markers that may be predictive of response or resistance to therapy. Molecular markers are used in drug development to test hypotheses and new agents in biologically defined cohorts of patients. The US Food and Drug Administration considers neoadjuvant trials for accelerated drug approval in early breast cancer, particularly for tumors with high risk of recurrence and unfavorable prognosis, recognizing the potential of this drug testing model [6]. pCR has been shown to be a surrogate marker for survival [7]; therefore, it has been postulated that patients treated with NAST might have better long-term outcomes because of early eradication of micrometastatic disease, but meta-analyses and large trials have failed to show this correlation.

It has been postulated that NAST might increase patient survival due to early eradication of micrometastatic disease as has been observed in animal models. In animal experiments, excision of primary tumors in mice leads to the release from the tumor to the circulation of growth-stimulating factors and of malignant cells that can form metastatic foci. When NAST was administered to tumor-carrying mice, both tumor cell proliferation and the release of growth-stimulating factors into the serum were suppressed, and survival was improved [8]. Thus, in mice, NAST appears to be associated with better local and systemic disease control although these results have never been shown in patients.

Standardized methodologies are therefore necessary to draw conclusions that impact long-term survival and immediate conclusions on drug and schedule development in clinical trials. There are standard surgical pathology measures of systemic therapy activity in the neoadjuvant setting. There are other imaging and histopathological measures that have been developed to study drug response before surgery. The standardization of such procedures used in NAST is the objective of this work.

2 Materials

2.1 Breast Biopsy

Breast core needle biopsy is the initial procedure for histopathological diagnosis of breast cancer. Core biopsy has replaced fine-needle aspiration (FNA) cytology due to the ability to perform histopathological studies, namely immunohistochemistry (IHC) studies that are critical in BC. A breast core needle biopsy should be performed in an experienced radiology department. The material requirement is a sterile biopsy tray that includes scalpel, generally a standard spring-loaded 14-gauge biopsy needle, probe guide, skin-cleansing sponge, syringes for administration of local anesthesia, tweezers, and gauze. Syringes contain 3 mL of 1 % lidocaine buffered with sodium bicarbonate (9:1 mixture) for skin and subcutaneous anesthesia, 10 mL of 1 % lidocaine with epinephrine buffered with sodium bicarbonate (9:1 mixture) for deep breast parenchyma anesthesia, and 10 mL of 1 % lidocaine with epinephrine buffered with sodium bicarbonate (9:1 mixture) for deep breast parenchyma anesthesia to be administered via the probe (if possible) [9] (*see* **Note 1**).

2.2 Diagnostic Pathology Report

Biopsy of the primary tumor is mandatory. All patients should have a histopathological characterization of the tumor and histological grading by an experienced breast pathologist, quantitative ER, PgR, HER2, and Ki67 staining by immunohistochemistry. The requirement is a hospital-based pathology lab with expertise in breast pathology [10] (*see* **Note 2**).

2.3 Conventional Radiology

Patients with breast cancer that are candidates for NAST should always have performed bilateral mammography, ultrasonography of the breasts, and axilla. The requirement is a hospital-based radiology department with expertise in mammary radiology [11] (*see* **Note 3**).

2.4 Axillary Biopsy

All radiologically suspect involved axillary nodes should be evaluated with an image-guided biopsy. In clinically detectable unequivocally involved axillary nodes biopsy is not mandatory, except in research settings, but it is increasingly common. The material requirement is a sterile biopsy tray that includes scalpel, generally a standard spring-loaded 14-gauge biopsy needle, probe guide, skin-cleansing sponge, syringes for administration of local anesthesia, tweezers, and gauze. Syringes contain 3 mL of 1 % lidocaine buffered with sodium bicarbonate (9:1 mixture) for skin and subcutaneous anesthesia, 10 mL of 1 % lidocaine with epinephrine buffered with sodium bicarbonate (9:1 mixture) for deep axillary anesthesia, and 10 mL of 1 % lidocaine with epinephrine buffered with sodium bicarbonate (9:1 mixture) for deep axillary anesthesia to be administered via the probe (if possible) [9, 12, 13] (*see* **Note 4**).

2.5 Skin Biopsy

All patients with equivocal skin involvement (TNM stage cT4b) should have a skin biopsy performed by an experienced breast surgeon. The material requirement is a skin biopsy tray that includes a skin antiseptic like isopropyl alcohol, povidone-iodine solution or chlorhexidine, an anesthetic like lidocaine, 1 or 2 % with epinephrine, and the circular cutting instrument—a trephine—that exists in 2–10 mm sizes [14] (*see* **Note 5**).

2.6 Magnetic Resonance Imaging

Magnetic resonance imaging (MRI) should be performed before any decision on NAST. MRI results impact the decision of NAST versus surgery as well as the surgical approach. The requirement is a hospital-based radiology department with expertise in mammary MRI. The equipment itself in terms of field strength should be 1.5 T or higher with double-breast coil with bilateral coverage acquisitions with slice thickness of 3 mm [15, 16] (*see* **Note 6**).

2.7 Positron Emission Tomography

Positron emission tomography (PET) scan is not mandatory in routine clinics, although it is informative, namely on detecting internal mammary gland, and supraclavicular and mediastinal lymph node disease. In clinical trials FDG-PET staging is done per protocol but it is increasingly demanded [17]. The requirement is a hospital-based nuclear medicine department with expertise in FDG-PET (*see* **Note 7**).

2.8 Multidisciplinary Team

A group of BC-treating physicians has to make NAST decisions in the multidisciplinary team (MDT). The professionals present should be the radiologist, the pathologist, the nuclear medicine physician, the surgeon, the oncologist and the radiotherapist, the medical geneticist, the psycho-oncologist, and the breast nurse [18] (*see* **Note 8**).

2.9 NAST   The choice of therapy has to do with the clinical context. In clinical trials, specific drugs are being tested; in daily practice, NAST is quite well defined. In patients that are too frail and old or have contraindications for CT, ET can be an option in inoperable cancer. In patients who have contraindications to BC surgery ET is also the chosen option, if biologically possible [19–21]. In fit patients who are good candidates for NACT the two types of CT that should be administered are anthracyclines and taxanes [22]. In patients that do not respond to either chemotherapy groups there is conflicting data on the best conduct. Although controversial, there are clinical trial and anecdotal evidence that generally the exposure to platinum salts, vinca alkaloids, or antimetabolites such as 5FU might rescue some non-responding. The choice of NACT according to biological subtypes is shown in Figs. 1, 2, 3, 4, and 5. All patients with tumors that have amplification of HER2 gene should be treated with regimens containing trastuzumab; lapatinib has not been validated for this purpose [23–36]. The logistic requirement to perform NACT is a well-trained breast oncologist in a medical oncology department of a hospital (*see* **Note 9**).

2.10 Tumor Labeling   NAST can, in selected cases, totally eradicate the tumor. This is, on one hand, favorable for the patient because pCR correlates with survival, and on the other unfavorable for the surgeon who does not

Miller-Payne System	Histopathologic findings
Grade 1	No change or some alteration to individual malignant cells, but no reduction in overall cellularity
Grade 2	A minor loss of tumor cells, but overall cellularity is still high; up to 30% loss
Grade 3	Between and estimated 30% and 90% reduction in tumor cells
Grade 4	A marked disappearance of tumor cells such that only small clusters or widely dispersed individual cells remain, more than 90% loss of tumor cells
Grade 5	No malignant cells identifiable in sections from the site of the tumor; only vascular fibroelastotic stroma remains, often containing macrophages; however, ductal carcinoma in situ may be present

Fig. 1 The Miller-Payne response criteria classification

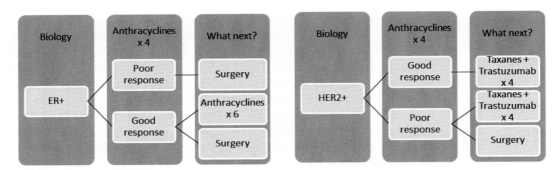

Fig. 2 The choice of NACT according to biological subtypes. Decision matrix based on clinical, epidemiological, radiological, pathological, biological, and response data collected during NAST NACT for ER+ and HER2+ carcinomas

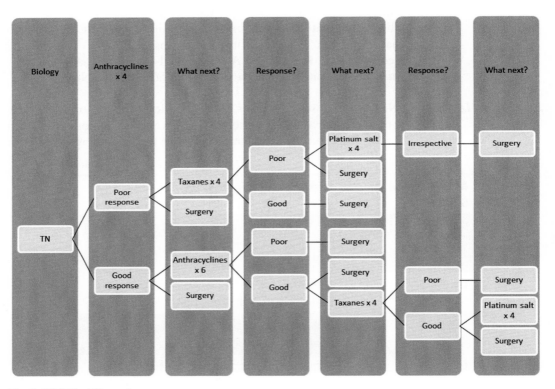

Fig. 3 NACT for TN carcinomas

detect the tumor and therefore cannot perform breast-conserving surgery (BCS). To overcome this, very small titanium clips or carbon particles to delineate the tumor contour are used [12]. This procedure can be done before or during NACT if very good response is observed to the initial two anthracycline cycles. The requirement is a hospital-based pathology lab with expertise in breast pathology and the material is the same as for breast biopsy (*see* **Note 10**).

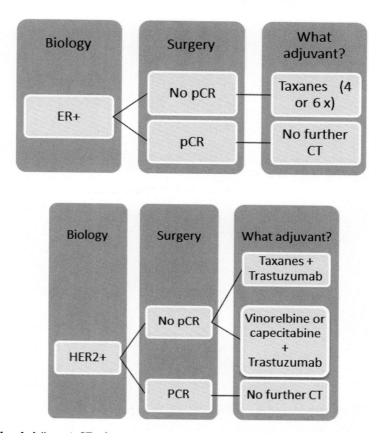

Fig. 4 Adjuvant CT after surgery in ER+ and HER2+ carcinomas that were treated with NACT

2.11 Biopsy at 2 Weeks

A biopsy at 2 weeks is never a clinical requirement. It will always be a research requirement to assess molecular biomarkers of early response. Standard methodologies should be implemented to enable the evaluated molecules to be used to test hypotheses and new agents [6]. This material can be used to extract RNA for PCR and cDNA microarray gene expression profiling. The gene expression profiling generally requires frozen material. The requirement is a hospital-based pathology lab with expertise in breast pathology (*see* **Note 11**).

2.12 Clinical Response Assessment

Clinical observation by an experimented oncologist in conducting NAST is invaluable in response assessment and therefore resistance. Perpendicular diameters of the lesion, if palpable, should be recorded, and caliber measurements are most adequate [37]. Photographs, if possible with drawn pictures in the skin and a ruler next to the lesion, are desirable. The logistic requirement to perform NACT is a well-trained breast oncologist in a medical oncology department of a hospital (*see* **Note 14**).

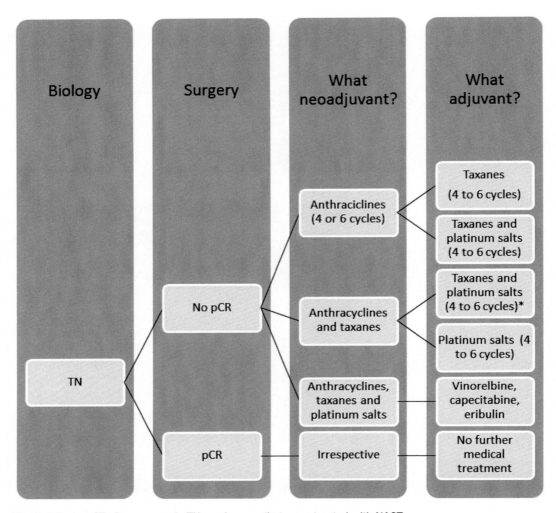

Fig. 5 Adjuvant CT after surgery in TN carcinomas that were treated with NACT

2.13 Radiological Response Assessment

In clinical practice this is not mandatory. Sonography may be requested in equivocal cases where non-response is equated. In the clinical trial setting it is generally mandatory to assess radiological response after a short course of 2 months, especially when targeted agents are being tested without classical CT. The stringent radiological response evaluation criteria are in place because targeted agents in this setting are not standard of care and if the lesion is not responding, clinical trial withdrawal is obligatory [11, 15, 37]. The requirement is a hospital-based radiology department with expertise in mammary sonography and MRI (*see* **Note 12**).

2.14 PET Response Assessment

PET can help monitor NAST response, specifically in the clinical trial setting, where, as explained, when testing nonstandard agents one must be extremely careful not to harm patients that could otherwise have a curative treatment [38]. The requirement is a hospital-based nuclear medicine department with expertise in FDG-PET (*see* **Note 13**).

2.15   Surgery

In the last three decades we have accumulated evidence supporting that a BCS followed by radiotherapy is as effective as mastectomy in terms of local control and disease-free and overall survival as first therapeutic approach in newly diagnosed breast cancer patients [39, 40]. However, the evidence is less clear regarding the role of BCS following NA therapy. Nevertheless, data points out that such patients managed with BCS have an acceptably low rate of loco-regional recurrences [19–24, 41]. Therefore BCS with immediate reconstruction, sentinel node biopsy (SNB), and axillary lymph node dissection (ALND) are all validated and can be performed in patients undergoing NAST [42] (*see* **Note 15**).

2.16   Surgical Pathology

The most important assessment of surgical pathology in breast cancer patients treated with NAST is the histological grading of pCR. pCR, as has been said, correlates with survival. Another crucial determinant of long-term survival is the number of involved lymph nodes after NAST. Frequently there is variation of tumor biological parameters—tumors lose ER expression or HER2 expression—this is a reflection of clonal evolution of the tumor and shows us the dying versus the surviving clones. Several, very large, 3000 patient trials have consistently seen that those patients obtaining pCR have excellent overall survival, irrespective of initial staging [4, 7, 41, 43–45]. This observation is also seen in clinical practice. We are very confident, by looking at our own patients, that pCR correlates with OS. The material obtained in the surgical specimen is also valuable to study biomarkers of response to NAST; such single-gene approaches are generally done with PCR and the multilane approaches are performed with high-throughput data methods such as gene expression profiling (*see* **Notes 16** and **17**).

2.17   Adjuvant Therapy

Adjuvant treatment will be prescribed according to all clinical, epidemiological, radiological, pathological, biological, and response data collected during NAST. All features are taken into account and it is a quite daunting task to reproduce our decision matrix which is attempted in the figures [26, 28–36] (Fig. 2).

2.18   Survival

Breast cancer is not a very acute oncological disease; results of interventions take years to mature, in the majority of cases. Clinical trial data has showed that BC survival has not been affected by NAST, as was initially postulated. The hypothesis put forward was that NAST would kill circulating tumor cells or microscopic distant organ metastasis prematurely, and, therefore, increase survival. This has unfortunately not been shown by the large NAST trials.

Mature survival data is only available at 5 years, and, even at 10- and 15-year time points, very interesting outcome data is collected, namely on late recurrences and second primaries. Breast cancer patients should be followed indefinitely if the institutional profile can adapt to such clinical burden [4, 7, 41, 43–45] (*see* **Note 18**).

3   Methods

3.1   *Breast Biopsy*

Stereotactic biopsy can be performed using a traditional prone stereotactic biopsy table or an add-on stereotactic biopsy unit, attachable to the mammography as well as tomosynthesis equipment. Attachable biopsy units allow the conversion of the mammography or tomosynthesis units into a biopsy guidance system and allow the patient to sit upright or be recumbent, but not prone, during the biopsy. Allowing the patient to sit during the procedure may cause more anxiety due to visualization of the biopsy needle. The patient has to sit still during 30 min. Before the biopsy, the patient's mammograms or tomosynthesis images are reviewed to determine the approach that allows the lesion's best visualization, shortest distance from the skin to the lesion, and avoidance of intervening vessels. The best approach is also determined by the patient's breast size, compressed breast thickness, and lesion's depth and location. The needle should hit the target inferior or superior to the lesion, especially if the lesion is small, so that the lesion is visible adjacent to the needle aperture, on the images post-firing to permit confirmation of accurate targeting. Once the lesion coordinates have been calculated and confirmed to be adequate for biopsy, the skin is cleansed. The skin and breast parenchyma are anesthetized, and a skin incision is made. The biopsy needle is then introduced through the skin incision into the breast parenchyma, to the pre-firing position just proximal to the target. Pre-firing stereotactic images are obtained to confirm the expected needle trajectory. Retargeting may be performed before "firing," if necessary. The needle is then fired and post-firing stereotactic images are obtained to confirm optimal final-needle aperture position. Repositioning can be performed if the probe aperture is not adjacent to the targeted lesion. It is generally not necessary to create another skin incision when repositioning. When the new target has been selected and transmitted to the equipment, simply retract the needle until the tip is just beneath the skin, select the new target, move the needle to the newly targeted pre-fire position, and then "fire" the probe. After the post-firing images confirm accurate targeting, tissue sampling in a 360° fashion may be performed. More accurately, the sampling is performed mainly at the clock positions where the lesion is located, as identified on the post-firing images. For example, if the lesion is now located superior to the probe aperture, only the areas from 9- to 3-o'clock position, going through the 12-o'clock position, need to be sampled. Sometimes, selective sampling of the 10-, 12-, and 2-o'clock positions may be sufficient, depending on how large the lesion is. The use of vacuum-assisted biopsy (VAB) needle devices is now the standard of care for stereotactic biopsy. These needles are powered with suction and have a rotating cutter. They can obtain multiple samples in a 360° fashion without the need to remove the needle from the lesion. The vacuum draws the breast tissue into an aperture in the probe, where the tissue is cut; the tissue is then transported to the specimen port for collection. VAB needles have been shown to be more accurate and allow larger volume tissue sampling than automated biopsy needles. In addition, VAB offers a higher calcification retrieval rate and a lower rate of targeting errors with decreased rebiopsy and underestimation rates. The needle size ranges from 11 to 7 gauge. Depending on the type of lesion and size of the biopsy needle, on average, 6–12 samples are obtained. For a small cluster of calcifications, a small number of samples may be sufficient for accurate diagnosis. However, an area of architectural distortion generally requires a larger number of core samples to avoid missing any underlying malignancy and to ensure accurate diagnosis of a radial scar.

3.2 Pathology Report

The pathology report should be read by an experienced breast pathologist. The histological subtype should be reported as well as histological grade. Immunohistochemistry to detect ER, PgR, HER2, and Ki67 is obligatory. The following primary antibodies are used: anti-Ki67 (clone MIB-1; Dako cat. 7240, 1:300), estrogen receptor (clone SP1; Ventana Roche cat. 790-4324), and progesterone receptor (clone 1E2; Ventana Roche cat. 790-2223). After incubation with the primary antibodies, immunohistochemical studies are carried out using the Autostainer link 48 (DakoCytomation, Carpinteria, CA). Slides are incubated with Polymer Kit-Envision (HRP) DAKO 150 µl for 30 min, washed, and sequentially stained with diaminobenzidine, Mayer's hematoxilin, and ethanol. Nuclei are counterstained with Mayer's hematoxylin. Images are acquired on a Leica DM5500 microscope. HER2 gene amplification and Ki67 quantitative staining should be assessed by methods that use genomic probes to detect polysomy in equivocal cases on immunohistochemical staining. HER2 expression and/or amplification is evaluated by pathway HER-2/neu (clone 4B5; Ventana Roche cat. 780-001) in Ventana BENCHMARK ULTRA instrument. The probes used are as follows: centromere enumeration probe 17, generally labeled in green, and locus-specific identifier ERBB2 probe, generally labeled in orange (Vysis-Abbott, Downers Grove, IL). Slides are prepared according to the manufacturer's instructions for paraffin sections. A positive result is defined as an ERBB2 gene/chromosome 17 ratio of ≥ 2.2. Images are visualized on a fluorescence microscope and captured on a workstation (MetaSystems, Altlussheim, Germany). A minimum of 100 nuclei are counted per case. The Ki-67 positivity level is defined as percentage of stained cells. The quantitative ER and PgR evaluation is established with the percentage of stained cells. The paraffin-embedded sections are stained by hematoxylin–eosin (H&E) for light microscopic examination. Tumor histology is assessed by analyzing the morphological features. All tumors are graded by a pathologist according to the Elston–Ellis histological grading. Tumors showing loss of E-cadherin plasma membrane staining are lobular carcinomas. The antibody used to study E-cadherin is clone NCH-38. It is of great interest to predict response to NACT in order not to use such a modality in tumors that will be chemoresistant. A classification system that predicts pathological complete response pCR is the breast cancer index (BCI). This index has been developed to assist prognosis in adjuvant patients [46]. The index tests by RT-PCR the presence of mRNA of nine genes: two genes HOXB13 and IL17BR had been previously tested as a ratio, HOXB13/IL17BR; two other genes are HER2 and ER and the last five represent histological grade and cellular differentiation and are BUB1B, CENPA, NEK2, RACGAP1, and RRM2. This test in the initial biopsy has been shown to be predictive of response to NACT [47]. Other proteins such as *tau* protein a microtubule stabilizer have been associated with less response to taxanes but results are conflicting [48].

3.3 Conventional Radiology

Conventional radiology entails mammography and ultrasonography. Mammography is able to detect radiopaque formations. It is frequent for cells of solid tumors to become necrotic due to lack of blood supply. Necrotic cell debris frequently appear as aggregates of microcalcifications, a radiological hallmark of cancer. Ultrasonography is able to detect solid lesions that do not have calcifications. Young women and BRCA germline mutations are bad candidates for mammographic screening [11].

3.4 Axillary Biopsy

The biopsy should be performed from inferolateral to superomedial locations toward the target to avoid major vessels and muscles. To flatten the axilla, a wedge pillow is used to rotate the patient's body and elevate the targeted area. The radiologist should visualize the target lymph node clearly at prebiopsy scanning and use color Doppler ultrasonography (US) to determine if there are large vessels around the target. After the puncture site and optimal needle approach have been determined, a local anesthetic is placed both superficially and deeply. During deep placement of anesthetic, the needle can be used as a probe to detect any sensitive nerves around the target and simulate the best approach. After the patient has been anesthetized, a small skin incision is made with a #11 scalpel blade. The biopsy needle is then advanced manually under real-time US guidance. Note that the outer cannula remains uncocked during manual needle advancement. In some cases, a tougher fascial layer under the superficial tissues requires a deeper incision or the use of a diamond-tipped guide cannula, and advancing the needle after cocking the outer cannula leaves a relatively thin and flexible portion of the stylet to withstand the insertion through the fascial tissues, potentially compromising placement accuracy or even bending the needle at the collection trough, the thinnest portion. One should advance the entire uncocked needle—including the outer, more rigid cutting cannula—to or through the cortex of the abnormal lymph node and then cocking the needle to expose the inner collection trough. The needle tip should be visualized at all times, especially when there are large vessels around the target. Liberal use of color Doppler US is encouraged, both to avoid major vessels and to select a portion of the target node that is less likely to bleed. Somewhat surprisingly, targeting the central hilar area, even when considerable vascularity is present, has neither resulted in any significant hematoma nor compromised sampling or pathologic interpretation. At present, Doppler US at low flow rates (approximately 4 cm/s) is highly sensitive for imaging blood flow and tends to lead to overestimation of the potential for vascular injury. When the needle tip is just at or within the target, the gun is cocked once, which opens the trough without advancing the needle tip. After manually adjusting the position of the trough to center on the target, the radiologist presses the button, thereby releasing the outer cutting cannula and closing the trough. The radiologist inserts the needle with the bevel facing up to facilitate penetration of the superficial tissues and advance the needle with the plastic flanges used to cock the needle optimally positioned so as not to compromise the angle of advancement. The open trough can be positioned with exquisite accuracy just prior to sampling using this technique. It is recommended that a portion of the trough bridges the cortex and surrounding fat, which helps the pathologist see the interface between the target tissue and surrounding normal tissue and thus more readily identify the target sample as being from a lymph node. Unlike with most other US core needle biopsy procedures, the needle is placed through the lymph node manually rather than by firing the trocar and the cutting cannula automatically. The fatty tissue of the axilla and the relatively soft consistency of both normal and abnormal lymph nodes allows the firing of just the outer cutting cannula to produce a very good core specimen. Use of this procedure prevents inadvertent needle damage, since the needle does not move after placement and the cannula samples only tissue that has already been traversed by the specimen trocar. After sampling and documentation of the needle throw position are achieved under US guidance, the needle is withdrawn, and the radiologist should apply manual pressure on the biopsy site to minimize bleeding. The radiologist should evaluate the specimen visually both before and after placing the sample into a 10 % formalin solution. The specimen should have a white (generally pathologic) or brown tan (lymph node tissue) component, possibly with a yellow component representing adjacent fatty tissue. Because the trough of the needle (19 mm) is usually longer than the target lymph node, some adjacent fat (density, 0.9 g/mL) will often accompany the sampled lymph node tissue (density, 1.04–1.07 g/mL), causing the sample to float or partially sink. A lymph node replaced by tumor will usually produce a sample that sinks to the bottom of the container, sometimes after losing a few air bubbles that accompany the core initially. As mentioned earlier, we usually obtain two samples, one through the middle of cortex and the other at the periphery. If the samples appear not to contain lymph node tissue (cortex) or tumor at visual assessment, additional sampling is advised. On average, we obtain about 2.2 samples per case. Occasionally, we will target more than one lymph node if another suspicious node is in the immediate vicinity, but this is not common.

3.5 Skin Biopsy

The site of the biopsy is chosen according to the presence of peau d'orange induration and infiltrated subcutaneously with 2–5 ml of 1 % plain lignocaine. A nipple biopsy usually requires injection of the anesthesia at the base of the nipple to produce a "nipple block." A single disk of full-thickness skin with a layer of underlying adipose tissue is removed and multiple biopsies are performed if required. Pressure is applied to the biopsy site for a few minutes until hemostasis is achieved.

3.6 MRI

Generally the breast MRI protocol corresponds to a 3D gradient-echo, dynamic imaging before and after intravenous contrast administration (gadolinium dimeglumine pentetate) and after a subtraction technique. Regions of interest (ROIs) are drawn around tumor areas. Wash in and wash out in the same point of the tumor is measured to quantify vascular permeability. Tumors are measured in their two longest perpendicular axes and reassessed at day 15 reported using RECIST [15].

3.7 PET

In investigational protocols, two FDG-PET examinations are performed, on days 1 and 15. Imaging is performed on a PET (Siemens, Medical Solutions, USA, IC) 1 h after iv injection of 370 MBq 18F-fluorodeoxyglucose. The PET scan is equipped with a lutetium oxyorthosilicate crystal (64 crystals/block), with an axial extent of 162 mm and a spatial resolution of 6 mm at 1 cm. Standardized uptake value (SUV) measurement of glucose metabolism is analyzed. Response is defined as a reduction of 20 % or more (maximum values of SUV) in all lesions present in the baseline time point (t1).

3.8 MDT

The MDT should discuss all operable BC patients in order to select adequate candidates that may benefit from NA approach. Contraindications to NAT are those patients not suited to be treated with CT and those patients with operable tumors not sensitive to CT. These are patients with strongly ER-positive tumors, simultaneously PgR positive and with low proliferation indexes (histological grade and Ki67 labeling). Generally, patients with HER2-positive tumors benefit from NAT, attaining pCR levels of 50 %. Patients with TN tumors due to the heterogeneity present in this disease have a more uncertain response pattern. However, pCR rates of 50 % have been obtained in certain well-selected patients with TNBC. Patients with ER-positive tumors should be treated only if BCS is unfeasible and the tumors, despite expressing ER, are expected to respond to NAT. Frail and extremely old patients with ER-positive tumors may be treated with NAET [49].

3.9 PST

Anthracyclines for four cycles every 3 weeks should be administered. Patients with strongly ER-positive tumors with low proliferation markers and low histological grade might not respond to NACT; therefore early stopping and surgery are most of the times performed after 12 weeks, or even 9 weeks of NACT. If patients with ER-positive tumors are responding well, with manageable toxicity, the full course of anthracyclines (six cycles) should be administered. Patients with triple-negative tumors (tumors that do not express ER, PgR, or HER2) have very heterogeneous sensitivity to NACT. These tumors are either highly sensitive or highly resistant; this finding has been termed the "TNBC paradox." Patients with TN carcinomas responding well to four cycles of anthracyclines should continue NACT to attain pCR. There is an option between immediate surgery in operable patients or pursuing therapy. The drawback of pursuing therapy is that one can turn a responder into a non-responder and patients should be surgically treated at best response. On the other hand, administering full-course NACT might give the best chance to obtain pCR. NACT should be used with care beyond four cycles of anthracyclines in patients with ER-positive and TN tumors that are not responding well [43, 50]. Regarding patients with TN tumors that respond well to chemotherapy or patients with HER2-positive disease, the conduct is quite different. These two entities have a pCR rate of over 50 %, so the ideal management should be to treat with four or six cycles of anthracyclines and four cycles of taxanes which corresponds to 12 weekly administrations, if this is the preferred modality. Patients with TNBC should, after the results of the GeparSixto trial, be offered carboplatin in addition to the current third-generation anthracycline and taxane backbone because it has increased pCR from 37 to 53 % in this subset of patients that relapse early when pCR is not attained [51]. In case of patients with HER2-positive tumors trastuzumab should be added to taxanes, not to anthracyclines for safety reasons, due to increased potential for causing cardiotoxicity (Figs. 1 and 2) [26, 28–36].

3.10   Tumor Labeling

The primary tumor, if expected to disappear, should be marked with metal or carbon clips; disappearance of the lesion depends on the biological subtype. Overall 50 % of primary lesions can disappear; a great percentage of these lesions are TN or HER2 positive. Once the specimen retrieval is completed, a localizing post-biopsy marker clip, usually a 3 mm titanium clip, is placed at the biopsy site. To ensure that the clip is deployed at the biopsy site, an image of the biopsied area is obtained, while the breast is still under compression on the biopsy unit. Because clip migration has been reported in up to 20 % of cases, 2-view mammography, including a cranio-caudal and a medio-lateral or later-omedial view, should be performed after the breast is decompressed to assess the final position of the deployed clip. Regarding metal clips and the metal used, titanium clips that measure 3 mm should be employed; these can only be placed with a 10–14-gauge needle. These are expensive and if not removed surgically will be permanent and radio-opaque. Carbon-coated microparticles are easier to insert, cheaper, and much smaller, 100–1000 μm, that are injected with a hypodermic needle and will last from 5 to 8 weeks (http://www.google.com/patents/US6394965) [12].

3.11   Biopsy at 2 Weeks

The main assessments at the early time point are always for research purposes. They include Ki67 change and other molecular biomarkers of early response depending on the drugs being studied. Nowadays material should be frozen in –80 °C refrigerator for further study, namely protein or nucleic acid extraction. Alternatively, RNA or protein extraction should be done immediately after the procedure and harvested at –20 °C. If no frozen material is obtained, FFPE material should be archived indefinitely in all cases. RNA extraction can be performed from FFPE samples [6].

3.12   Radiological Response Assessment

In radiological response assessment the two perpendicular diameters of the lesions are recorded and a comparison is made with the baseline image. A 10–30 % reduction in the greatest diameter of the primary tumor is defined as minor response (MR). A clinical partial response (cPR) is considered when the decrease in the greatest diameter of the primary tumor is ≥30 %. Progression (PD) is established if the primary tumor increased in size or new tumor lesions are observed. Otherwise, we consider stable disease (SD) [49].

3.13   PET Response Assessment

In response assessment using PET, delta SUV is the relevant indicator [17, 49].

3.14   Clinical Response Assessment

Patients are evaluated for clinical response and toxicity before each NAST cycle. Tumors are measured by caliper in two perpendicular diameters and the greatest diameter is considered for evaluation. A 10–30 % reduction in greatest diameter of the primary tumor is defined as minor response (MR). A clinical partial response (cPR) is considered when the decrease in greatest diameter of the primary tumor was ≥30 %. Progression (PD) is established if the primary tumor increased in size or new tumor lesions are observed. Otherwise, we consider stable disease (SD).

3.15 Surgery	SLNB after NAST has been criticized for wide variations in the false-negative rates, whereas pre-PST sentinel lymph node biopsy commits many patients to a completion ALND on the basis of pretreatment nodal positivity, thereby negating the PST downstaging benefits [52]. The adoption of axillary lymphatic mapping and sentinel lymph node biopsy (SLNB) in stage I/II, clinically N0 patients has promoted the judicious use of SLNB in selected patients who have undergone NACT, if the nodes are "downstaged" and are clinically negative at the completion of NACT. SLNB in these patients remains highly controversial, as does the application of NACT in patients with smaller (T1, N1, or T2, N01) cancers. ALND should be performed in all patients with involved immediate reconstruction. The amazing results of NAST prompted investigations of breast conservation for selected patients, generally very good responders. Criteria for BCT in patients that undergo NAST are (1) patient desire for breast conservation, (2) absence of multicentric disease which means tumors in multiple quadrants of the breast, (3) absence of diffuse microcalcifications on mammogram, (4) absence of skin involvement consistent with inflammatory breast cancer, and (5) residual tumor mass amenable to a margin-negative lumpectomy resection. LABC was traditionally considered a contraindication for immediate breast reconstruction (IBR). Therefore we perform immediate breast reconstruction with implants or with muscles from the transverse rectus abdominus myocutaneous (TRAM) or the latissimus dorsi flap [53].
3.16 Surgical Pathology	Several types of response are seen, either concentric diminution of the initial mass or a primary tumor that leaves scattered foci in the remaining parenchyma. The most important measure is the residual cancer burden (RCB) [54]. Residual disease (RD) assessment has strong outcome correlation. RD is determined by bidimensional diameters of the primary tumor bed (d1 and d2) that contains invasive carcinoma (f inv), the number of axillary lymph nodes containing metastatic carcinoma (LN), and the diameter of the largest metastasis in an axillary lymph node (d met). D prim is the square root of d1 times d2. The proportion of invasive carcinoma (f inv) within the cross-sectional area of the primary tumor bed is estimated from the overall percent of carcinoma (%CA) and then corrected for the component of in situ carcinoma (%CIS):

$$F\,inv = \left(1 - \left(\%CIS/100\right)\right) \times \left(\%CA/100\right)$$

To develop the RCB index the four parameters of residual tumor (d prim, f inv, LN, and d met) are individually associated. To calculate a single index of RCB first there is information on the primary tumor bed (RCB prim = f inv × d prim) and next on regional metastasis (RCB met = 4(1 − 0.75 e LN) d met). Finally the RCB index applies a power transformation to avoid skewness of the distribution:

$$RCB = 1.4 (f\ inv \times d\ prim)\, e\, 0.17 + [4(1\ 0.75\, e\, LN)\, d\ met]\, e\, 0.17$$

Besides RCB other more classic studies are done on the surgical pathology specimen. The Miller-Payne response criteria classification, shown in Fig. 1, was widely used before RCB was defined [55]. The tumor cells are evaluated for dissociation, dyscohesion, and loss of organization of the tumor cells and necrobiotic changes such as necrosis, vacuolation of nucleus and cytoplasm, karyorrhexis, pyknosis, and karyolysis. Any change in pattern or type of carcinoma should be recorded. The stroma is examined for host response including fibrosis, elastosis, and collagenization, and infiltration by lymphocytes, plasma cells, fibroblasts, histiocytes, and giant cell formation was observed. Similar changes in tumor cells and stroma are observed in the lymph nodes. Epithelial-to-stromal ratio is calculated as the mean of readings in all sections and viable-to-nonviable tumor cell ratios are calculated in both pretreatment biopsies and post-NAST specimens, viability being defined as distinct nuclear chromatin with intact nuclear and cytoplasmic membrane in the absence of the criteria of necrosis (karyorrhexis, karyolysis, pyknosis) [56]. A third system developed in the Netherlands is called the neoadjuvant response index (NRI). NRI is calculated by adding a breast response score (a number from a five-point scale) to an axillary response score (a number from a three-point scale) and dividing this by the score that would have been obtained in case of a pCR in both breast and axilla. Consequently, the NRI is a number between 0 (representing no response) and 1 (a pCR of both breast and axilla) [57].

Lymphocytic response is graded as: grade 1, scattered lymphocytes between tumor cells; grade 2, formation of microaggregates of lymphocytes; and grade 3, dense infiltration of lymphocytes destroying tumor cells or forming masses [58]. The presence of lymphovascular embolization and in situ disease/cancerization of ducts are separately noted.

For drug development, RNA extraction, and cDNA synthesis from paraffin-embedded blocks (FFPE), archived breast cancer surgical blocks in FFPE are cut into tissue sections of 5 μm. These are deparaffinized and counterstained with Mayer's hematoxylin and eosin. Cancer-enriched areas are needle microdissected under the breast pathologist's guidance. Total RNA is extracted with the RNeasy FFPE kit, according to the manufacturer's instructions with a slight modification: proteinase K cell lysis at 56 °C is performed overnight. The RNase-Free DNase Set "on column" DNA digestion procedure is included. Each extracted RNA is reverse-transcribed with the First-Strand cDNA Synthesis kit, using a 1:1 mixture of random primers (pd(N)6) and oligo-dT primers (NotI-d(T)18). cDNA from control samples (high-quality RNA from HCT-116 and a primary skin fibroblast cell line) is synthesized from 3 µg of total RNA.

3.17 RT-qPCR

RNA concentration and integrity cannot be assessed using standard methods due to known FFPE degradation issues and to the small amounts of extracted samples. Thus, to indirectly check the amount of each isolated total RNA FFPE sample and its RT-qPCR downstream performance, we prepare two standard dilution series using cDNA from the two control cell lines, corresponding to 100, 10, 1, 0.1, 0.01, 0.001, and 0.0001 ng of the original total RNA. These series are subsequently used to calculate a RT-qPCR standard curve for the non-differentially expressed gene MAPKAPK2 (Lods = −2.7). Primer sets were designed with the NCBI Primer-BLAST tool [59], to work at 59 °C and with an amplicon length of 70–100 bp. Duplicates of each breast cancer sample are analyzed by RT-qPCR using SsoFast™ EvaGreen Supermix reagent in 10 L of reaction mixture containing template (2 L, ~200 pg/L) and primers (0.5 M each). Samples are processed in a CFX96 Touch Real-Time PCR Detection System according to the cycling program: 95 °C for 60 s, 50 cycles of 95 °C for 10 s, and 59 °C for 15 s. Fluorescence data collection occurs at 59 °C. Relative differential expression analysis of target genes by RT-qPCR is based on the 2−∆∆Ct methodology from [60] using mean quantification cycle of duplicates as cycle threshold.

3.18 Immuno histochemistry

Immunohistochemistry (IHC) is performed in breast cancer samples (3 µm thick tissue sections), and according to standard protocols. Primary antibodies to whichever genes one is studying are diluted in Bond Primary Antibody Diluent plus background-reducing components at the most adequate dilutions, generally 1:100 or 1:150. Evaluation of tumor biopsies and surgical specimen evaluation are performed by an investigator blinded to clinical data, generally a breast pathologist. Three biological samples from each patient at the various time points studied are studied by histopathology. Surgical samples are evaluated according to ypTNM. Tumor samples were classified according to standard pathology criteria for histological subtype, grade of differentiation, Ki67 staining, hormonal receptor expression, and HER2 expression and/or amplification. For sample collection for molecular studies tumor tissue is obtained by image-guided biopsy prior to treatment, after 14 days of the studied molecule, and from the definitive surgical specimen. Sometimes a second biopsy on treatment might be required. Tissue is flash-frozen in liquid nitrogen and stored at −80 °C. Hematoxylin- and eosin-stained sections of formalin-fixed paraffin-embedded specimen are used to evaluate histological characteristics. For microarray data analysis, total RNA is extracted from frozen material with RNeasy Mini Kit with RNase-Free DNase Set "on column" DNA digestion, according to the manufacturer's instructions. RNA concentration is determined on an RNA quantity and quality, respectively, examined with a Nano Drop ND-1000 UV-vis Spectrophotometer and an RNA 6000 Nano Assay Kit in a Bioanalyser 2100. Only high-quality total RNA samples with a 260/280 ratio >1.8 and containing at least 100 ng are further processed. For gene expression profiling, ample labeling and GeneChip processing are performed at an Affymetrix Core Facility. Total RNA (100 ng) is used to generate cDNA with the Ambion WT Expression Kit for Affymetrix GeneChip Whole Transcript WT Expression Arrays. Biotin-labeled cRNAs, produced with the Affymetrix GeneChip WT Terminal Labeling kit according to the Hybridization User Manual, are hybridized to Human Gene 1.0 ST arrays. Liquid handling steps are performed with GeneChip Fluidics Station 450 and arrays scanned with GeneChip Scanner 3000 7G using Command Console v1.1. Raw data should be publicly available at GEO under a series number (GSE *xxxxxxxx*). Microarray data analysis is performed with R Statistical Computing software complemented with Bioconductor packages. Microarrays are generally pre-processed using aroma. affymetrix v2.4.0 package (http://aroma-project.org/). Heatmaps are plotted using "gplots" or any other bioconductor package. Differential expression analysis is obtained with limma package for the several comparisons: comparison 1 is generally diagnosis versus the first 2 weeks on therapy, both with biopsy-collected samples. Comparison 2 is done between non-responding and responding tumor samples. Finally, comparison 3 is performed between on-therapy biopsy versus surgically collected samples. Selection of differentially expressed genes (DEGs) is based on limma output parameters LODS >0 (log-odds) and log2-ratio ≥+0.58 or ≤0.58. Gene set enrichment analysis (GSEA) of GO biological processes (GO-BP) among filtered DEGs is analyzed with GSEA tools from InnateDB (http://www.innatedb.ca/).

3.19 Adjuvant Systemic Therapy

In patients with pCR no other systemic chemotherapy is necessary. Targeted therapy with trastuzumab or antiestrogens is mandatory. For patients that do not have pCR taxanes may be continued or maintained and there is data for platinum salts in TNBC and for capecitabine or vinorelbine in non-responding patients [61, 62]. All adjuvant treatment data is displayed in the figures (Figs. 2, 3, 4, 5, and 6).

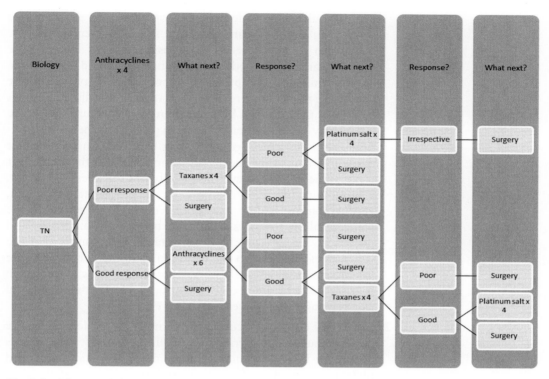

Fig. 6 Decision matrix based on clinical, epidemiological, radiological, pathological, biological and response data collected during NAST.

3.20 Survival

Patients that undergo PST and obtain pCR have better survival. Patients with residual disease after pCR, specially of with TNBC, have poor survival [45]. It is also known that stage-matched data from operated patients comparing ypTNM with pTNM also shows worse survival in the ypTNM patients showing that for those that have residual tumor after PST prognosis is worse than for stage-matched primarily operated patients [5].

4   Notes

1. Breast biopsy: Stereotactic biopsy is challenging in patients with a small or thin breast, superficial lesion close to the skin, deep lesion adjacent to the chest wall, or lesion located in the very superior portion of the breast (inner or outer quadrants). Dedicated maneuvers generally allow any lesion visible under mammography or tomosynthesis to be successfully biopsied. Image quality of the small breast or anterior breast may be poor if the breast does not fill the opening of the compression paddle. A malleable putty or aluminum foil may be placed adjacent to the breast to decrease any "air gap" and to improve

image quality. When the patient's compressed breast thickness is less than 3 cm on mammography, special maneuvers may be needed to permit stereotactic biopsy. First, the breast tissue may be manually pushed inward on both sides and up from the nipple, while slow compression of the breast is applied with the paddle, so that more breast tissue gathers centrally and bulges anteriorly through the biopsy rectangular aperture. Then the breast may be subsequently taped in place with the aid of additional sponge, gauze, or saline bag to ensure that the breast tissue maintains this thickness during biopsy. If this maneuver is not sufficient to achieve adequate breast thickness for biopsy, a "double-paddle" approach may be taken. With this approach, another biopsy paddle may be placed in front of the image plate, to allow the breast tissue to bulge both anteriorly and posteriorly. In all cases, it is important to remember that the lesion does not need to be at the center of the needle aperture for successful sampling. With VAB probes, as long as the lesion is at the edge of the needle aperture, successful tissue retrieval can generally be achieved.

2. Pathology report: All cutoff points must be predefined before the correlations are carried out, in all of the cases. Pathologists should be blind to clinical, epidemiological, radiological, and pharmacologic data. E-cadherin staining might be helpful to diagnose lobular BC because it is insensitive to CT.

3. Conventional imagiology: One should not use necessarily only the maximum tumor diameters in the standard cranio-caudal and medio-lateral oblique views. Mammographic findings associated with malignancy include nodular densities with poorly defined or spiculated margins, clustered or pleomorphic calcifications, or architectural distortion. Real-time sonography findings associated with malignancy include hypoechoic solid masses with poorly defined margins, shadowing, or acoustical enhancement.

4. Axillary biopsy: It might bring interesting information, namely on clonal evolution. Increasingly this biopsy should be mandatory due to pCR in the axilla; this biopsy is the only unequivocal confirmation of axillary disease. The whole maneuver described in Subheading 3 requires some practice and physical strength but is easily accomplished with experience.

5. Skin biopsy: After a skin biopsy documenting skin involvement, because it changes surgical conduct, namely nipple-sparing mastectomies and BCS should be rediscussed and generally avoided.

6. MRI: MRI is also useful to study the contralateral breast. Decisions to surgically explore the contralateral breast are taken in 10–20 % of cases. In our experience breast MRI

should be performed before any NAST. Accreditation and standardization of breast MRI procedures have not yet been internationally decided and to date there have been few established guidelines for image acquisition in breast MRI.

7. PET: In several NAST clinical trials PET-CT to evaluate early (at 15 days) or more delayed response has been incorporated. In daily clinical practice, PET-CT evaluation is not routinely used and the accuracy in detecting regional lymph node infiltration is still debated because strong specificity (97 %) but poor sensitivity (20 %) has been reported mainly because of high-dimensional detection threshold (1 cm) [17]. Initial baseline SUV uptake is higher in grade 3, high Ki67, ER-negative tumors. Change in SUV is greatest in TN- and HER2-positive tumors but this change in TN tumors has not been shown to be predictive of path CR as is the case with HER2-positive tumors [38]. These results are in accordance to the heterogeneity in path CR rates observed in TN tumors.

8. MDT: For decisions on PST the interaction between surgeons and oncologists in the MDT is crucial. As in all other team-work, the team should respect one another and work very closely. Breast oncologists and breast surgeons talk daily as well as pathologists.

9. PST: The current use of NACT for LOBC should be limited to TNBC and HER2-positive BC and arguably some poorly differentiated tumors that express low level of steroid hormone receptors and have very high ki67 labeling. In LABC and IBC the use of NACT is mandated due to a staging imperative not a biological rationale. There is a reduced amount of data, data regarding hundreds of patients, that dose-dense NACT might be of interest in some biological subtypes. Dose-dense chemotherapy might be an option in young fit patients with TN breast cancer; otherwise there is increased toxicity with small incremental benefit in path CR [63].

10. Tumor labeling is performed as has been explained with metal clips or carbon microparticles. The most important caveat of labeling is that one should always remember to label tumors that are responding because if unlabeled a mastectomy might be performed in a patient that would otherwise have been treated with breast conservation. A newer approach with iodine isotope 125 has been developed that enables labeling of multifocal and multicentric lesions with the objective of performing BCS; this newer method should be further explored because up to now any multifocal lesion is a strong indication for mastectomy [64].

11. Biopsy at 2 weeks Ki67 change: The current ki67 change results do not support using short-term changes in Ki67 to guide treatment in individual patients. This clinical model would be useful,

however, for assessing new agents as a screening tool before conducting large-scale trials. Molecular biomarkers of early response: Ki67 assessment is considered a reproducible method while the additional role of other proliferative markers such as cyclin D1 and E is uncertain. In the opinion of most panelists, published data are strong enough to state that Ki67 assessed after primary endocrine therapy is a marker of long-term clinical outcome and warrants validation as a surrogate endpoint. Conversely, the question of whether an early increase in Ki67 expression during treatment should be considered a sign of treatment failure that may lead to a change of the therapy has divided the panelists. The identification of early changes in the expression of molecular biomarkers after treatment makes PST the ideal approach for planning the so-called window-of-opportunity studies that have a molecular change as primary endpoint. Standard guidelines for study design are required however. Changes in soluble biomarkers after PST do not provide complementary information with respect to tissue biomarkers. Similarly, circulating tumor cell counts and circulating tumor cell molecular characterization before and after treatment have not provided up to now sufficient data to consider this method useful in the clinical management of patients having PST. Further studies are required to define its role in this setting. Antiangiogenic agents are increasingly being used in PST regimens in prospective clinical trials. However, their role is being debated after concerns were raised regarding antiangiogenics in the metastatic setting. Commonly adopted response criteria may not be the best way to monitor the activity of these drugs. Several methods have been proposed, and MRI and PET scans have provided the most interesting data. The majority of panelists suggest MRI ± PET scan a useful approach, but a clear consensus to formulate a recommendation was not obtained.

12. Radiological response assessment: Mammography is not performed during NAST for response assessment because it is not predictive of response. Sonography namely with elastography is a useful technique. Ultrasound elastography is a dynamic imaging technique that has been established as a promising modality to identify relative tissue by measuring the degree of strain-related distortion under the application of an external force. It consists of applying quasi-static pressure on the examined tissue surface periodically and estimating the induced strain distribution by tracking tissue motion. The likelihood of path CR when sonography demonstrates no residual disease is 80 % [65].

13. PET response assessment: In our experience PET was more useful than MRI in predicting NAST response. Delta SUV obtained by PET after two cycles can predict path CR and potentially identify a subgroup of non-responding patients for

whom continuation of toxic and ineffective chemotherapy should be avoided and immediate breast surgery performed.

14. Clinical response assessment: Up to 10 % of patients progress upon NAT; generally between 10 and 20 % do not respond but have stable disease. These patients have a worse long-term outcome; however to treat them with NAST is not deleterious as the worse outcome is unrelated to the timing of ST, be it adjuvant or NA. A residual mass after NACT is generally considered a contraindication to BCS. In a retrospective analysis of a large NA trial the patients with residual clinically palpable mass were less likely to be treated with breast conservation.

15. Surgery SNB is a controversial but feasible approach. It should be more explored because there is more than 50 % axillary downstaging and SNB can avoid the morbidity of axillary clearance [42]. ALND, immediate reconstruction. Multifocal or multicentric tumors should generally be treated with mastectomy, as well as T3 tumors, i.e., that measure 5 cm or more in their greatest diameter.

16. Surgical pathology: grading response, biology, Ki67: pCR is an important predictor of survival in all BC biological subtypes: those that express steroid hormone receptors, HER2-amplified BC, and TNBC. On the contrary, failure to achieve pCR is only clearly associated with worse outcomes in TNBC and HER2-positive BC; it is not observed for the majority of hormone receptor-positive BC. In this subtype of BC path CR rates are invariably low because these patients arguably receive the most efficacious therapy, endocrine therapy, as adjuvant treatment. Molecular biomarkers at surgery: The most relevant biomarker is tumor cell viability because it has long-term survival impact. Tumor cell viability has been well classified in the residual cancer burden scoring system as has been detailed.

17. Adjuvant chemotherapy: In patients treated with anthracyclines in the neoadjuvant setting that have had suboptimal pathological response taxanes have been used [66]. In patients already treated with both regimens, the choice of adjuvant therapy is less clear and less backed up by data. There is German data using capecitabine, vinorelbine, and platinum salts, but no regimen is standard of care. All the data and experience gathered have shown that non-responders to anthracyclines and taxanes have a very unfavorable long-term prognosis.

18. Survival: NAST takes place during a year: four to six months of NAST, surgery, and adjuvant therapy. Then patients are followed and breast cancer long-term survival and local and distant relapse data are collected. Long-term data in BC implies 5–10-year follow-up. It is not difficult to imagine that the majority of women with breast cancer will become candidates

for NACT as more information about tumor response and outcome data are accumulated. Nowadays, increasingly we are treating patients with 2 cm tumors (T1c) with negative axillas with NAST; although there is no data for better outcome there is data for equivalence [67].

References

1. Siegel R, Naishadham D, Jemal A (2013) Cancer statistics, 2013. CA Cancer J Clin 63:11–30

2. Giordano SH (2003) Update on locally advanced breast cancer. Oncologist 8:521–530

3. Rea D, Francis A, Hanby AM, Speirs V, Rakha E, Shaaban A, Chan S, Vinnicombe S, Ellis IO, Martin SG, Jones LJ, Berditchevski F, UK Inflammatory Breast Cancer Working group (2015) Inflammatory breast cancer: time to standardise diagnosis assessment and management, and for the joining of forces to facilitate effective research. Br J Cancer 112:1613–1615

4. Fisher B, Bryant J, Wolmark N, Mamounas E, Brown A, Fisher ER, Wickerham DL, Begovic M, DeCillis A, Robidoux A, Margolese RG, Cruz AB Jr, Hoehn JL, Lees AW, Dimitrov NV, Bear HD (1998) Effect of preoperative chemotherapy on the outcome of women with operable breast cancer. J Clin Oncol 16:2672–2685

5. Lee NK, Shin KH, Park IH, Lee KS, Ro J, Jung SY, Lee S, Kim SW, Kim TH, Kim JY, Kang HS, Cho KH (2012) Stage-to-stage comparison of neoadjuvant chemotherapy versus adjuvant chemotherapy in pathological lymph node positive breast cancer patients. Jpn J Clin Oncol 42:995–1001

6. Bardia A, Baselga J (2013) Neoadjuvant therapy as a platform for drug development and approval in breast cancer. Clin Cancer Res 19:6360–6370

7. Cortazar P, Zhang L, Untch M, Mehta K, Costantino JP, Wolmark N, Bonnefoi H, Cameron D, Gianni L, Valagussa P, Swain SM, Prowell T, Loibl S, Wickerham DL, Bogaerts J, Baselga J, Perou C, Blumenthal G, Blohmer J, Mamounas EP, Bergh J, Semiglazov V, Justice R, Eidtmann H, Paik S, Piccart M, Sridhara R, Fasching PA, Slaets L, Tang S, Gerber B, Geyer CE Jr, Pazdur R, Ditsch N, Rastogi P, Eiermann W, von Minckwitz G (2014) Pathological complete response and long-term clinical benefit in breast cancer: the CTNeoBC pooled analysis. Lancet 384:164–172

8. Fisher B, Gunduz N, Coyle J, Rudock C, Saffer E (1989) Presence of a growth-stimulating factor in serum following primary tumor removal in mice. Cancer Res 49:1996–2001

9. Huang ML, Adrada BE, Candelaria R, Thames D, Dawson D, Yang WT (2014) Stereotactic breast biopsy: pitfalls and pearls. Tech Vasc Interv Radiol 17:32–39

10. Fitzgibbons PL, Page DL, Weaver D, Thor AD, Allred DC, Clark GM, Ruby SG, O'Malley F, Simpson JF, Connolly JL, Hayes DF, Edge SB, Lichter A, Schnitt SJ (2000) Prognostic factors in breast cancer. College of American Pathologists Consensus Statement 1999. Arch Pathol Lab Med 124:966–978

11. Londero V, Bazzocchi M, Del Frate C, Puglisi F, Di Loreto C, Francescutti G, Zuiani C (2004) Locally advanced breast cancer: comparison of mammography, sonography and MR imaging in evaluation of residual disease in women receiving neoadjuvant chemotherapy. Eur Radiol 14:1371–1379

12. Corsi F, Sorrentino L, Bossi D, Sartani A, Foschi D (2013) Preoperative localization and surgical margins in conservative breast surgery. Int J Surg Oncol 2013:793819

13. Kronenwett U, Huwendiek S, Castro J, Ried T, Auer G (2005) Characterisation of breast fine-needle aspiration biopsies by centrosome aberrations and genomic instability. Br J Cancer 92:389–395

14. AlGhamdi KM, AlEnazi MM (2011) Versatile punch surgery. J Cutan Med Surg 15:87–96

15. Lobbes MB, Prevos R, Smidt M, Tjan-Heijnen VC, van Goethem M, Schipper R, Beets-Tan RG, Wildberger JE (2013) The role of magnetic resonance imaging in assessing residual disease and pathologic complete response in breast cancer patients receiving neoadjuvant chemotherapy: a systematic review. Insights Imaging 4:163–175

16. Prevos R, Smidt ML, Tjan-Heijnen VC, van Goethem M, Beets-Tan RG, Wildberger JE, Lobbes MB (2012) Pre-treatment differences and early response monitoring of neoadjuvant chemotherapy in breast cancer patients using magnetic resonance imaging: a systematic review. Eur Radiol 22:2607–2616

17. Monzawa S, Adachi S, Suzuki K, Hirokaga K, Takao S, Sakuma T, Hanioka K (2009) Diagnostic performance of fluorodeoxyglucose-positron emission tomography/computed tomography of breast cancer in detecting axillary lymph node metastasis: comparison with ultrasonography and contrast-enhanced CT. Ann Nucl Med 23:855–861

18. Boileau J-F, Simmons C, Clemons M, Gandhi S, Lee J, Chia SK, Basik M, Provencher L, Untch M, Brackstone M (2012) Extending neoadjuvant care through multi-disciplinary collaboration: proceedings from the fourth annual meeting of the Canadian Consortium for Locally Advanced Breast Cancer. Curr Oncol 19(2):106–114

19. Dowsett M, Smith IE (2011) Presurgical progesterone in early breast cancer: so much for so little? J Clin Oncol 29:2839–2841

20. Dowsett M, Ebbs SR, Dixon JM, Skene A, Griffith C, Boeddinghaus I, Salter J, Detre S, Hills M, Ashley S, Francis S, Walsh G, Smith IE (2005) Biomarker changes during neoadjuvant anastrozole, tamoxifen, or the combination: influence of hormonal status and HER-2 in breast cancer—a study from the IMPACT Trialists. J Clin Oncol 23:2477–2492

21. Smith IE, Dowsett M, Ebbs SR, Dixon JM, Skene A, Blohmer J-U, Ashley SE, Francis S, Boeddinghaus I, Walsh G (2005) Neoadjuvant treatment of postmenopausal breast cancer with anastrozole, tamoxifen, or both in combination: the Immediate Preoperative Anastrozole, Tamoxifen, or Combined with Tamoxifen (IMPACT) multicenter double-blind randomized trial. J Clin Oncol 23:5108–5116

22. Deo SVS, Bhutani M, Shukla NK, Raina V, Rath GK, Purkayasth J (2003) Randomized trial comparing neo-adjuvant versus adjuvant chemotherapy in operable locally advanced breast cancer (T4b N0-2 M0). J Surg Oncol 84:192–197

23. Baselga J, Bradbury I, Eidtmann H, Di Cosimo S, de Azambuja E, Aura C, Gomez H, Dinh P, Fauria K, Van Dooren V, Aktan G, Goldhirsch A, Chang TW, Horvath Z, Coccia-Portugal M, Domont J, Tseng LM, Kunz G, Sohn JH, Semiglazov V, Lerzo G, Palacova M, Probachai V, Pusztai L, Untch M, Gelber RD, Piccart-Gebhart M, Neo AST (2012) Lapatinib with trastuzumab for HER2-positive early breast cancer (NeoALTTO): a randomised, open-label, multicentre, phase 3 trial. Lancet 379:633–640

24. Mittendorf EA, Wu Y, Scaltriti M, Meric-Bernstam F, Hunt KK, Dawood S, Esteva FJ, Buzdar AU, Chen H, Eksambi S, Hortobagyi GN, Baselga J, Gonzalez-Angulo AM (2009) Loss of HER2 amplification following trastuzumab-based neoadjuvant systemic therapy and survival outcomes. Clin Cancer Res 15:7381–7388

25. Miolo G, Muraro E, Martorelli D, Lombardi D, Scalone S, Spazzapan S, Massarut S, Perin T, Viel E, Comaro E, Talamini R, Bidoli E, Turchet E, Crivellari D, Dolcetti R (2014) Anthracycline-free neoadjuvant therapy induces pathological complete responses by exploiting immune proficiency in HER2+ breast cancer patients. BMC Cancer 14:954

26. von Minckwitz G, Kummel S, Vogel P, Hanusch C, Eidtmann H, Hilfrich J, Gerber B, Huober J, Costa SD, Jackisch C, Loibl S, Mehta K, Kaufmann M, German Breast Group (2008) Intensified neoadjuvant chemotherapy in early-responding breast cancer: phase III randomized GeparTrio study. J Natl Cancer Inst 100:552–562

27. Qi M, Li JF, Xie YT, Lu AP, Lin BY, Ouyang T (2010) Weekly paclitaxel improved pathologic response of primary chemotherapy compared with standard 3 weeks schedule in primary breast cancer. Breast Cancer Res Treat 123:197–202

28. Chang HR, Glaspy J, Allison MA, Kass FC, Elashoff R, Chung DU, Gornbein J (2010) Differential response of triple-negative breast cancer to a docetaxel and carboplatin-based neoadjuvant treatment. Cancer 116:4227–4237

29. Bear HD, Anderson S, Smith RE, Geyer CE Jr, Mamounas EP, Fisher B, Brown AM, Robidoux A, Margolese R, Kahlenberg MS, Paik S, Soran A, Wickerham DL, Wolmark N (2006) Sequential preoperative or postoperative docetaxel added to preoperative doxorubicin plus cyclophosphamide for operable breast cancer: National Surgical Adjuvant Breast and Bowel Project Protocol B-27. J Clin Oncol 24:2019–2027

30. Byrski T, Huzarski T, Dent R, Marczyk E, Jasiowka M, Gronwald J, Jakubowicz J, Cybulski C, Wisniowski R, Godlewski D, Lubinski J, Narod SA (2014) Pathologic complete response to neoadjuvant cisplatin in BRCA1-positive breast cancer patients. Breast Cancer Res Treat 147:401–405

31. Guiu S, Gauthier M, Coudert B, Bonnetain F, Favier L, Ladoire S, Tixier H, Guiu B, Penault-Llorca F, Ettore F, Fumoleau P, Arnould L (2010) Pathological complete response and survival according to the level of HER-2 amplification after trastuzumab-based neoadjuvant therapy for breast cancer. Br J Cancer 103:1335–1342

32. Giacchetti S, Porcher R, Lehmann-Che J, Hamy AS, de Roquancourt A, Cuvier C, Cottu PH, Bertheau P, Albiter M, Bouhidel F, Coussy F, Extra JM, Marty M, de The H, Espie M (2014) Long-term survival of advanced triple-negative breast cancers with a dose-intense cyclophosphamide/anthracycline neoadjuvant regimen. Br J Cancer 110: 1413–1419

33. Gerber B, von Minckwitz G, Eidtmann H, Rezai M, Fasching P, Tesch H, Eggemann H, Schrader I, Kittel K, Hanusch C, Solbach C, Jackisch C, Kunz G, Blohmer JU, Huober J, Hauschild M, Nekljudova V, Loibl S, Untch M (2014) Surgical outcome after neoadjuvant chemotherapy and bevacizumab: results from the GeparQuinto study (GBG 44). Ann Surg Oncol 21:2517–2524

34. Gerber B, Loibl S, Eidtmann H, Rezai M, Fasching PA, Tesch H, Eggemann H, Schrader I, Kittel K, Hanusch C, Kreienberg R, Solbach C, Jackisch C, Kunz G, Blohmer JU, Huober J, Hauschild M, Nekljudova V, Untch M, von Minckwitz G, German Breast Group Investigators (2013) Neoadjuvant bevacizumab and anthracycline-taxane-based chemotherapy in 678 triple-negative primary breast cancers; results from the geparquinto study (GBG 44). Ann Oncol 24:2978–2984

35. de Azambuja E, Holmes AP, Piccart-Gebhart M, Holmes E, Di Cosimo S, Swaby RF, Untch M, Jackisch C, Lang I, Smith I, Boyle F, Xu B, Barrios CH, Perez EA, Azim HA Jr, Kim SB, Kuemmel S, Huang CS, Vuylsteke P, Hsieh RK, Gorbunova V, Eniu A, Dreosti L, Tavartkiladze N, Gelber RD, Eidtmann H, Baselga J (2014) Lapatinib with trastuzumab for HER2-positive early breast cancer (NeoALTTO): survival outcomes of a randomised, open-label, multi-centre, phase 3 trial and their association with pathological complete response. Lancet Oncol 15:1137–1146

36. Untch M, Rezai M, Loibl S, Fasching PA, Huober J, Tesch H, Bauerfeind I, Hilfrich J, Eidtmann H, Gerber B, Hanusch C, Kuhn T, du Bois A, Blohmer JU, Thomssen C, Dan Costa S, Jackisch C, Kaufmann M, Mehta K, von Minckwitz G (2010) Neoadjuvant treatment with trastuzumab in HER2-positive breast cancer: results from the GeparQuattro study. J Clin Oncol 28:2024–2031

37. Shenkier T, Weir L, Levine M, Olivotto I, Whelan T, Reyno L, Steering Committee on Clinical Practice Guidelines for the Care and Treatment of Breast Cancer (2004) Clinical practice guidelines for the care and treatment of breast cancer: 15. Treatment for women with stage III or locally advanced breast cancer. CMAJ 170:983–994

38. Humbert O, Berriolo-Riedinger A, Riedinger JM, Coudert B, Arnould L, Cochet A, Loustalot C, Fumoleau P, Brunotte F (2012) Changes in 18F-FDG tumor metabolism after a first course of neoadjuvant chemotherapy in breast cancer: influence of tumor subtypes. Ann Oncol 23:2572–2577

39. Fisher B, Anderson S, Bryant J, Margolese RG, Deutsch M, Fisher ER, Jeong JH, Wolmark N (2002) Twenty-year follow-up of a randomized trial comparing total mastectomy, lumpectomy, and lumpectomy plus irradiation for the treatment of invasive breast cancer. N Engl J Med 347:1233–1241

40. Veronesi U, Cascinelli N, Mariani L, Greco M, Saccozzi R, Luini A, Aguilar M, Marubini E (2002) Twenty-year follow-up of a randomized study comparing breast-conserving surgery with radical mastectomy for early breast cancer. N Engl J Med 347:1227–1232

41. Beriwal S, Schwartz GF, Komarnicky L, Garcia-Young JA (2006) Breast-conserving therapy after neoadjuvant chemotherapy: long-term results. Breast J 12:159–164

42. Schwartz GF, Tannebaum JE, Jernigan AM, Palazzo JP (2010) Axillary sentinel lymph node biopsy after neoadjuvant chemotherapy for carcinoma of the breast. Cancer 116:1243–1251

43. Carey LA, Dees EC, Sawyer L, Gatti L, Moore DT, Collichio F, Ollila DW, Sartor CI, Graham ML, Perou CM (2007) The triple negative paradox: primary tumor chemosensitivity of breast cancer subtypes. Clin Cancer Res 13:2329–2334

44. Esserman LJ, Berry DA, DeMichele A, Carey L, Davis SE, Buxton M, Hudis C, Gray JW, Perou C, Yau C, Livasy C, Krontiras H, Montgomery L, Tripathy D, Lehman C, Liu MC, Olopade OI, Rugo HS, Carpenter JT, Dressler L, Chhieng D, Singh B, Mies C, Rabban J, Chen Y-Y, Giri D, van't Veer L, Hylton N (2012) Pathologic complete response predicts recurrence-free survival more effectively by cancer subset: results from the I-SPY 1 TRIAL—CALGB 150007/150012, ACRIN 6657. J Clin Oncol 30:3242–3249

45. Liedtke C, Mazouni C, Hess KR, André F, Tordai A, Mejia JA, Symmans WF, Gonzalez-Angulo AM, Hennessy B, Green M, Cristofanilli M, Hortobagyi GN, Pusztai L (2008) Response to neoadjuvant therapy and long-term survival in patients with triple-negative breast cancer. J Clin Oncol 26:1275–1281

46. Ma XJ, Salunga R, Dahiya S, Wang W, Carney E, Durbecq V, Harris A, Goss P, Sotiriou C, Erlander M, Sgroi D (2008) A five-gene molecular grade index and HOXB13:IL17BR are

complementary prognostic factors in early stage breast cancer. Clin Cancer Res 14:2601–2608

47. Mathieu MC, Mazouni C, Kesty NC, Zhang Y, Scott V, Passeron J, Arnedos M, Schnabel CA, Delaloge S, Erlander MG, Andre F (2012) Breast Cancer Index predicts pathological complete response and eligibility for breast conserving surgery in breast cancer patients treated with neoadjuvant chemotherapy. Ann Oncol 23:2046–2052

48. Guarneri V, Barbieri E, Conte P (2011) Biomarkers predicting clinical benefit: fact or fiction? J Natl Cancer Inst Monogr 2011:63–66

49. Berruti A, Generali D, Kaufmann M, Puztai L, Curigliano G, Aglietta M, Gianni L, Miller WR, Untch M, Sotiriou C, Daidone M, Conte P, Kennedy D, Damia G, Petronini P, Di Cosimo S, Bruzzi P, Dowsett M, Desmedt C, Mansel RE, Olivetti L, Tondini C, Sapino A, Fenaroli P, Tortora G, Thorne H, Bertolini F, Ferrozzi F, Danova M, Tagliabue E, de Azambuja E, Makris A, Tampellini M, Dontu G, Van't Veer L, Harris AL, Fox SB, Dogliotti L, Bottini A (2011) International expert consensus on primary systemic therapy in the management of early breast cancer: highlights of the fourth Symposium on primary systemic therapy in the management of operable breast cancer, Cremona, Italy (2010). J Natl Cancer Inst Monogr 2011:147–151

50. Purushotham A, Pinder S, Cariati M, Harries M, Goldhirsch A (2010) Neoadjuvant chemotherapy: not the best option in estrogen receptor-positive, her2-negative, invasive classical lobular carcinoma of the breast? J Clin Oncol 28:3552–3554

51. von Minckwitz G, Puglisi F, Cortes J, Vrdoljak E, Marschner N, Zielinski C, Villanueva C, Romieu G, Lang I, Ciruelos E, De Laurentiis M, Veyret C, de Ducla S, Freudensprung U, Srock S, Gligorov J (2014) Bevacizumab plus chemotherapy versus chemotherapy alone as second-line treatment for patients with HER2-negative locally recurrent or metastatic breast cancer after first-line treatment with bevacizumab plus chemotherapy (TANIA): an open-label, randomised phase 3 trial. Lancet Oncol 15:1269–1278

52. Vitug AF, Newman LA (2007) Complications in breast surgery. Surg Clin North Am 87:431–451, x

53. Criscitiello C, Azim HA Jr, Agbor-tarh D, de Azambuja E, Piccart M, Baselga J, Eidtmann H, Di Cosimo S, Bradbury I, Rubio IT (2013) Factors associated with surgical management following neoadjuvant therapy in patients with primary HER2-positive breast cancer: results from the NeoALTTO phase III trial. Ann Oncol 24:1980–1985

54. Symmans WF, Peintinger F, Hatzis C, Rajan R, Kuerer H, Valero V, Assad L, Poniecka A, Hennessy B, Green M, Buzdar AU, Singletary SE, Hortobagyi GN, Pusztai L (2007) Measurement of residual breast cancer burden to predict survival after neoadjuvant chemotherapy. J Clin Oncol 25:4414–4422

55. von Minckwitz G, Eidtmann H, Rezai M, Fasching PA, Tesch H, Eggemann H, Schrader I, Kittel K, Hanusch C, Kreienberg R, Solbach C, Gerber B, Jackisch C, Kunz G, Blohmer JU, Huober J, Hauschild M, Fehm T, Muller BM, Denkert C, Loibl S, Nekljudova V, Untch M, German Breast Group, Arbeitsgemeinschaft Gynäkologische Onkologie–Breast Study Groups (2012) Neoadjuvant chemotherapy and bevacizumab for HER2-negative breast cancer. N Engl J Med 366:299–309

56. Ogston KN, Miller ID, Payne S, Hutcheon AW, Sarkar TK, Smith I, Schofield A, Heys SD (2003) A new histological grading system to assess response of breast cancers to primary chemotherapy: prognostic significance and survival. Breast 12:320–327

57. Rodenhuis S, Mandjes IA, Wesseling J, van de Vijver MJ, Peeters MJ, Sonke GS, Linn SC (2010) A simple system for grading the response of breast cancer to neoadjuvant chemotherapy. Ann Oncol 21:481–487

58. Melichar B, Studentova H, Kalabova H, Vitaskova D, Cermakova P, Hornychova H, Ryska A (2014) Predictive and prognostic significance of tumor-infiltrating lymphocytes in patients with breast cancer treated with neoadjuvant systemic therapy. Anticancer Res 34:1115–1125

59. Ye J, Coulouris G, Zaretskaya I, Cutcutache I, Rozen S, Madden T (2012) Primer-BLAST: a tool to design target-specific primers for polymerase chain reaction. BMC Bioinformatics 13:134

60. Livak KJ, Schmittgen TD (2001) Analysis of relative gene expression data using real-time quantitative PCR and the $2-\Delta\Delta CT$ method. Methods 25:402–408

61. von Minckwitz G, Blohmer JU, Costa SD, Denkert C, Eidtmann H, Eiermann W, Gerber B, Hanusch C, Hilfrich J, Huober J, Jackisch C, Kaufmann M, Kummel S, Paepke S, Schneeweiss A, Untch M, Zahm DM, Mehta K, Loibl S (2013) Response-guided neoadjuvant chemotherapy for breast cancer. J Clin Oncol 31:3623–3630

62. Telli ML (2013) Insight or confusion: survival after response-guided neoadjuvant chemotherapy in breast cancer. J Clin Oncol 31:3613–3615

63. Untch M, Fasching PA, Konecny GE, von Koch F, Conrad U, Fett W, Kurzeder C, Lück H-J, Stickeler E, Urbaczyk H, Liedtke B, Salat C, Harbeck N, Müller V, Schmidt M, Hasmüller S, Lenhard M, Schuster T, Nekljudova V, Lebeau A, Loibl S, von Minckwitz G, Arbeitsgemeinschaft Gynäkologische Onkologie PREPARE investigators (2011) PREPARE trial: a randomized phase III trial comparing preoperative, dose-dense, dose-intensified chemotherapy with epirubicin, paclitaxel and CMF versus a standard-dosed epirubicin/cyclophosphamide followed by paclitaxel ± darbepoetin alfa in primary breast cancer—results at the time of surgery. Ann Oncol 22:1988–1998

64. Gobardhan PD, de Wall LL, van der Laan L, ten Tije AJ, van der Meer DCH, Tetteroo E, Poortmans PMP, Luiten EJT (2013) The role of radioactive iodine-125 seed localization in breast-conserving therapy following neoadjuvant chemotherapy. Ann Oncol 24: 668–673

65. Keune JD, Jeffe DB, Schootman M, Hoffman A, Gillanders WE, Aft RL (2010) Accuracy of ultrasonography and mammography in predicting pathologic response after neoadjuvant chemotherapy for breast cancer. Am J Surg 199:477–484

66. Bines J, Earl H, Buzaid AC, Saad ED (2014) Anthracyclines and taxanes in the neo/adjuvant treatment of breast cancer: does the sequence matter? Ann Oncol 25:1079–1085

67. Bottini A, Berruti A, Generali D, Dogliotti L (2011) The treatment of individual patients is more than a trial. J Natl Cancer Inst Monogr 2011:53–54

Chapter 16

Proteomics in the Assessment of the Therapeutic Response of Antineoplastic Drugs: Strategies and Practical Applications

Vukosava Milic Torres, Lazar Popovic, Fátima Vaz, and Deborah Penque

Abstract

Uncovering unknown pathological mechanisms and body response to applied medication are the driving forces toward personalized medicine. In this post-genomic era, all eyes are turned to the proteomics field, searching for answers and explanations by investigating the gene end point functional units—proteins and their proteoforms. The development of cutting-edge mass spectrometric technologies and bioinformatics tools have allowed the life-science community to discover disease-specific proteins as biomarkers, which are often concealed by high sample complexity and dynamic range of abundance. Currently, there are several proteomics-based approaches to investigate the proteome. This chapter focuses on gold standard proteomics strategies and related issues toward candidate biomarker discovery, which may have diagnostic/prognostic as well as mechanistic utility in cancer drug resistance.

Key words Cancer, Chemotherapeutic response, Proteomics, Discovery-based proteomics, Target-proteomics

---

1 Introduction

In spite of great progress in the management of oncological diseases, there is still too many patients with malignancy who are underdiagnosed or not responsive to a given therapy. Therefore, there is a urgent need to find predictive biomarkers which can identify causes of therapy response and progress of disease and recognize timing of the emergence of drug-resistant cancer cells and the need to switch to an appropriate alternative therapy. The emergence of proteomics, which examines a larger number of proteins in certain time/condition, allows researchers to perform a comprehensive analysis of the proteome in health and complex diseases such as cancer. Through proteomics, innovative tests based on protein biomarkers/signature can be developed to detect cancer and predict whether a patient will respond to a particular therapeutic drug, thereby leading to personalized medicine. However, to

José Rueff and António Sebastião Rodrigues (eds.), *Cancer Drug Resistance: Overviews and Methods*, Methods in Molecular Biology, vol. 1395, DOI 10.1007/978-1-4939-3347-1_16, © Springer Science+Business Media New York 2016

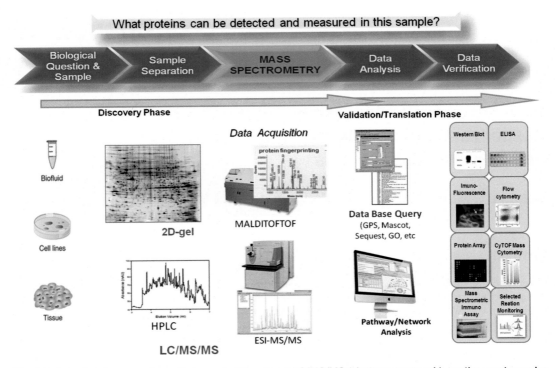

Fig. 1 Proteomics discovery-based approach: 2D-gel and LC/MS/MS (shotgun approach) are the most popular proteomics technological strategy to identify and quantify proteins toward biomarker discovery

accomplish such proteomic promises is still quite a challenge, requiring enormous efforts similar to drug development by pharmaceutics industry [1]. Nevertheless, with the rapid advances in the field, findings from proteomics research have begun to be translated into clinical applications [2]. The main goal of the first phase of proteomic discovery platforms is to deliver the most promising biomarker candidates which could have future clinical utility. Two main proteomics approaches can be used to explore the proteome toward candidate biomarker identification: discovery proteomics-based approach and target proteomics-based approach.

1.1 Proteomics Strategies

1.1.1 Discovery Proteomics Approach

The discovery proteomics approach operates in unbiased mode uncovering theoretically every potential candidate biomarkers expressed in a certain sample associated with a given time and/or condition (Fig. 1). In other words, proteins associated with a specific disease or state (for example: chemotherapeutic effects of a specific anticancer drug) can be identified and quantified in a proteome of a specific biological specimen (body fluids/cells/tissues).

The main challenge in this analytical approach is to know how much of this proteome do we need to explore to find out the best potential biomarker candidates and how deep in the proteome can we delve with available technologies [3]. The complete protein composition and proportion of any specific fluid/cell/tissue are still unknown, since no methods currently available enable detection of all the protein components of a biological system. The number and location of proteins, proteoforms, and type and degrees of posttranslational modifications (an important information for a comprehensive functional description of a proteome) are very dynamic and thus difficult to ascertain even for plasma that has been extensively evaluated by proteomics in the last decades.

Moreover, proteins exhibit an enormous dynamic range of concentrations in biological systems, defined as the ratio of concentration of the most abundant to the concentration of the least abundant proteins, which is $10^5/10^6$ for cells/tissues and 10^{10} for serum [4], making identification and quantification problematic specially for the proteins present at low abundance in a complex sample.

Knowing that protein quantities cannot be simply amplified prior to measurement as in the case of DNA, ultrasensitive techniques are needed to analyze and measure proteins of interest, often present as low-abundant complex mixture.

Multiple approaches for simplifying complex proteome prior to analysis as well as different methods and technologies for separation and identifications of proteins have been described and are in continuous improvements (see below).

1.2 Gold Standards in Discovery Proteomics

Two-dimensional-difference in gel electrophoresis (2D-DIGE)—is an improved modification of the traditional two-dimensional polyacrylamide gel electrophoresis (2D-PAGE) that when associated with mass spectrometry (MS) identification has become the most popular and versatile method in proteomics [5]. In 2D-DIGE, proteins are pre-labeled (minimal/saturated) with CyDye DIGE Fluor dyes prior to 2D electrophoresis for difference analysis that enables multiplexing within the same 2D gel. The use of fluorescent CyDyes (CyDye2, CyDye3, or Cydye5) with different excitation and emission wavelengths allows parallel runs up to three different samples (internal standard, control and case study) in a single gel running. This strategy overcomes some limitations of the original 2D technique by increasing reproducibility between gel to gel and sensitivity in protein detection by up to five orders of magnitude [5]. The expression of thousands of proteins can be visualized for each CyDye-labeled condition. The corresponding images are captured by high-quality fluorescence scanners for further specific image analysis to detect and normalize statistically significant changes in protein abundance. Protein spots of interest are excised from the gel and prepared for further MS identification. Although the CyDyes are expensive, the advantages of 2D-DIGE over standard 2DE means that 2D-DIGE is the recommended 2D gel-based method for biomarker discovery. Nevertheless, as classical 2D-PAGE, 2D-DIGE method is a non-high-throughput method involving multiple steps. The limitations associated with

2D-PAGE's inability to resolve high-hydrophobic proteins, high acid, or high-basic proteins exceeding the IPG range capacity or proteins with upper or lower weight size of the gel limits remain in 2D-DIGE. Guidelines together with an optimized protocol for 2D-DIGE were recently reported [6, 7].

Shotgun/bottom-up profiling—In the shotgun methodology, proteins are extracted from samples using biochemical methods, eventually involving fractionation or enrichment (see below sample preparation), and a protease (typically trypsin) is used to digest the proteins into peptides at specific amino acid residues (trypsin predominantly cleaves at the carboxyl side of lysine and arginine except when either is bound to a C-terminal proline) [8]. The produced peptides are separated by liquid chromatography (LC) such as strong anion exchange combined with reverse-phase chromatography, and the fractions analyzed by electrospray ionization to volatize and ionize the peptides coupled to tandem mass spectrometry, a method also known by LC/MS/MS. Alternatively, in order to improve protein coverage, proteins are pre-separated by 1D SDS-PAGE and stained, protein bands along the gel lane at different molecular weights are cut off from the gel and digested, and the peptides injected into LC for tandem MS analysis. Tandem (MS/MS) mass spectrometer measures the mass-to-charge ratio (m/z) of ionized peptides and produces spectra (singular spectrum) for each peptide sequence, which correspond to the number of ions at each m/z value. The derived peptide tandem mass spectra are matched to a library spectra generated from in silico digestion of a protein database to infer the peptide sequences and, by extension, the protein identification (bottom-up approach). This process involves sophisticated bioinformatics tools and careful analysis of the data since peptides can be either uniquely assigned to a single protein or be shared by several proteins and therefore it needs further weight by score or *P*-values as input to protein inference/identification. The correct mapping of peptides to proteins, particularly for redundant proteins and proteoforms, can be challenging and is largely related to the complexity of databases of proteins that have been increasing continuously. A number of bioinformatics tools and guidelines have been thus proposed to address this relevant issue [8–16].

The proteome is extremely complex with genomic variants and more than 400 different possibilities of posttranslational modifications (PTMs). Bottom-up proteomics often do not distinguish proteoforms (e.g., PTMs, mutated, truncated) of the same protein. Also, it does not provide accurate information on the site location of PTMs and even if the site location is well determined the quantitative value cannot be directly linked to quantitative values of protein states. In a simple example with two modifications located on two different tryptic peptides, four protein states exist but the quantitative data obtained on the modified and unmodified peptides cannot be directly used to infer the quantitative values of the four protein states [17].

The shotgun approach was initially designed for full proteome profiling but soon extended to quantitative and comparative studies in biomarker discovery by the introduction of label and label-free quantitative approaches. Stable isotope labeling strategies like isotope-coded affinity tag (ICAT) [18–24], stable isotope (*in vivo*) labeling with amino acids in cell culture (SILAC) [25], and more recently isobaric tags (iTRAQ) [26] and tandem mass tags (TMT) [27, 28] are some of the approaches for relative and absolute quantification. In general, these reagents use "tags" of different masses to label control and disease peptides, respectively, which are then pooled and simultaneously measured by MS to identify and quantify respective peptides and hence proteins. The limitations are related to incomplete labeling, variability in chromatographic runs, complex sample preparation, need for a higher protein concentration, and, not less important, higher costs.

Label-free quantitative approaches have become an attractive technique by showing superior information, particularly when coverage was taken into account and more than one peptide was required for identification [29]. The label-free quantification approach is based on either spectrometric signal intensity approach [19, 20] or spectral counting approach [21]. The spectrometric intensity approach implies the measurement of chromatographic peak areas of peptide precursor ions that correlates linearly in a wide range with the corresponding protein abundance. Several experimental and technical aspects have to be considered to obtain reliable results [19, 20]. The spectral counting, which is widely used, implies a counting and comparison of the number of fragment-ion spectra (MS/MS) acquired for peptides of a given protein based on the empiric observation that higher abundance of a particular protein in a sample is related to higher number of corresponding peptide MS/MS spectra [21]. Modified approaches of spectral counting have been reported in order to normalize and eliminate variances between replicate measurements and to increase reproducibility and significance of quantification [22–24, 29]. Although label or label-free LC/MS/MS methods still suffer from some issues with reproducibility and dynamic range, these methods have been shown to provide increased proteome coverage compared to 2DE gels. Nevertheless, comparative studies between 2D/MS and LC/MS/MS have shown limited overlap among proteome identifications suggesting the complementary nature of these approaches.

1.3 Target Proteomics

Target proteomics is a biased approach testing whether a given protein or a set of proteins is measurable in a particular sample. Target proteomics focuses measurements on a subset of proteins (typically <100 proteins per analysis) relevant to a particular

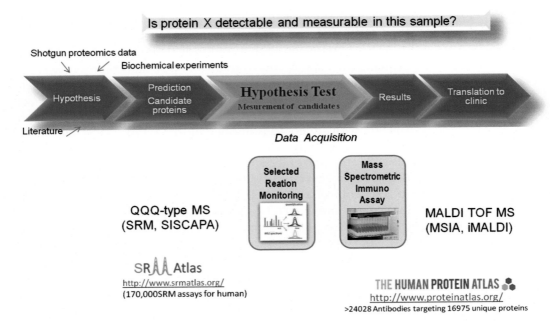

Fig. 2 Target proteomics workflow. Selected reaction monitoring (SRM), stable isotope standards and capture by anti-peptide antibodies (SISCAPA) or mass spectrometric immunoassay (MSIA) are the most popular in target proteomics

hypothesis based on prior information or intuition (Fig. 2). It attempts to reproducibly and accurately quantify sets of pre-defined potential biomarker candidates across multiple samples to verify and/or validate their clinical value. In contrast to discovery-proteomics, where there is no guess about which candidates will be identified by profiling as many proteins as possible in a sample or sub-sample, target-proteomics by predefining the group of candidates to investigate based on a prior hypothesis, a more specific and sensitive assay to identify and quantify these particular proteins can be built. Nevertheless, as for discovery-proteomics, sample complexity and protein dynamic range can also be a technical issue in target-proteomics approach (see below).

1.3.1 Frequently Used Methods in Target-Based Proteomics

Selected or multiple reaction monitoring (SRM/MRM) method— Using a triple quadrupole (QQQ)-based MS instrumentation, which operates as a dual-mass filter, the molecular ions corresponding to the peptide sequences unique to the target protein are pre-selected, based on their m/z ratio, for relative/absolute quantification by tandem MS analysis. The pick area for SRM transitions, a combination of intact ions (precursor) and resulting specific fragment ions (products), is integrated as measures of peptide abundance, which correlates with the surrogate protein abundance in a sample [30–32]. To achieve absolute quantification, heavy isotope-labeled reference peptides are used for spiking the sample.

Web-based tools such as SRMAtlas [33] have been created to facilitate selection of peptide transitions and design of reference peptides that are unique to a given target protein. In order to support ultrahigh-throughput multiplexed analysis, a next generation of SRM assay has been developed such as the sequential window acquisition of all theoretical mass spectra (SWATH), a method that relies on data-independent acquisition, in which all sequential peptides are selected for fragmentation without regard to its signal intensity.

Immunoaffinity-based target methods—Immunoaffinity enrichment of target peptides before MS analysis, a method known by stable isotope standards and capture by anti-peptide antibodies (SISCAPA), was developed to overcome the need for fractionation of complex samples and/or depletion of most abundant proteins before SRM or LC/MS/MS analysis [34]. SISCAPA and iMALDI (another version using MALDI spectrometry [35]) are modified approaches from mass spectrometric immune assay (MSIA) pioneered in mid-1990s by Nelson's group [36, 37]. In contrast to SISCAPA or iMALDI, in which tryptic peptides from a digested sample are antibody-based captured using a magnetic bead protocol, MSIA was designed to capture intact proteins from samples by using pipette tips with immobilized antibodies before full-length MALDI MS characterization and relative quantification. Analyzing full-length proteins (top-down approach) by MSIA, genetic and post-translational proteoforms for a single protein can be revealed and some of them can be specifically associated with a disease state [37]. Knowing that any immuno-based approach is limited by the availability of antibodies, the Human Protein Atlas project (www.proteinatlas.org) has been produced and thousands of antibodies made available and a database with millions of images showing the spatial distribution of proteins in different normal/cancer human cells/tissues. Very recently, a combination between the high-throughput target enrichment capability of MSIA and the absolute quantification of SRM method has been proposed to provide a high-throughput approach for monitoring all unique proteoforms of a target protein at near the low picomolar range, opening up new unexplored strategy in the search for novel biomarkers [38, 39].

1.4 Verification Phase

Hundreds to thousands of biomarker candidates can be generated but few of them may warrant assessment in a time-consuming and costly large-scale clinical validation trial. In a biomarker development pipeline, experimental design, prioritization, and verification are crucial early stages in the selection of the most promising candidates for clinical validation and translation to a specific clinical application [40, 41]. It would help to develop meaningful criteria for prioritizing candidates for further verification stage. One strategy is a better understanding of the disease state and/or the drug-target pathways, of which putative proteins could be involved, taking into account their functional relevance to the cell/tissue or biofluid where they were identified as dysregulated during the discovery phase [41]. Methods based on immunoaffinity such as western blotting, ELISA, immunocyto/histochemistry, and more recently SRM-based methods are the most popular in the verification stage (also referred to as assessment of performance when any candidates are verified by an orthogonal method, i.e., other than the one used in the discovery phase). The relevance of the verification stage is reflected by the fact that for many proteomics journals of high impact, it is a mandatory requirement for reporting of clinical biomarker studies. A statistical framework for the design of biomarker discovery and verification studies was recently proposed by Skates et al. [42], whose ultimate goal is to set clinical relevance and appropriate biospecimen sample size for discovery and verification stages prior to clinical validation stage. This approach will certainly contribute to make the journey of a protein biomarker from the bench to the clinic shorter and successful. However, there is still a need for sophisticated computational methods and algorithms to allow for consistent analysis and interpretation of proteomic data using statistical principles.

The translational phase dedicated to the development and testing of promising candidate biomarkers to clinical utility is a long journey requiring concerted efforts and necessary steps involving regulatory procedures [1, 43]. However, the discussion of this issue is out of the scope of this chapter.

1.5 Which Proteome to Investigate?

Whatever the proteomic platform approach is chosen for a particular biological question, it is important to know which biospecimen to investigate and how this should be prepared to uncover specific biomarkers for a particular disease/condition.

Peripheral blood is the biospecimen of election for biomarker discovery since blood circulates inside the body to basically all locations to deliver, collect, or exchange molecular information with cells, tissues or organs. Moreover, peripheral blood is easy to obtain by minimal invasive procedures from patients. Therefore, plasma/serum and blood cells are very attractive for disease biomarker discovery and/or development of new clinical tests in monitoring the effectiveness and side effects of chemotherapeutic drugs. The Human Proteome Organization recommends the use of plasma over serum because it is a more reproducible sample [44, 45]. More than 3000 different protein components have been identified in plasma [46]. However, only 22 proteins are the most abundant (on an mg/mL basis), constituting more than 99 % of the mass of the total plasma proteins. The remaining 1 %, the low-abundance component, is composed of thousands of proteins that may represent the most interesting subset of the proteome to search for biomarkers [46]. To access low-abundance proteins in plasma, selective depletion and/or selective equalization of the most abundant interfering proteins have been proved useful strategies by extending the range of profiling to up to fourfold lower concentrations. Nevertheless, none of these methods guarantee efficient discovery of low-abundance disease-specific biomarkers in plasma [47].

Saliva and urine are gaining much attention as easier and non-invasive alternatives to blood in biomarker discovery to determine health or disease status [48–50]. Both are complex and dynamic biological fluids containing a wide range of compounds. Saliva is mainly composed of secretions from salivary glands and gingival crevicular fluid participating in oral and teeth protection and ingestion and digestion of food and that mediate taste sensations. The composition of saliva changes with the physiological states of the human body thus being a good indicator of the plasma/serum levels of various substances such as hormones and drugs [51].

Urine is mainly composed by water but also contains inorganic salts and organic compounds, including proteins, peptides, hormones, and a wide range of body metabolic waste and bacterial by-products [52–55]. Combined analytical procedures have been developed based on solvent precipitation, ultracentrifugation, vesicle separation or combinatorial peptide ligand libraries to maximize sensitivity and reproducibility of both saliva and urinary proteome analysis [53]. Up to date about 1500 proteins in saliva and 3400 proteins in urine have been identified as potential source of information on both normal and pathological human body physiology [55–58]. However, as for any other biofluid/cell/tissue, the exact number of proteins that are present in saliva or urine is still unknown and depending on the procedure used, different groups of proteins can be detected and analyzed.

Human cancer-derived cell lines cultured either in traditional 2D monolayer or in 3D matrices to mimic tissue structure have proven to be valuable for cancer biomarker research and therapeutic drug discovery and evaluation [59–61]. To better address this challenge, larger human cancer cell line panels, such as CMT100 platform and Cancer Cell Line Encyclopedia, were recently created providing a detailed genetic characterization of more than 1000 human cancer cell lines, including pharmacological profile of many anticancer drugs across many of these cell lines [62–64]. However, not all cancer lines have equal value as tumor model and significant changes in cellular characteristics are highly probable due to long periods of cell culture and culture conditions; therefore precautions in their use and interpretation of data must be taken into account [65]. Primary tumor cultures, which are maintained for a relatively short period of time, could be an alternative to the cell line models [66]. Proteomics studies of xenografts and genetically engineered mouse models for some human cancer aspects have also shown promising results in the evaluation of cancer characteristics and drug resistance in a context that reflects tumor heterogeneity and the organism environment influencing tumor development and progression, which is not possible with individual cell lines [67].

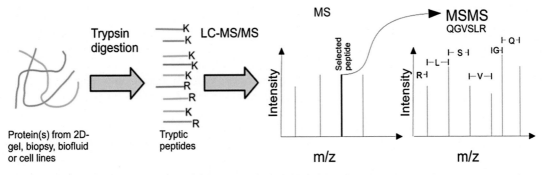

Fig. 3 Simplified scheme of a typical MS-based strategy to identify potential biomarkers in the discovery phase

1.6   Mass Spectrometry and Considerations When Starting Proteomics-Based Biomarker Study

Mass spectrometer measures the mass-to-charge ratio (m/z) of ionized molecules such as proteins/peptides and produces spectra (singular spectrum), which correspond to the number of ions at each m/z value. Matrix-assisted LASER desorption/ionization (MALDI) and electrospray ionization (ESI) are the two most popular techniques to volatize and ionize the proteins or peptides for subsequent mass spectrometric analysis. MALDI-MS is mostly used to analyze relatively simple peptide mixtures (e.g., gel protein spots). Trypsin-digested samples are deposited at MALDI plate, and let to dry with a crystalline matrix to be ionized via LASER pulses. In contrast, ESI ionizes the analytes out of a solution coming from an LC system. Integrated LC system with ESI-MS systems (LC-MS) are preferentially used for analysis of complex samples. There are different types of mass analyzers, with different advantages/disadvantages, such as time-of-flight (TOF), ion trap, quadrupole, Fourier transform ion cyclotron (FT-MS), and orbitrap [4].

MS strategies for protein studies roughly fall into three categories: (1) bottom up, (2) middle down, and (3) top down. Figure 3 outlines bottom-up, the most popular MS strategy used for identifying potential biomarkers. First extracted proteins are digested with trypsin followed by liquid chromatography separation and mass analysis. The mass spectrometer typically analyzes both the mass of the intact tryptic peptides in the MS scans and of smaller peptide fragments generated inside the mass spectrometer by for example colliding a peptide with a collision gas such as nitrogen. Collision-induced dissociation typically causes breakage of the peptide bond along the peptide sequence with different frequencies (the most abundant ions are b- and y-ion series), thereby generating series of ions for which mass differences can be associated with amino acid residues. Partial or full peptide sequence can be obtained from the assigned residues in a given ion series. Typically peptide sequences are assigned to spectra by matching the spectra against a sequence database also called database-dependent search. The direct output from such searches is a list of peptides assigned to spectra. Frequently, it is the protein identifications that are of interest and the next step is therefore to infer which proteins are in the sample based on the identified peptides. Botton-up protein inference is often complicated by the fact that many tryptic peptides are shared among proteins (see discussion in next section).

The current status of proteomics-based biomarker discovery can generate many potential biomarkers of which very few make it to clinical practice. There are multiple reasons for this situation such as poorly funded studies, lack of standardization of sample collection and storage, statistical problems, and also technology and data analysis problems [68–70].

Computational analysis of the MS data can also lead to flawed or non-optimal results. Furthermore, although data analysis software is mostly correct they do make mistakes. These problems are often ignored or overlooked although they can have immense consequences for biomarker studies and is therefore the focus of the discussion below.

Protein inference: Protein inference is a problem that is largely ignored in bottom-up MS experiments. Bottom-up proteomics provides a list of peptides which is weighted by score or *P*-values which serves as input to protein inference. Many database-dependent search engines have a simplistic approach to protein inference where for example the score of all none redundant peptides matching a protein is summed to calculate a protein score. However, this approach leads to the unfortunate result that a protein can be ranked with a high score just because it shares many peptides with other proteins. To this end we divide proteins and peptides into five evidence groups which provide a more transparent view of the reliability of the identified proteins and peptides [9, 10]. We have also developed computational methods to explore protein connected by peptide spectra matches by alignment which overlay identified peptides and can display the network of protein to peptide spectra assignments of a connected group of proteins. These visual tools can be used to design additional experiments to resolve the potential protein inference problem for a group of proteins [10].

Protein states: The proteome is extremely complex with genomic variants and more than 200 posttranslational modifications (PTMs). Bottom-up proteomics often do not provide accurate information on the site location of PTMs and even if the site location is well determined the quantitative value cannot be directly linked to quantitative values of protein states. In a simple example with two modifications located on two different tryptic peptides four protein states exist but the quantitative data obtained on the modified and unmodified peptides cannot directly be used to infer the quantitative values of the four protein states [17].

Basic search parameters: Finally the mass accuracy used for database-dependent search, the specific sequence database used for the search, the filtering criteria such as minimum number of peptides identified, and false discovery rate (FDR) cutoff can dramatically change the proteins and peptides in the final report and can be a reason for lost information and low reproducibility when validation is performed in another laboratory that uses a different data analysis pipeline.

1.7 Proteomics in Practice: Chemotherapeutic Response

Multidrug resistance (MDR) is the main obstacle which hampered development of the new targeted therapies [71]. Biomarkers of MDR and individual response on the therapy are of crucial importance in personalized approach of cancer treatment management. Oncotype-Dx and Mammaprint are commercially available genetic tests primarily designed to estimate a woman's recurrence risk of early-stage breast cancer. Besides prognostic information, these tests estimate how likely women with breast cancer will benefit from certain types of chemotherapy and radiotherapy [72]. However, expression of oncogenes is not a unique MDR mechanism and it is impossible to have single-modulation drug-resistance marker to overcome resistance phenotype in cancer. Drug resistance is a complex process containing multiple and overlapping routes [73–77]. For better understanding of underlying mechanisms of drug response in its microenvironment, complementary technology is needed, the one that can assess signaling pathways and provide information which is not appeared at genomic and transcriptomic level. Gene expression profiling has limitations, because it does not capture posttranslational modifications (PTMs) which influence protein functionality and stability [72]. Studies showed that data of gene and protein expression have poor concordance when directly compared. In other words, there is no certainty that gene expression can predict protein expression levels. The functional components of signal transduction pathways at the subcellular level by nature are posttranslational and involve complex interplay between phosphorylation, dephosphorylation, oxidation, alkylation, protein degradation through ubiquitination, etc. [78]. These rapid and dynamic changes are impossible to capture in real time through the gene expression. With appearance of proteomics as science in 1994 [79] started post-genomic era, and eyes of the life science turn to look at the proteins as a final functional units, searching for the information about protein abundance, and their modifications along with interacting pathways and networks in order to study cellular processes. Subdisciplines in proteomics among others are clinical proteomics, oncoprotoemics, pharmacoproteomics, and toxoproteomics which are of special interest for further development of medicine. Because of complex nature of the disease, proteomics become an indispensable tool for studying cancer, particularly to identify biosignatures of the disease and markers responsible for the diagnosis, prognosis, and prediction of therapeutic response. Proteomics-based biomarkers are truly step ahead to personalized medicine.

Recent achievements in therapy response and acquired antineoplastic resistance using proteomics approach revealed some putative candidate biomarkers as Transgelin (TAGLN2) [77], stathmin/oncoprotein 18 [80], hnRNP A2, GDI2 [81], DEFA, and MAP2 [82] in resistance to neoadjuvant taxane therapy. Work of Fling and co-workers provides strong evidence that stress hormones modulate breast cancer cell sensitivity to paclitaxel [83]. Dramatic overexpression of two protein disulfide isomerases PDIA4 and PDIA6 discovered by 2D-DIGE/MS was found in cisplatin-resistant cell models as well in the patients with lung adenocarcinoma [84]. Proteomic analysis of epithelial ovarian carcinoma (EOC) cell secretoma revealed multiplexed upregulation of COL11A1 protein which made it an excellent candidate with clinical utility for prediction of response to cisplatin therapy. A direct correlation of COL11A1 was found with overall and progressive-free survival in patients with advanced EOC [85]. The strategy of MALDI imaging followed by liquid chromatography–tandem mass spectrometry (LC–MS/MS) applied for discovery of protein fingerprint in cisplatin-responsive cells of esophageal adenocarcinoma revealed that reduced cytochrome c oxidase prior to treatment exhibited favorable clinical outcome to neoadjuvant chemotherapy [86]. It is known that positive HER-2neu (also known as ErbB) status of primary breast cancer has been associated to poor outcome [87]. Gold standard for testing Her-2neu biomarker, based on immunohistochemistry (IHC) and in situ hybridization techniques (FISH, CISH), has estimated a 20 % level of uncertainty [88]. Tests based on proteomics approach, using HPLC-MS/MS technique, undoubtedly can identify and quantify expression of ErbB in archived breast cancer tissue. These tests are in the clinical validation stage with just one step away from implementation into routine oncological diagnostics [89, 90].

2 Conclusions

While genome sequencing has great potential to propel the understanding of diseases, by itself it is insufficient for full description of disease phenotypes and their dynamic interaction/modulation with the environment. Proteome information provides a broad picture of patient phenotype, rather than indicating unchanging probabilistic risks that can be inferred from sequence analysis of genomic DNA. As it can be seen options for analyzing proteome are numerous and mainly dependent on the type of information we would like to receive, sample type, and equipment available in the lab. Step-by-step protocols for sample preparation: cleanup, separation (1D, 2D), and image and MS analysis are numerous and well documented elsewhere [91–99]. However, there is no established gold standard for global proteome analysis. Quantities of data that can be obtained by shotgun analysis are rapidly increasing as mass spectrometry instruments and bioinformatics tools develop. Expansion of new MS technologies and supporting algorithms for data analysis still have not reached asymptotic limit causing a continuous evolution in the field of proteome analysis. Recently, some proteomic assays have reached clinical routine. Parallel validation of new protein biomarkers and development of user-friendly low-cost benchtop mass spectrometers of high resolution will result in the establishment of standardized protocols and implementation of proteomic tools into clinical diagnostic laboratories.

Acknowledgements

The authors would like to acknowledge funding support from FCT grants (HMSP-ICJ/0022/2011, Poly-Annual Funding Program, and FCT-fellowship SFRH/BPD/43365/2008) and FEDER/Saúde XXI Program (Portugal) and Rune Matthiesen for comments on the chapter.

References

1. Fuzery AK, Levin J, Chan MM, Chan DW (2013) Translation of proteomic biomarkers into FDA approved cancer diagnostics: issues and challenges. Clin Proteomics 10:13

2. Li D, Chan DW (2014) Proteomic cancer biomarkers from discovery to approval: it's worth the effort. Expert Rev Proteomics 11:135–136

3. Patterson SD (2004) Proteomics: beginning to realize its promise? Arthritis Rheum 50:3741–3744

4. Aebersold R, Mann M (2003) Mass spectrometry-based proteomics. Nature 422:198–207

5. Penque D (2009) Two-dimensional gel electrophoresis and mass spectrometry for biomarker discovery. Proteomics Clin Appl 3:155–172

6. Gao W (2014) Analysis of protein changes using two-dimensional difference gel electrophoresis. In: Keohavong P, Grant SG (eds) Molecular toxicology protocols, vol 1105, Methods in molecular biology. Humana, New York, NY, pp 17–30, doi:10.1007/978-1-62703-739-6_2

7. Dautel F, Kalkhof S, Trump S, Lehmann I, Beyer A, Martin VB (2011) Large-scale 2-D DIGE studies – guidelines to overcome pitfalls and challenges along the experimental procedure. J Integr OMICS 1:9

8. Zhang Y, Fonslow BR, Shan B, Baek MC, Yates JR 3rd (2013) Protein analysis by shotgun/bottom-up proteomics. Chem Rev 113:2343–2394

9. Matthiesen R (2013) Algorithms for database-dependent search of MS/MS data. Methods Mol Biol 1007:119–138

10. Matthiesen R, Prieto G, Amorim A, Aloria K, Fullaondo A, Carvalho AS, Arizmendi JM (2012) SIR: deterministic protein inference from peptides assigned to MS data. J Proteomics 75:4176–4183

11. Carr S, Aebersold R, Baldwin M, Burlingame A, Clauser K, Nesvizhskii A (2004) The need for guidelines in publication of peptide and protein identification data: working group on publication guidelines for peptide and protein identification data. Mol Cell Proteomics 3:531–533

12. Nesvizhskii AI, Aebersold R (2005) Interpretation of shotgun proteomic data: the protein inference problem. Mol Cell Proteomics 4:1419–1440

13. Latterich M (2006) Publishing proteomic data. Proteome Sci 4:8

14. Wilkins MR, Appel RD, Van Eyk JE, Chung MC, Gorg A, Hecker M, Huber LA, Langen H, Link AJ, Paik YK, Patterson SD, Pennington SR, Rabilloud T, Simpson RJ, Weiss W, Dunn MJ (2006) Guidelines for the next 10 years of proteomics. Proteomics 6:4–8

15. Tabb DL (2008) What's driving false discovery rates? J Proteome Res 7:45–46

16. Binz PA, Barkovich R, Beavis RC, Creasy D, Horn DM, Julian RK Jr, Seymour SL, Taylor CF, Vandenbrouck Y (2008) Guidelines for reporting the use of mass spectrometry informatics in proteomics. Nat Biotechnol 26:862

17. Matthiesen R, Azevedo L, Amorim A, Carvalho AS (2011) Discussion on common data analysis strategies used in MS-based proteomics. Proteomics 11:604–619

18. Gygi SP, Rist B, Gerber SA, Turecek F, Gelb MH, Aebersold R (1999) Quantitative analysis of complex protein mixtures using isotope-coded affinity tags. Nat Biotechnol 17: 994–999

19. Bondarenko PV, Chelius D, Shaler TA (2002) Identification and relative quantitation of protein mixtures by enzymatic digestion followed by capillary reversed-phase liquid chromatography-tandem mass spectrometry. Anal Chem 74:4741–4749

20. Chelius D, Bondarenko PV (2002) Quantitative profiling of proteins in complex mixtures using liquid chromatography and mass spectrometry. J Proteome Res 1:317–323

21. Liu H, Sadygov RG, Yates JR 3rd (2004) A model for random sampling and estimation of relative protein abundance in shotgun proteomics. Anal Chem 76:4193–4201

22. Lundgren DH, Hwang SI, Wu L, Han DK (2010) Role of spectral counting in quantitative proteomics. Expert Rev Proteomics 7:39–53

23. Neilson KA, Ali NA, Muralidharan S, Mirzaei M, Mariani M, Assadourian G, Lee A, van Sluyter SC, Haynes PA (2011) Less label, more free: approaches in label-free quantitative mass spectrometry. Proteomics 11:535–553

24. Wu WW, Wang G, Baek SJ, Shen RF (2006) Comparative study of three proteomic quantitative methods, DIGE, cICAT, and iTRAQ, using 2D gel- or LC-MALDI TOF/TOF. J Proteome Res 5:651–658

25. Ong SE, Blagoev B, Kratchmarova I, Kristensen DB, Steen H, Pandey A, Mann M (2002) Stable isotope labeling by amino acids in cell culture, SILAC, as a simple and accurate approach to expression proteomics. Mol Cell Proteomics 1:376–386

26. Ross PL, Huang YN, Marchese JN, Williamson B, Parker K, Hattan S, Khainovski N, Pillai S, Dey S, Daniels S, Purkayastha S, Juhasz P, Martin S, Bartlet-Jones M, He F, Jacobson A, Pappin DJ (2004) Multiplexed protein quantitation in Saccharomyces cerevisiae using amine-reactive isobaric tagging reagents. Mol Cell Proteomics 3:1154–1169

27. Thompson A, Schafer J, Kuhn K, Kienle S, Schwarz J, Schmidt G, Neumann T, Johnstone R, Mohammed AK, Hamon C (2003) Tandem mass tags: a novel quantification strategy for comparative analysis of complex protein mixtures by MS/MS. Anal Chem 75:1895–1904

28. Dayon L, Hainard A, Licker V, Turck N, Kuhn K, Hochstrasser DF, Burkhard PR, Sanchez JC (2008) Relative quantification of proteins in human cerebrospinal fluids by MS/MS using 6-plex isobaric tags. Anal Chem 80:2921–2931

29. Megger DA, Bracht T, Meyer HE, Sitek B (2013) Label-free quantification in clinical proteomics. Biochim Biophys Acta 1834: 1581–1590

30. Anderson L, Hunter CL (2006) Quantitative mass spectrometric multiple reaction monitoring assays for major plasma proteins. Mol Cell Proteomics 5:573–588

31. Picotti P, Aebersold R (2012) Selected reaction monitoring-based proteomics: workflows, potential, pitfalls and future directions. Nat Methods 9:555–566

32. Liebler DC, Zimmerman LJ (2013) Targeted quantitation of proteins by mass spectrometry. Biochemistry 52:3797–3806

33. Deutsch EW, Lam H, Aebersold R (2008) PeptideAtlas: a resource for target selection for emerging targeted proteomics workflows. EMBO Rep 9:429–434

34. Anderson NL, Anderson NG, Haines LR, Hardie DB, Olafson RW, Pearson TW (2004) Mass spectrometric quantitation of peptides and proteins using stable isotope standards and capture by anti-peptide antibodies (SISCAPA). J Proteome Res 3:235–244

35. Warren EN, Elms PJ, Parker CE, Borchers CH (2004) Development of a protein chip: a MS-based method for quantitation of protein expression and modification levels using an immunoaffinity approach. Anal Chem 76: 4082–4092

36. Nelson RW, Hutchens TW (1992) Mass spectrometric analysis of a transition-metal-binding peptide using matrix-assisted leaser-desorption time-of-flight mass spectrometry. A demonstration of probe tip chemistry. Rapid Commun Mass Spectrom 6:4–8

37. Nelson RW, Borges CR (2011) Mass spectrometric immunoassay revisited. J Am Soc Mass Spectrom 22:960–968

38. Yassine H, Borges CR, Schaab MR, Billheimer D, Stump C, Reaven P, Lau SS, Nelson R (2013) Mass spectrometric immunoassay and MRM as targeted MS-based quantitative approaches in biomarker development: potential applications to cardiovascular disease and diabetes. Proteomics Clin Appl 7:528–540

39. Niederkofler EE, Phillips DA, Krastins B, Kulasingam V, Kiernan UA, Tubbs KA,

Peterman SM, Prakash A, Diamandis EP, Lopez MF, Nedelkov D (2013) Targeted selected reaction monitoring mass spectrometric immunoassay for insulin-like growth factor 1. PLoS One 8, e81125

40. Rifai N, Gillette MA, Carr SA (2006) Protein biomarker discovery and validation: the long and uncertain path to clinical utility. Nat Biotechnol 24:971–983

41. Paulovich AG, Whiteaker JR, Hoofnagle AN, Wang P (2008) The interface between biomarker discovery and clinical validation: the tar pit of the protein biomarker pipeline. Proteomics Clin Appl 2:1386–1402

42. Skates SJ, Gillette MA, LaBaer J, Carr SA, Anderson L, Liebler DC, Ransohoff D, Rifai N, Kondratovich M, Tezak Z, Mansfield E, Oberg AL, Wright I, Barnes G, Gail M, Mesri M, Kinsinger CR, Rodriguez H, Boja ES (2013) Statistical design for biospecimen cohort size in proteomics-based biomarker discovery and verification studies. J Proteome Res 12:5383–5394

43. Ioannidis JP (2013) Biomarker failures. Clin Chem 59:202–204

44. Omenn GS (2004) The Human Proteome Organization Plasma Proteome Project pilot phase: reference specimens, technology platform comparisons, and standardized data submissions and analyses. Proteomics 4:1235–1240

45. Omenn GS (2007) THE HUPO Human Plasma Proteome Project. Proteomics Clin Appl 1:769–779

46. Nanjappa V, Thomas JK, Marimuthu A, Muthusamy B, Radhakrishnan A, Sharma R, Ahmad Khan A, Balakrishnan L, Sahasrabuddhe NA, Kumar S, Jhaveri BN, Sheth KV, Kumar Khatana R, Shaw PG, Srikanth SM, Mathur PP, Shankar S, Nagaraja D, Christopher R, Mathivanan S, Raju R, Sirdeshmukh R, Chatterjee A, Simpson RJ, Harsha HC, Pandey A, Prasad TS (2014) Plasma Proteome Database as a resource for proteomics research: 2014 update. Nucleic Acids Res 42:D959–D965

47. Tu C, Rudnick PA, Martinez MY, Cheek KL, Stein SE, Slebos RJ, Liebler DC (2010) Depletion of abundant plasma proteins and limitations of plasma proteomics. J Proteome Res 9:4982–4991

48. Wang Q, Yu Q, Lin Q, Duan Y (2014) Emerging salivary biomarkers by mass spectrometry. Clin Chim Acta 438:214–221

49. Schafer CA, Schafer JJ, Yakob M, Lima P, Camargo P, Wong DT (2014) Saliva diagnostics: utilizing oral fluids to determine health status. Monogr Oral Sci 24:88–98

50. Yakob M, Fuentes L, Wang MB, Abemayor E, Wong DT (2014) Salivary biomarkers for detection of oral squamous cell carcinoma – current state and recent advances. Curr Oral Health Rep 1:133–141

51. de Almeida Pdel V, Gregio AM, Machado MA, de Lima AA, Azevedo LR (2008) Saliva composition and functions: a comprehensive review. J Contemp Dent Pract 9:72–80

52. Husi H, Barr JB, Skipworth RJ, Stephens NA, Greig CA, Wackerhage H, Barron R, Fearon KC, Ross JA (2013) The Human Urinary Proteome Fingerprint Database UPdb. Int J Proteomics 2013:760208

53. Kiprijanovska S, Stavridis S, Stankov O, Komina S, Petrusevska G, Polenakovic M, Davalieva K (2014) Mapping and identification of the urine proteome of prostate cancer patients by 2D PAGE/MS. Int J Proteomics 2014:594761

54. Beretov J, Wasinger VC, Graham PH, Millar EK, Kearsley JH, Li Y (2014) Proteomics for breast cancer urine biomarkers. Adv Clin Chem 63:123–167

55. Santucci L, Candiano G, Petretto A, Bruschi M, Lavarello C, Inglese E, Righetti PG, Ghiggeri GM (2014) From hundreds to thousands: widening the normal human urinome (1). J Proteomics 112C:53–62

56. Rudney JD, Staikov RK, Johnson JD (2009) Potential biomarkers of human salivary function: a modified proteomic approach. Arch Oral Biol 54:91–100

57. Wilmarth PA, Riviere MA, Rustvold DL, Lauten JD, Madden TE, David LL (2004) Two-dimensional liquid chromatography study of the human whole saliva proteome. J Proteome Res 3:1017–1023

58. Denny P, Hagen FK, Hardt M, Liao L, Yan W, Arellanno M, Bassilian S, Bedi GS, Boontheung P, Cociorva D, Delahunty CM, Denny T, Dunsmore J, Faull KF, Gilligan J, Gonzalez-Begne M, Halgand F, Hall SC, Han X, Henson B, Hewel J, Hu S, Jeffrey S, Jiang J, Loo JA, Ogorzalek Loo RR, Malamud D, Melvin JE, Miroshnychenko O, Navazesh M, Niles R, Park SK, Prakobphol A, Ramachandran P, Richert M, Robinson S, Sondej M, Souda P, Sullivan MA, Takashima J, Than S, Wang J, Whitelegge JP, Witkowska HE, Wolinsky L, Xie Y, Xu T, Yu W, Ytterberg J, Wong DT, Yates JR 3rd, Fisher SJ (2008) The proteomes of human parotid and submandibular/sublingual gland salivas collected as the ductal secretions. J Proteome Res 7:1994–2006

59. Gillet JP, Varma S, Gottesman MM (2013) The clinical relevance of cancer cell lines. J Natl Cancer Inst 105:452–458

60. Zhang W, Dolan ME (2009) Use of cell lines in the investigation of pharmacogenetic loci. Curr Pharm Des 15:3782–3795

61. Paul D, Kumar A, Gajbhiye A, Santra MK, Srikanth R (2013) Mass spectrometry-based proteomics in molecular diagnostics: discovery of cancer biomarkers using tissue culture. Biomed Res Int 2013:783131

62. McDermott U, Sharma SV, Settleman J (2008) High-throughput lung cancer cell line screening for genotype-correlated sensitivity to an EGFR kinase inhibitor. Methods Enzymol 438:331–341

63. Sharma SV, Haber DA, Settleman J (2010) Cell line-based platforms to evaluate the therapeutic efficacy of candidate anticancer agents. Nat Rev Cancer 10:241–253

64. Barretina J, Caponigro G, Stransky N, Venkatesan K, Margolin AA, Kim S, Wilson CJ, Lehar J, Kryukov GV, Sonkin D, Reddy A, Liu M, Murray L, Berger MF, Monahan JE, Morais P, Meltzer J, Korejwa A, Jane-Valbuena J, Mapa FA, Thibault J, Bric-Furlong E, Raman P, Shipway A, Engels IH, Cheng J, Yu GK, Yu J, Aspesi P Jr, de Silva M, Jagtap K, Jones MD, Wang L, Hatton C, Palescandolo E, Gupta S, Mahan S, Sougnez C, Onofrio RC, Liefeld T, MacConaill L, Winckler W, Reich M, Li N, Mesirov JP, Gabriel SB, Getz G, Ardlie K, Chan V, Myer VE, Weber BL, Porter J, Warmuth M, Finan P, Harris JL, Meyerson M, Golub TR, Morrissey MP, Sellers WR, Schlegel R, Garraway LA (2012) The cancer cell line encyclopedia enables predictive modelling of anticancer drug sensitivity. Nature 483: 603–607

65. Kulasingam V, Diamandis EP (2008) Tissue culture-based breast cancer biomarker discovery platform. Int J Cancer 123:2007–2012

66. Centenera MM, Raj GV, Knudsen KE, Tilley WD, Butler LM (2013) Ex vivo culture of human prostate tissue and drug development. Nat Rev Urol 10:483–487

67. Terp MG, Ditzel HJ (2014) Application of proteomics in the study of rodent models of cancer. Proteomics Clin Appl 8:640–652

68. Regnier FE, Skates SJ, Mesri M, Rodriguez H, Tezak Z, Kondratovich MV, Alterman MA, Levin JD, Roscoe D, Reilly E, Callaghan J, Kelm K, Brown D, Philip R, Carr SA, Liebler DC, Fisher SJ, Tempst P, Hiltke T, Kessler LG, Kinsinger CR, Ransohoff DF, Mansfield E, Anderson NL (2010) Protein-based multiplex assays: mock presubmissions to the US Food and Drug Administration. Clin Chem 56:165–171

69. Diamandis EP (2010) Cancer biomarkers: can we turn recent failures into success? J Natl Cancer Inst 102:1462–1467

70. Cox J, Heeren RM, James P, Jorrin-Novo JV, Kolker E, Levander F, Morrice N, Picotti P, Righetti PG, Sanchez JC, Turck CW, Zubarev R, Alexandre BM, Corrales FJ, Marko-Varga G, O'Donovan S, O'Neil S, Prechl J, Simoes T, Weckwerth W, Penque D (2011) Facing challenges in proteomics today and in the coming decade: report of roundtable discussions at the 4th EuPA scientific meeting, Portugal, Estoril 2010. J Proteomics 75:4–17

71. Wind NS, Holen I (2011) Multidrug resistance in breast cancer: from in vitro models to clinical studies. Int J Breast Cancer 2011: 967419

72. Zanotti L, Bottini A, Rossi C, Generali D, Cappelletti MR (2014) Diagnostic tests based on gene expression profile in breast cancer: from background to clinical use. Tumour Biol 35:8461–8470

73. Pokharel D, Padula MP, Lu JF, Tacchi JL, Luk F, Djordjevic SP, Bebawy M (2014) Proteome analysis of multidrug-resistant, breast cancer-derived microparticles. J Extracell Vesicles 3

74. Hu T, To KK, Wang L, Zhang L, Lu L, Shen J, Chan RL, Li M, Yeung JH, Cho CH (2014) Reversal of P-glycoprotein (P-gp) mediated multidrug resistance in colon cancer cells by cryptotanshinone and dihydrotanshinone of Salvia miltiorrhiza. Phytomedicine 21:1264–1272

75. Jope T, Lammert A, Kratzsch J, Paasch U, Glander HJ (2003) Leptin and leptin receptor in human seminal plasma and in human spermatozoa. Int J Androl 26:335–341

76. Bogle R, Wilkins M (2007) Treating acute myocardial infarction: something in the wind? Lancet 370:1461–1462

77. Chen S, Dong Q, Hu S, Cai J, Zhang W, Sun J, Wang T, Xie J, He H, Xing J, Lu J, Dong Y (2014) Proteomic analysis of the proteins that are associated with the resistance to paclitaxel in human breast cancer cells. Mol Biosyst 10:294–303

78. Seo J, Lee KJ (2004) Post-translational modifications and their biological functions: proteomic analysis and systematic approaches. J Biochem Mol Biol 37:35–44

79. Wilkins M, Appel R (2007) Ten Years of the Proteome. In: Wilkins M, Appel R, Williams K, Hochstrasser D (eds) Proteome research. Principles and practice. Springer, Berlin, pp 1–13. doi:10.1007/978-3-540-72910-5_1

80. Balasubramani M, Nakao C, Uechi GT, Cardamone J, Kamath K, Leslie KL, Balachandran R, Wilson L, Day BW, Jordan MA (2011) Characterization and detection of cellular and proteomic alterations in stable stathmin-overexpressing, taxol-resistant BT549

breast cancer cells using offgel IEF/PAGE difference gel electrophoresis. Mutat Res 722:154–164

81. Lee DH, Chung K, Song JA, Kim TH, Kang H, Huh JH, Jung SG, Ko JJ, An HJ (2010) Proteomic identification of paclitaxel-resistance associated hnRNP A2 and GDI 2 proteins in human ovarian cancer cells. J Proteome Res 9:5668–5676

82. Bauer JA, Chakravarthy AB, Rosenbluth JM, Mi D, Seeley EH, De Matos Granja-Ingram N, Olivares MG, Kelley MC, Mayer IA, Meszoely IM, Means-Powell JA, Johnson KN, Tsai CJ, Ayers GD, Sanders ME, Schneider RJ, Formenti SC, Caprioli RM, Pietenpol JA (2010) Identification of markers of taxane sensitivity using proteomic and genomic analyses of breast tumors from patients receiving neoadjuvant paclitaxel and radiation. Clin Cancer Res 16:681–690

83. Flint MS, Kim G, Hood BL, Bateman NW, Stewart NA, Conrads TP (2009) Stress hormones mediate drug resistance to paclitaxel in human breast cancer cells through a CDK-1-dependent pathway. Psychoneuroendocrinology 34:1533–1541

84. Tufo G, Jones AW, Wang Z, Hamelin J, Tajeddine N, Esposti DD, Martel C, Boursier C, Gallerne C, Migdal C, Lemaire C, Szabadkai G, Lemoine A, Kroemer G, Brenner C (2014) The protein disulfide isomerases PDIA4 and PDIA6 mediate resistance to cisplatin-induced cell death in lung adenocarcinoma. Cell Death Differ 21:685–695

85. Teng PN, Wang G, Hood BL, Conrads KA, Hamilton CA, Maxwell GL, Darcy KM, Conrads TP (2014) Identification of candidate circulating cisplatin-resistant biomarkers from epithelial ovarian carcinoma cell secretomes. Br J Cancer 110:123–132

86. Aichler M, Elsner M, Ludyga N, Feuchtinger A, Zangen V, Maier SK, Balluff B, Schone C, Hierber L, Braselmann H, Meding S, Rauser S, Zischka H, Aubele M, Schmitt M, Feith M, Hauck SM, Ueffing M, Langer R, Kuster B, Zitzelsberger H, Hofler H, Walch AK (2013) Clinical response to chemotherapy in oesophageal adenocarcinoma patients is linked to defects in mitochondria. J Pathol 230:410–419

87. Sjogren S, Inganas M, Lindgren A, Holmberg L, Bergh J (1998) Prognostic and predictive value of c-erbB-2 overexpression in primary breast cancer, alone and in combination with other prognostic markers. J Clin Oncol 16:462–469

88. Wolff AC, Hammond ME, Hicks DG, Dowsett M, McShane LM, Allison KH, Allred DC, Bartlett JM, Bilous M, Fitzgibbons P, Hanna W, Jenkins RB, Mangu PB, Paik S, Perez EA, Press MF, Spears PA, Vance GH, Viale G, Hayes DF (2013) Recommendations for human epidermal growth factor receptor 2 testing in breast cancer: American Society of Clinical Oncology/College of American Pathologists clinical practice guideline update. J Clin Oncol 31:3997–4013

89. Bateman NW, Sun M, Bhargava R, Hood BL, Darfler MM, Kovatich AJ, Hooke JA, Krizman DB, Conrads TP (2011) Differential proteomic analysis of late-stage and recurrent breast cancer from formalin-fixed paraffin-embedded tissues. J Proteome Res 10:1 323–1332

90. Nuciforo P, Thyparambil S, Garrido-Castro AC, Peg V, Prudkin L, Jimenez J, Hoos WA, Burrows J, Hembrough TA, Perez-Garcia JM, Cortes J, Scaltriti M (2014) Correlation of high levels of HER2 measured by multiplex mass spectrometry with increased overall survival in patients treated with anti-HER2-based therapy. J Clin Oncol 32:5 (suppl.; abstr. 649)

91. Protocol-Online Two dimensional SDS-PAGE. http://www.protocol-online.org/prot/Molecular_Biology/Protein/Protein_Electrophoresis/Two-Demensional_SDS-PAGE/

92. BIO-RAD (2014) 2-D electrophoresis for proteomics. http://www.pmf.colostate.edu/Protocols/BioRad%202D%20manual.pdf

93. Hannigan A, Burchmore R, Wilson JB (2007) The optimization of protocols for proteome difference gel electrophoresis (DiGE) analysis of preneoplastic skin. J Proteome Res 6:3422–3432

94. Marcus K, Joppich C, May C, Pfeiffer K, Sitek B, Meyer H, Stuehler K (2009) High-resolution 2DE. Methods Mol Biol 519: 221–240

95. May C, Brosseron F, Chartowski P, Meyer HE, Marcus K (2012) Differential proteome analysis using 2D-DIGE. Methods Mol Biol 893:75–82

96. Beckett P (2012) The basics of 2D DIGE. Methods Mol Biol 854:9–19

97. Scherp P, Ku G, Coleman L, Kheterpal I (2011) Gel-based and gel-free proteomic technologies. Methods Mol Biol 702:163–190

98. GE_Healthcare (2004) 2-D electrophoresis principles and methods. http://www.gelifesciences.com/gehcls_images/GELS/Related%20Content/Files/1335426794335/litdoc80642960_20140929120525.pdf

99. Rabilloud T, Lelong C (2011) Two-dimensional gel electrophoresis in proteomics: a tutorial. J Proteomics 74:1829–1841

Chapter 17

Managing Drug Resistance in Cancer: Role of Cancer Informatics

Ankur Gautam, Kumardeep Chaudhary, Rahul Kumar, Sudheer Gupta, Harinder Singh, and Gajendra P.S. Raghava

Abstract

Understanding and managing cancer drug resistance is the main goal of the modern oncology programs worldwide. One of the major factors contributing to drug resistance in cancer cells is the acquired mutations in drug targets. Advances in sequencing technologies and high-throughput screening assays have generated huge information related to pharmaco-profiling of anticancer drugs and revealed the mutational spectrum of different cancers. Systematic meta-analysis of this complex data is very essential to make useful conclusions in order to manage cancer drug resistance. Bioinformatics can play a significant role to interpret this complex data into useful conclusions. In this chapter, the use of bioinformatics platforms, particularly CancerDR, in understanding the cancer drug resistance is described.

Key words Drug resistance, Bioinformatics, CancerDR, Pharmaco-profiling

1 Introduction

Cancer is a deadly disease with a very high mortality rate and affects the life of millions of people every year [1]. Both developed and developing countries are in the grip of this deadly disease. Apart from surgery and radiotherapy, chemotherapy remains the principle mode of treatment for the cancer patients. Considerable efforts have been made over the last few decades to control this disease. Despite advances in targeted therapy, cancer is still a leading killer worldwide. One of the main reasons for the failure of cancer treatment is the frequent development of resistance in the cancer cells towards the current chemotherapeutic drugs [2]. There are many factors which are responsible for the development of resistance among which mutations in drug targets are one of the major contributing factors [2]. The knowledge of these mutations with respect to anticancer drug sensitivity is very important in order to understand and manage the cancer drug resistance.

José Rueff and António Sebastião Rodrigues (eds.), *Cancer Drug Resistance: Overviews and Methods*, Methods in Molecular Biology, vol. 1395, DOI 10.1007/978-1-4939-3347-1_17, © Springer Science+Business Media New York 2016

Due to advances in the next-generation sequencing (NGS) technologies, a great deal of attention has been made to reveal the genomic information of the cancers [3]. As a result of such efforts, the mutational landscapes of all the cancers are now available. At the same time, high-throughput assays have provided the drug sensitivity information for hundreds of anticancer drugs across almost whole range of cancer cell lines [4]. With such flood of information of cancer genomics and phramacogenomics, bioinformatics can play a major role in understanding the molecular basis of cancer drug resistance by interpreting this complex data. This understanding of relationship between drug resistance and mutations in drug targets will further lay the foundation for the personalized medicine in near future. This chapter sheds light on how bioinformatics can help in understanding the cancer drug resistance. The main emphasis is given on CancerDR, which is the first resource providing information on cancer drug resistance.

1.1 Cancer Drug Resistance

Despite the increasing arsenal of anticancer agents, development of drug resistance is a persistent problem in cancer treatment. Drug resistance refers to a condition when anticancer drugs are not able to exert their effects on cancer cells or simply when cancer cells do not respond to anticancer therapy. There are many factors contributing to the drug resistance in cancer cells, viz. altered membrane transport, alteration in drug targets, enhanced DNA repair, impaired apoptotic machinery, and increased drug metabolisms [2] (Fig. 1). There are many proteins known as ATP-binding cassette (ABC) proteins which regulate the transport of drugs in and outside the cancer cell. Differences in the expression levels of these proteins affect the drug transport significantly. Low influx and high efflux of drugs keep intracellular drug concentration below a cell-killing threshold and thus remains one of the reasons for the development of resistance. Chemotherapeutic drugs often have intracellular targets and thus to exert their effects these have to bind with their intracellular targets at desired concentration. This interaction between drug and the target is very specific and any sort of alteration in drug targets impairs the drug binding resulting in the low efficacy of the drugs. One of the other important factors contributing to the development of resistance is the increased metabolism of drug molecules, which is due to the enhanced expression of drug-metabolizing enzymes. Apart from these, the most common cause of cancer drug resistance is the acquired mutations in the drug targets or in transport proteins during the cancer progression, making these proteins/targets unable to do their function properly.

1.2 Cancer Informatics and Management of Drug Resistance

Over the last decade, a huge amount of pharmacogenomics, epigenomics, and transcriptomics data has been generated by different high-throughput and omics approaches, which provide a high resolution and global view of the cancer genome (Fig. 2).

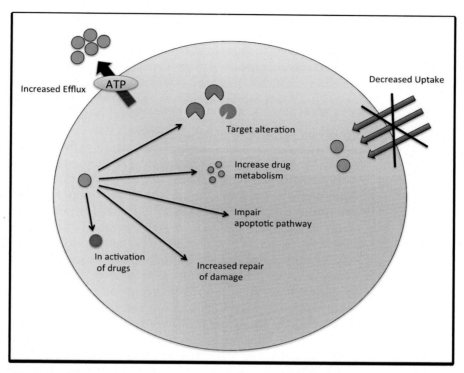

Fig. 1 Schematic representation of mechanisms of drug resistance in cancer

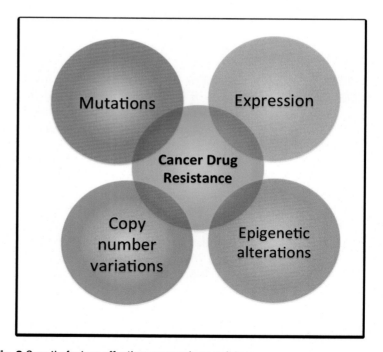

Fig. 2 Genetic factors affecting cancer drug resistance

This flood of information broadened our molecular understanding of cancer causation, progression, and pharmacological response to the drugs. In the past, many experimental and bioinformatic investigations have provided several predictive biomarkers for drug resistance in cancer. For example, Barretina et al. [5] proposed the MEK inhibitor efficacy associated with AHR expression in NRAS-mutant lines; and SLFN11 expression correlated with sensitivity to topoisomerase inhibitors. Moreover, resistance to tyrosin kinase inhibitors has been associated with EGFR mutation (T790M) [6]. Apart from genomics, epigenomics has also been associated with drug resistance in cancer; for example, TFAP2E hypermethylation is associated with clinical nonresponsiveness to 5-fluorouracil in colorectal cancer [7] and promoter hypermethylation of IGFBP-3 upholds cisplatin resistance in non-small-cell lung cancers [8]. As it can be seen that these high-throughput studies provided meaningful insights into cancer drug resistance. However, the present status of predictive biomarkers for drug responsiveness and personalized therapy of cancer is very primitive. With powerful bioinformatics tools and advances in NGS, the hidden aspects in the complex cancer genomics can be unveiled and useful information pertaining to drug resistance like expression of drug-metabolizing enzymes, expression of transport proteins, mutations in drug targets, expression of enzymes involved in apoptosis, and epigenomics can be extracted. A systematic meta-analysis of this information may help to handle cancer drug resistance in a better way. In the past, a few cancer informatics resources have been developed, which can be helpful to understand the cancer drug resistance. A brief description of these resources is as follows:

1.3 COSMIC and CCLE

Catalogue of somatic mutations in cancer (COSMIC) is a vast initiative of Wellcome Trust in the field of cancer. This catalogue houses the critical information related to the various aspects of cancer, e.g., mutation, copy number variation, and gene expression. Genomics of drug sensitivity in cancer (GDSC) [3] is one such project of COSMIC, which contains the crucial information about pharmacogenomic data of around 140 anticancer drugs on hundreds of cancer cell lines. This elegant study of GDSC revealed many associations between the genomic alterations (mutations, copy number variations, and gene expression) and drug sensitivity. A similar study was carried out by Barretina et al. [4] in 2012 at Broad Institute, which leads to the formation of Cancer Cell Line Encyclopedia (CCLE). This encyclopedia houses the genomic information of around 900 cancer cell lines and pharmacogenomic data of 24 anticancer drugs in nearly 500 cancer cell lines. This study also revealed the critical associations between genomic alterations (mutations, copy number variations, and gene expression) and drug sensitivity and some associations were common between these two studies, like mutations in *BRAF* and *NRAS* gene lead to the sensitivity of MEK inhibitors. Data generated from these two studies are highly enriched with the drug sensitivity information, which could be helpful in exploration and decoding of the drug resistance mechanism in cancer.

1.4   HerceptinR

Among several small molecular drugs and antibodies as therapeutics in cancer treatment, Herceptin antibody (Ab) has been a well-known drug, which is mainly efficient in Her2+ cancers such as breast cancer. This Ab binds to the extracellular domain of Her2 protein, present on the cell surface. Although there are several successful clinical cases for effectiveness of Herceptin, we still find reports of both *de novo* and acquired resistance. The molecular mechanisms leading to these two types of resistance are still not fully comprehended, which in turn leads to the administration of drug to unsuitable patients. In the absence of efficient biomarkers/diagnostic kits for Herceptin resistance, the hope of therapy costs both life and money. In order to investigate the possible mechanisms, biomarkers, and supplementary drugs for Herceptin-resistant breast cancers, a database namely HerceptinR is available [9]. HerceptinR houses the cell line-based assays performed with Herceptin, drugs supplemented to overcome resistance, and genomic properties of cancer cell lines which might be contributing towards resistance. The database also provides tools, which enable users to compare the mutation, expression, and copy number variation profile of Herceptin-resistant and -sensitive cell lines. The database is available at http://crdd.osdd.net/raghava/herceptinr/.

1.5   CancerDR

CancerDR is a cancer informatics platform, which has been developed with an aim to understand the cancer drug resistance [10]. In this database, information of pharmacological profiling of 148 anticancer drugs on 952 cancer cell lines in relation to the mutation status of the 116 drug target genes has been provided in a systematic manner. Users can understand the relationship between drug resistance and mutations in drug targets. For this, many user-friendly tools for analysis and data retrieval have been integrated. The users can make the best use of this resource in the following ways:

1.6   Alignment/ Mutation Module

In order to understand the relation between drug resistance in cancer cells with mutations in drug targets, it is imperative to know the spectrum of mutations in the target genes. Therefore, Alignment/Mutation module has been integrated in CancerDR to impart better understanding of mutations in drug targets at both sequence and structural level. Along with mutants, natural variants of a drug target can also be aligned simultaneously. In sequence alignment, wild-type sequence of drug target is aligned with other mutants/variants to look, mutation falling in conserved or non-conserved regions of the target. Mutation falling in conserved regions can destroy the normal functioning of the drug target, which could be advantageous for cancer growth. In structural alignment, wild-type structure of drug target is aligned with mutants/variants to see structural changes. Structural changes in conserved domains/motifs could be lethal for normal growth. Both types of alignments are crucial in revealing the impact of mutation on normal functioning of drug target and it can also change the fate of drug action, which could lead to drug resistance/sensitivity. This tool has been subdivided into four following modules:

1.6.1 Total Align

In total align module, user can choose any of the drug targets from the list of 116 drug targets to align either mutants or natural variants of selected drug target. By default alignment of all the mutants/variants comes in the Jalview window, but users also have the opportunity to align any of the mutant/variant separately by clicking the mutant given at the head of the Jalview window. At the bottom of the Jalview window a button named as "Start Jalview" is present. On clicking this button, a pop-up window of Jalview appears which shows sequence conservation, quality, and consensus in addition to the sequence alignment. In Fig. 3, we have shown the example of ERBB2 as drug target, which shows the alignment of 15 mutants of ERBB2 protein.

1.6.2 Custom Align

Custom align module helps the user to figure out association of drug target mutation with the IC_{50} of anticancer drug, if any. After selecting some drug target from the table provided, user gets the options of different drugs. Viability assays have been done on cancer cell lines in which selected target is mutated using the shown drugs. After clicking any of the drugs provided, a table appears showing the IC_{50} in different cell lines in increasing order. Scale of IC_{50} of different cell lines can reveal the resistance or sensitivity of that cell lines for particular drug (Fig. 4). Further, user can choose any number of mutants present in table to see their sequence alignment. This module also provides the opportunity to the user for alignment of user-defined sequence with other wild/mutant/variant sequences.

1.6.3 Mutants

Mutants module enlists all the mutations reported for drug targets at amino acid, codon, and cDNA level.

1.6.4 Structure Alignment

This module facilitates the user to do the structural alignment of mutants/variants of drug targets (Fig. 5). All the structural alignments have been done by using the MUSTANG software and Jmol applet is incorporated to view 3D alignment. User can also download the PDB file of the aligned 3D structure. Along with the structural alignment, MUSTANG also provides the sequence alignment

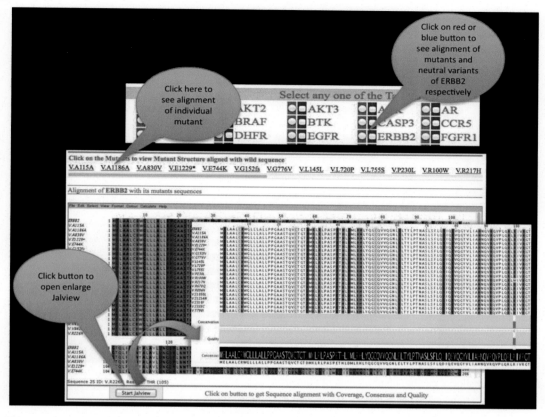

Fig. 3 Screenshot of total align module of CancerDR

based on structural alignment. If some mutants/variants have extremely variable length as compared to the other partners, they have been removed to obtain proper structural alignment.

1.7 Target Structure Module

A single mutation in drug target gene sometimes may have very significant effect on the functionality of drug target, if it is in the drug-binding pocket. Therefore, understanding of the effect of the mutations on the target's structure and crucial active sites/drug-binding sites is very important to understand the drug resistance. In this context, the target structure module allows users to view and compare the 3D structure of the target and their mutants. It helps in identification of structural changes due to variation in the protein sequence or mutation. Visualization of target and its mutant helps in identification of region, which causes significant changes in the 3D structure. This module together with structure alignment module will help users in identification of potential drug-sensitive and drug resistance mutation.

The target structure module of CancerDR consists of five different submodules. The first two modules allow users to view the

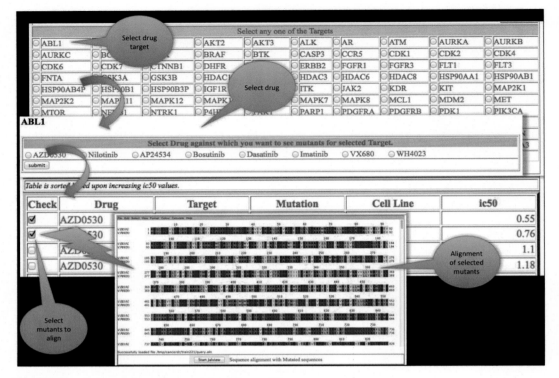

Fig. 4 Screenshot of custom align module of CancerDR

tertiary and secondary structures of the target and their known mutants. The tertiary structure module displays the predicted 3D structure of the selected target along with structure similarity details. It also allows user to download the PDB of the target, and view the Ramachandran map and the details of homologous structure in PDB. All the mutants of the selected target are displayed along the 3D structure of the target. Clicking on the name of the mutant will display the 3D structure with structural similarity and quality details of the mutant structure. The secondary structure module displays the predicted secondary structure of the selected target and its mutant. The module displays the DSSP 8 state secondary structure information for detailed analysis.

The next two modules allow users to compare the 3D structure and homology-based modeling of unknown mutants. The compare module allows users to compare the 3D structure of two mutants or one mutant with a user-specified structure in PDB format. The output page displays the aligned 3D structure of the selected mutants or the uploaded structure with a mutant. The display shows the overlapped and non-overlapped regions of different structures. The user sequence module models the 3D structure of unknown sequence based upon the 3D structure of the selected target. The last module

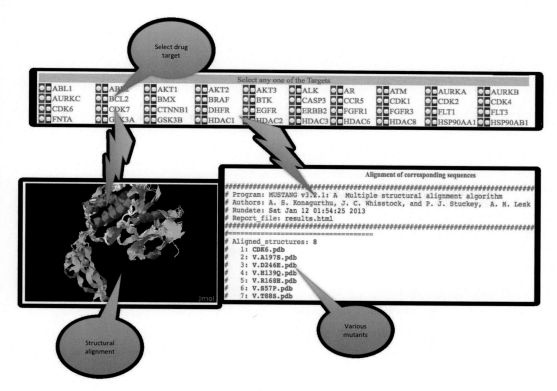

Fig. 5 Screenshot of structure alignment module of CancerDR

structures in PDB display the solved 3D structure of all the 148 targets. Since different regions of a target are solved by different studies, multiple PDBs exist for a single target.

1.8 Clusters/ Groups Module

In the pursuit of mechanisms contributing to drug resistance in cancer, our group explored the drug assay profile of all the cancer cell lines belonging to different tissue types along with an overview at mutational profile of all the drug targets. In this regard, the clusters/groups module has been developed, which has been further divided into submodules as follows:

1.8.1 Clustering of Mutant Drug Target Sequences (Target Cluster)

In order to investigate the positional conservation of amino acid residues and hot spots for drug sensitivity or resistance in the drug targets, we have devised this submodule. In this submodule, we have incorporated targets of 148 drugs in their mutated amino acid sequence form. Since each target has different mutational profile among 952 cancer cell lines, we aligned all the possible mutants of each target through Web-based Jalview software (Fig. 6). The result can be seen as precise alignment with wild-type sequence or in the form of distance tree.

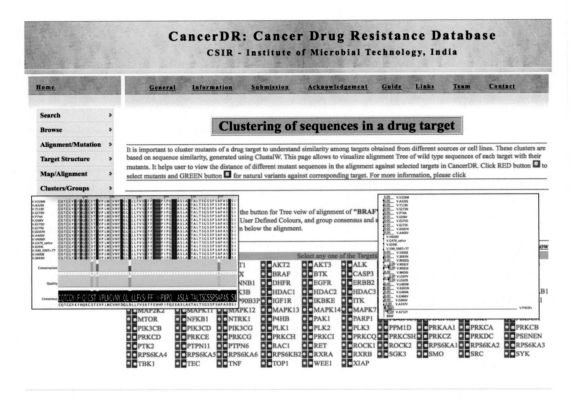

Fig. 6 Screenshot of target cluster module of CancerDR

1.8.2 Clustering of Cancer Cell Lines Assayed on Different Drugs (Cell Line Cluster)

This study analyzes pharmacological profiles of 952 cancer cell lines belonging to about 28 tissue types, where each of the tissue types or cell lines has their defined mutational pattern. Since the particular set of genomics leads to particular type of response towards drug, we clustered the cell lines assayed on particular drug and distributed the cell lines in different range of IC_{50s}, for example, a range of $IC_{50} >= Ref$ and $IC_{50} >= 3 \times Ref$, where Ref and 3×Ref represent the lowest IC_{50} and three times of lowest IC_{50}, respectively. In addition to this, we also provided the absolute ranges of IC_{50} such as R1: 0–0.001 μM, R2: 0.001–0.005 μM up to higher IC_{50} like >390 μM. The higher ranges of IC_{50} provide clusters of cell lines, which might harbor some common genomic features, which might be contributing towards drug resistance. For example, while selecting pancreatic cancer in cell line clustering (Fig. 7), the 17AAG drug shows one extremely resistant cell line (MZ1PC), which lies in the range of 250 times of reference IC_{50}. Such pharmacological profile-based distribution of large number of cell lines assayed over 148 anticancer drugs gives a broad overview of resistant cell lines for each of the drugs and their genomics.

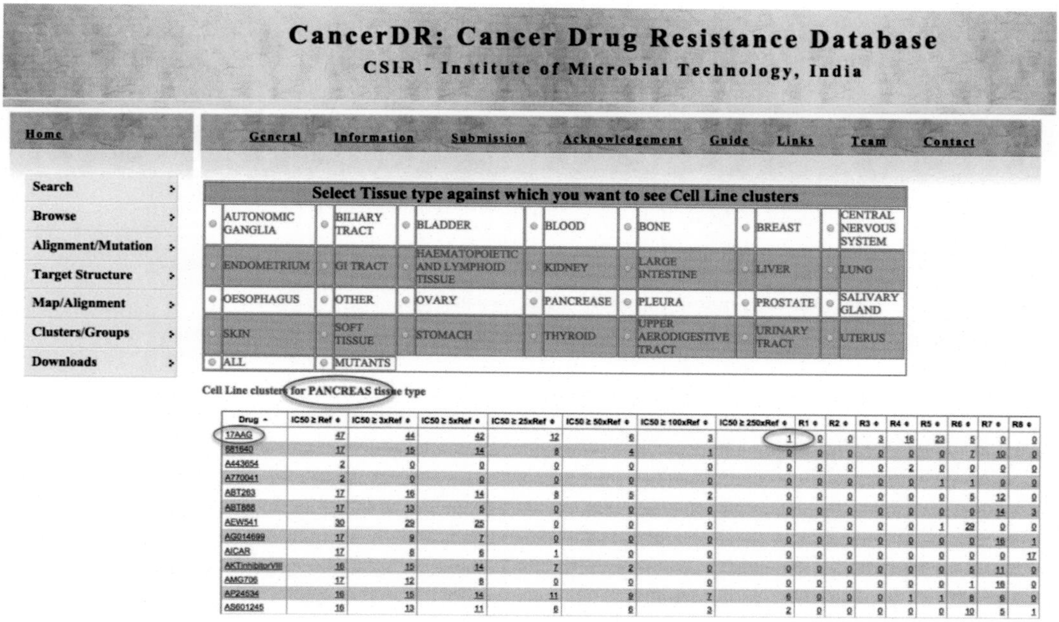

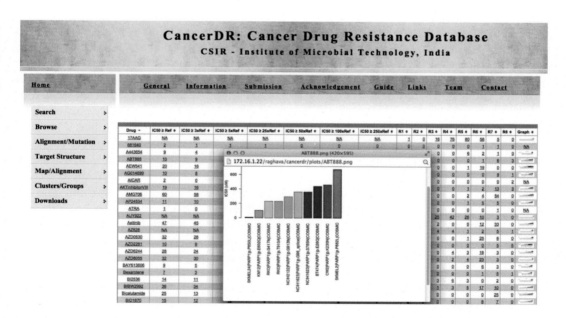

Fig. 7 Screenshot of cell line cluster module of CancerDR

Fig. 8 Screenshot of cell line module showing mutational profile of all the targets of 148 drugs

As a miscellaneous and additional section in this submodule, we have added mutational profile of all the targets of 148 drugs and tried to investigate the relation between mutations in target and its effect on drug resistance (Fig. 8). The profiles can be seen in the form of column charts in the table.

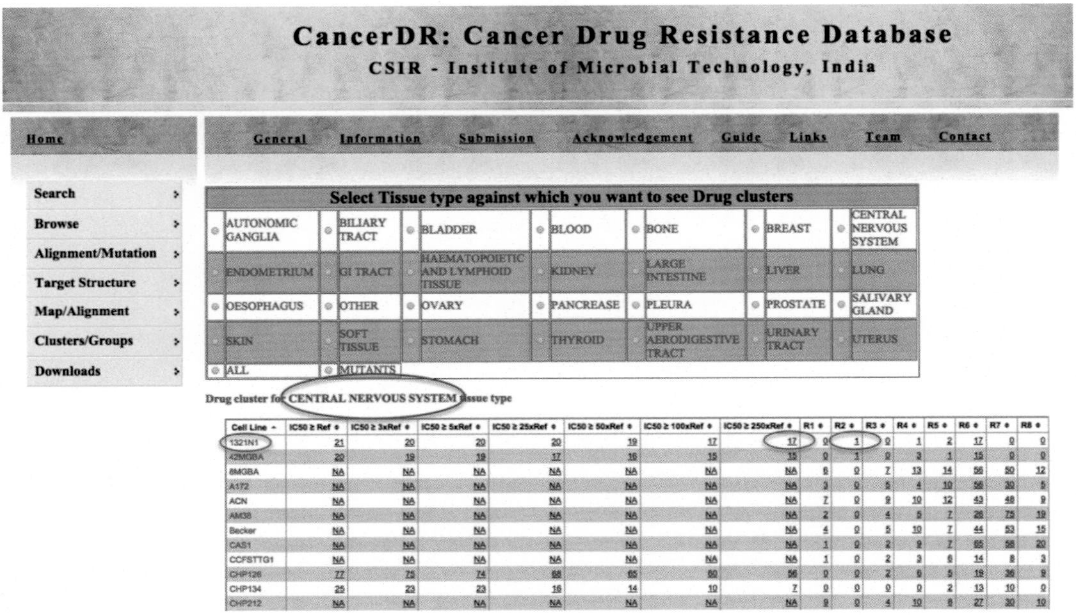

Fig. 9 Screenshot of drug cluster module of CancerDR

1.8.3 Clustering of Anticancer Drugs for Different Cancer Cell Lines (Drug Cluster)

As discussed above that we have analyzed pharmacological profiles of 952 cancer cell lines, each cell line has been assayed on a panel of drugs. Among panel of drugs, any particular cell line shows sensitivity for particular set of drugs while showing resistance towards other. In order to examine the distribution of different drugs for each of the cell lines for its resistance or sensitivity, we clustered the drugs as we discussed in the above section of cell lines. This sub-module works exactly same as cell line cluster except for the clustering of drugs here in place of cell lines. For example, in case of central nervous system, among all the drugs assayed on 1321 N1 cell line, there are 17 drugs, which are highly resistant and 21 drugs are fairly sensitive (Fig. 9). This module also presents one drug paclitaxel lying in the range of R2: 0.001–0.005 µM, which shows that any cell line/sample having genomic property like 1321 N1 may respond pharmacologically similar.

1.9 Map/Alignment Module

With the advent of NGS technology, it becomes imperative to take care of high-throughput data to deal with. This monstrous data, which is the integral part of, nowadays, almost every other cancer study contains underlying information, and hence is taken care with this module when it comes to drug resistance.

1.9.1 Short Reads

The raw output of NGS experiments in the form of reads is aligned to the desired/all drug targets in cancer. It is quite useful if there is any unusual perturbation in the genome in terms of variation/mutation. Tablet viewer (a Java-based applet) is quite handy in visualizing aligned raw reads of a patient against the well-known cancer drug targets. This module is flexible for taking single-end reads as well as paired-end reads and output is mailed to the user. Any variation/mutation even at a single point is clearly visible with this viewer.

1.9.2 Contigs

To move a step further, genomics or transcriptomics contigs (raw reads assembled by the well-known software) from patient are taken as input and are subjected downstream to gene annotation software, Augustus, to predict genes. These predicted genes are subjected to BLAST hit against well-known cancer drug targets to get inference. BLAST hits are done at both gene and protein level. This module finally facilitates the user to analyze any variation of predicted genes as compared to the cancer targets.

1.9.3 Sequences

If a user directly wants to compare (in terms of sequence alignment) the gene/protein with the drug targets and their mutants in CancerDR database, then this module is quite useful as it gives BLAST alignment options at gene as well as protein level. All variants of BLAST (BLASTN, BLASTP, BLASTX, TBLASTN, TBLASTX) have been incorporated for flexible alignment options and results are downloadable.

1.10 Limitations and Future Prospects

Despite considerable efforts to improve the cancer treatment, still drug resistance is an unsolved problem in cancer drug therapy. However, advances in sequencing technologies have unveiled novel alterations in cancer genome and subsequently increased our understanding of molecular mechanisms of drug resistance in cancer cells. Furthermore, the pharmacogenomics studies have provided novel insights into drug resistance in cancer cells but these studies are restricted to only cancer cell lines. Pharmacogenomics data from tumor tissues and patients are needed which would give a better picture and thus would be more helpful to understand the cancer drug resistance.

Since there are many factors responsible for drug resistance in cancer cells, to date, most of the studies have been carried out to understand the effect of only one factor at a time, like correlation between mutations in drug targets and drug sensitivity. A holistic analysis of all the factors like expression of transport proteins, mutations in drug targets, effects of epigenetic alterations, and copy number variations is essential to understand drug resistance problem. In the era of personalized medicine, it can be expected that in near future more personalized studies covering all these aspects would be carried out and will help to overcome the cancer drug resistance.

References

1. Siegel R, Naishadham D, Jemal A (2013) Cancer statistics, 2013. CA Cancer J Clin 63:11–30

2. Housman G, Byler S, Heerboth S, Lapinska K, Longacre M, Snyder N, Sarkar S (2014) Drug resistance in cancer: an overview. Cancers (Basel) 6:1769–1792

3. Garnett MJ, Edelman EJ, Heidorn SJ, Greenman CD, Dastur A, Lau KW, Greninger P, Thompson IR, Luo X, Soares J, Liu Q, Iorio F, Surdez D, Chen L, Milano RJ, Bignell GR, Tam AT, Davies H, Stevenson JA, Barthorpe S, Lutz SR, Kogera F, Lawrence K, McLaren-Douglas A, Mitropoulos X, Mironenko T, Thi H, Richardson L, Zhou W, Jewitt F, Zhang T, O'Brien P, Boisvert JL, Price S, Hur W, Yang W, Deng X, Butler A, Choi HG, Chang JW, Baselga J, Stamenkovic I, Engelman JA, Sharma SV, Delattre O, Saez-Rodriguez J, Gray NS, Settleman J, Futreal PA, Haber DA, Stratton MR, Ramaswamy S, McDermott U, Benes CH (2012) Systematic identification of genomic markers of drug sensitivity in cancer cells. Nature 483:570–575

4. Barretina J, Caponigro G, Stransky N, Venkatesan K, Margolin AA, Kim S, Wilson CJ, Lehar J, Kryukov GV, Sonkin D, Reddy A, Liu M, Murray L, Berger MF, Monahan JE, Morais P, Meltzer J, Korejwa A, Jane-Valbuena J, Mapa FA, Thibault J, Bric-Furlong E, Raman P, Shipway A, Engels IH, Cheng J, Yu GK, Yu J, Aspesi P Jr, de Silva M, Jagtap K, Jones MD, Wang L, Hatton C, Palescandolo E, Gupta S, Mahan S, Sougnez C, Onofrio RC, Liefeld T, MacConaill L, Winckler W, Reich M, Li N, Mesirov JP, Gabriel SB, Getz G, Ardlie K, Chan V, Myer VE, Weber BL, Porter J, Warmuth M, Finan P, Harris JL, Meyerson M, Golub TR, Morrissey MP, Sellers WR, Schlegel R, Garraway LA (2012) The Cancer Cell Line Encyclopedia enables predictive modelling of anticancer drug sensitivity. Nature 483:603–607

5. Wilks C, Cline MS, Weiler E, Diehkans M, Craft B, Martin C, Murphy D, Pierce H, Black J, Nelson D, Litzinger B, Hatton T, Maltbie L, Ainsworth M, Allen P, Rosewood L, Mitchell E, Smith B, Warner J, Groboske J, Telc H, Wilson D, Sanford B, Schmidt H, Haussler D, Maltbie D (2014) The Cancer Genomics Hub (CGHub): overcoming cancer through the power of torrential data. Database (Oxford) 2014

6. Yun CH, Mengwasser KE, Toms AV, Woo MS, Greulich H, Wong KK, Meyerson M, Eck MJ (2008) The T790M mutation in EGFR kinase causes drug resistance by increasing the affinity for ATP. Proc Natl Acad Sci U S A 105:2070–2075

7. Ebert MP, Tanzer M, Balluff B, Burgermeister E, Kretzschmar AK, Hughes DJ, Tetzner R, Lofton-Day C, Rosenberg R, Reinacher-Schick AC, Schulmann K, Tannapfel A, Hofheinz R, Rocken C, Keller G, Langer R, Specht K, Porschen R, Stohlmacher-Williams J, Schuster T, Strobel P, Schmid RM (2012) TFAP2E-DKK4 and chemoresistance in colorectal cancer. N Engl J Med 366:44–53

8. Ibanez de Caceres I, Cortes-Sempere M, Moratilla C, Machado-Pinilla R, Rodriguez-Fanjul V, Manguan-Garcia C, Cejas P, Lopez-Rios F, Paz-Ares L, de CastroCarpeno J, Nistal M, Belda-Iniesta C, Perona R (2010) IGFBP-3 hypermethylation-derived deficiency mediates cisplatin resistance in non-small-cell lung cancer. Oncogene 29:1681–1690

9. Ahmad S, Gupta S, Kumar R, Varshney GC, Raghava GP (2014) Herceptin resistance database for understanding mechanism of resistance in breast cancer patients. Sci Rep 4:4483

10. Kumar R, Chaudhary K, Gupta S, Singh H, Kumar S, Gautam A, Kapoor P, Raghava GP (2013) CancerDR: cancer drug resistance database. Sci Rep 3:1445

INDEX

A

ABCB1............................7, 8, 12, 22, 76, 77, 79, 82, 87–101, 140, 143, 146, 147, 155, 218
Acquired resistance and adaptive compensatory...............6–7
Acquired resistance and tumor micro-heterogeneity............4–6
Apoptosis....................................4, 6, 10, 27–29, 33, 40, 107, 115, 116, 127, 146, 149–151, 163, 167, 168, 174, 209, 212–214, 216, 221, 232, 233, 302

B

BCR-ABL
    Imatinib... 76, 77, 79
    Resistance in CML 76, 77, 79
Bioinformatics....................................210, 284, 294, 300, 302
Breast cancer (BC) 3, 6, 7, 11, 13, 24, 25, 30, 40, 76, 88, 105–128, 140, 144, 147, 148, 151–153, 189–203, 213, 214, 218–220, 233, 251–276, 292, 293, 303

C

Cancer 1–15, 19, 56, 69, 76, 87, 164–165, 229, 241, 281, 299
Cancer stem cells (CSCs).......................... 10, 139, 153–156, 165, 209, 241–248
CancerDR 300, 303, 305–311
Cell lines
    HepG2 ....57, 58, 60, 64, 67, 212, 218, 219
    K562 ....77–80, 83, 84, 140, 142, 147
Chemoresistance...............................39, 56, 57, 88, 148, 155, 168, 175, 211, 212, 216, 242, 248
Chemotherapeutic response 148, 292–293
Chemotherapy 6, 7, 9, 10, 14, 25–26, 29, 33, 39, 57, 69, 88–90, 105, 106, 114, 117, 119–121, 128, 139, 146, 151, 154, 165–174, 190, 209, 213, 231–233, 241, 242, 252, 255, 265, 270, 273, 275, 292, 293, 299
Chronic myelogenous leucemia (CML) 8, 76–80
Clonogenic assay 41–42, 46, 49
Comet assay.. 40–41, 43, 45
CYP3A..56–58
Cytokinetics..27–30

D

Discovery-based proteomics..282
DNA damage 9, 10, 26, 39, 40, 43–45, 50–52, 139, 149, 150, 172, 209, 212, 215, 216, 233
DNA repair and resistance 9–10, 139, 149
DPBS buffer...71
Drug resistance 1–15, 22, 24, 30–33, 39, 40, 70, 76–79, 137–156, 164, 165, 191, 192, 207–223, 229–237, 289, 292, 299–311
Drug transporters 12, 25, 75–84, 91, 139

E

Efflux transporters..69–74
Electron microscopy...217
Endocrine therapy (ET)....................11, 105–107, 109, 116, 121, 220, 252, 274, 275
Epigenetic 1, 2, 5–9, 11, 12, 14, 139, 152, 156, 164, 165, 208–211, 215, 231, 311
Epigenomics and Resistance 11–12

F

FFPE. *See* Formaldehyde Fixed-Paraffin Embedded (FFPE)
Flow cytometry..92–95, 99, 100, 243
Formaldehyde Fixed-Paraffin Embedded
    (FFPE)............... 175–179, 181, 189–203, 266, 268, 269

G

β-Galactosidase staining..42, 50
Gene expression..................................7, 8, 30, 40, 56–58, 64, 66, 67, 78, 83, 106, 108, 109, 138, 139, 147, 154, 164, 165, 174, 190, 208, 211, 215, 221, 223, 256, 257, 259, 270, 292, 302
Gene regulation..................................... 8, 138, 207, 218, 222

H

γ-H2AX 40, 41, 44–46, 51
Head and neck squamous cell carcinoma
    (HNSCC)...................140, 143, 145, 146, 155, 241–248
Hepatocytes....................................57, 58, 60–61, 65, 66, 213
HNSCC. *See* Head and neck squamous cell carcinoma
    (HNSCC)

José Rueff and António Sebastião Rodrigues (eds.), *Cancer Drug Resistance: Overviews and Methods*, Methods in Molecular Biology, vol. 1395, DOI 10.1007/978-1-4939-3347-1, © Springer Science+Business Media New York 2016

I

Immune checkpoints 29
Immunohistochemistry (IHC) 127, 144, 192,
193, 196–199, 253, 261, 269–270, 293
Inflammation...57, 58, 175
Intrinsic resistance and pharmacogenetic 3–4

K

Ki67 labeling 265, 273

L

Ligand independent ER activation...........................106–108
Long noncoding RNAs................................ 12, 164, 207–223

M

MDR. *See* Multi-drug resistance (MDR)
MDR1. *See* Multidrug resistance gene (MDR1)
Metabolic response... 30, 56, 144
microRNAs 4, 9, 11–12, 137–156, 163–181,
189–203, 207, 230
miRNAs. *See* microRNAs
Multi-drug resistance (MDR)............. 7, 8, 20, 24, 69, 70, 76–78,
82, 87–92, 95, 97–101, 146, 148, 209, 217, 218, 292
Multidrug resistance gene (MDR1) 22, 88, 91,
95, 96, 146, 218

N

NanoSight microscope 234
Neoadjuvant systemic therapy (NAST)....................251–276
Noncoding RNAs.........................12, 150, 164, 190, 207–223

O

Orospheres 242, 248
Overall survival..............26, 56, 114, 122, 151, 155, 166, 252, 259

P

Pancreatic cancer 19, 153, 154, 163–181,
213, 214, 217, 219, 308
Pathologic complete response (pCR)....................... 252, 255,
259, 262, 265, 268, 270, 272, 275

P-glycoprotein (P-gp) 22, 76, 88–89, 171,
218, 232
Pharmaco-profiling 252, 289, 303,
308, 310
Plasma 3, 21–24, 76, 88, 89, 119,
123, 165, 170, 172, 180, 181, 191, 219–220, 230, 232,
234–237, 262, 268, 283, 288, 289
Pre-metastatic niche 229–231
Prognosis 8, 88, 107, 123, 168, 171, 213,
214, 229, 251, 252, 262, 271, 275, 292
Proteomics..15, 281–294

R

Radiologic response 258–259,
266, 274
Real-time fluorimetry...93, 98
Relapse 4, 19, 56, 106, 108–111, 114,
191, 209, 211, 251, 265, 275
Repopulation 20
Resistance 1–15, 19–33, 39–52, 55, 69,
76, 87, 165, 231, 241, 292, 299
RT-qPCR.. 195, 199, 269

S

Self-renewal................................14, 154–156, 212, 213, 242
Stop and Lysis solutions 71

T

Target-proteomics 282, 285–286
TEER values 74
Transwells..74
Tumor-derived exosomes..231
Tumor microenvironment..229–233
Tumor microenvironment and resistance....................12–13
Tyrosine kinase inhibitors (TKIs)......................... 76, 78, 108,
110, 115, 116, 120, 126, 147, 173

U

Uptake 4, 7–8, 24, 31, 32, 57, 66,
69–74, 76, 99, 147, 231, 264, 273
Uptake and efflux transporters in resistance 69–74